Materiality and Organizing

Materiality and Organizing

Social Interaction in a Technological World

Edited By
Paul M. Leonardi, Bonnie A. Nardi, and
Jannis Kallinikos

Great Clarendon Street, Oxford, OX2 6DP,
United Kingdom

Oxford University Press is a department of the University of Oxford.
It furthers the University's objective of excellence in research, scholarship,
and education by publishing worldwide. Oxford is a registered trade mark of
Oxford University Press in the UK and in certain other countries

First Edition published 2012

Impression: 2

British Library Cataloguing in Publication Data
Data available

Library of Congress Cataloguing in Publication Data
Data available

ISBN 978–0–19–966405–4 (Hbk)
978–0–19–966406–1 (Pbk)

Printed in Great Britain by
MPG Books Group, Bodmin and King's Lynn

Table of Contents

Acknowledgments

The editors would like to thank the authors for stimulating discussion at the workshop held at Northwestern University in April 2011, and for a collection of thoughtful papers spanning a range of issues concerning materiality. Working with this group of authors was unusually rewarding, imparting a sense of the creative ferment that marks current understandings of materiality. Northwestern University's generosity in hosting the workshop and supporting the development of the volume is gratefully acknowledged. Specifically, we deeply appreciate the efforts of Dean Barbara O'Keefe who made all this possible. In addition to providing funding for our workshop, Dean O'Keefe kicked off the meting with a personal appearance, taking time out of her very busy schedule to give us a great start. Important financial support for the workshop was also provided by the Allen K. and Johnnie Cordell Breed fund at Northwestern. The authors received feedback from colleagues as they labored on their papers, and we thank these industrious reviewers who contributed to the quality of the "materials" that constitute this volume. We are fortunate in the guidance and support of our editor David Musson at Oxford University Press, as well as the efficient and helpful support staff.

List of Figures

List of Tables

Notes on the Contributors

Chad Anderson is Assistant Professor of Information Systems at the University of Nevada, Reno. He received his Ph.D. in Computer Information Systems from Georgia State University in 2011. His dissertation examined how the materiality of an electronic medical records system was influencing nurse's work practices on hospital patient care units. His work has been published in *MIS Quarterly*, *Communications of the AIS*, and the *Journal of Organizational and End User Computing*.

Bijan Azad is Associate Professor at the Olayan School of Business, American University of Beirut and Co-Director of Darwazah Center for Innovation Management and Entrepreneurship. His research is on technology enactment in organizations. He earned his Ph.D. from MIT. He has published in *Journal of Strategic Information Systems*, *European Journal of Information Systems*, and *Information Systems Journal*. He was recipient of the OCIS Division Best Conference Paper Award at the Academy of Management (2008).

Albert Borgmann is Regents Professor of Philosophy at the University of Montana, Missoula, where he has taught since 1970. His special area is the philosophy of society and culture. Among his publications are *Technology and the Character of Contemporary Life* (University of Chicago Press, 1984), *Crossing the Postmodern Divide* (University of Chicago Press, 1992), *Holding on to Reality: The Nature of Information at the Turn of the Millennium* (University of Chicago Press, 1999), *Power Failure: Christianity in the Culture of Technology* (Brazos Press, 2003), and *Real American Ethics* (University of Chicago Press, 2006).

Jenna Burrell is Assistant Professor in the School of Information at University of California-Berkeley. She is the author of *Invisible Users: Youth in the Internet Cafes of Urban Ghana* (MIT Press, 2012). Her interests span many research topics including theories of materiality, user agency, transnationalism, post-colonial relations, and digital representation. Recent fieldwork in rural and urban sites in Ghana and Uganda has examined the appropriation of various digital technologies by individuals and social groups.

François Cooren is Professor at the Department of Communication of the Université de Montréal. His interests lie in organizational communication, language and social interaction, and communication theory. He is the author of more than forty peer-reviewed articles published in international journals as well as twenty book chapters. He authored two books (*The Organizing Property of Communication* (2000) and *Action and Agency in Dialogue* (2010), both published by John Benjamins) and edited two volumes

(*Communication as Organizing* (2006) and *Interacting and Organizing* (2007), both published by Lawrence Erlbaum).

Christiane Demers is Professor in the Department of Management at HEC Montréal where she teaches courses on organizational change and strategy. Her research focuses on organizational transformation and evolution, with a particular emphasis on strategic dynamics and communication processes in professional organizations. She is the author of three books, including *Organizational Change Theorie: A Synthesis* (Sage, 2007). She is the author of several publications, including recent articles in *Journal of Organizational Change Management* and *Organization Science.*

Hamid Ekbia is Associate Professor of Information Science and Cognitive Science, and the Director of Center for Research on Mediated Interaction at the School of Library and Information Science, Indiana University, Bloomington. His research is focused on how technologies mediate interactions among people, organizations, and communities. He has written about technology and mediation in organizations, design and scholarly communities, and local populations. His recent book, *Artificial Dreams: The Quest for Non-Biological Intelligence* (Cambridge University Press, 2008), is a critical–technical analysis of Artificial Intelligence.

Gail T. Fairhurst is Professor of Organizational Communication at the University of Cincinnati, USA. Her research and writing interests are in organizational communication, leadership, and organizational discourse analysis. She has several publications in leading communication and management journals and is the author of three books, including *Discursive Leadership: In Conversation with Leadership Psychology* (Sage, 2007). She is Distinguished Scholar of the National Communication Association, Fellow of the International Communication Association, and a Fulbright Scholar.

Samer Faraj holds the Canada Research Chair in Technology, Management & Healthcare at the Desautels Faculty of Management at McGill University. He is interested in complex collaboration in settings as diverse as health care organizations, knowledge teams, and online communities and how technology is transforming organizations and allowing new forms of organizing to emerge. He is currently Senior Editor at *Organization Science* and at *Information Systems Research.*

Philip Faulkner is a Fellow and Senior College Teaching Officer in Economics at Clare College, Cambridge, and a Fellow of the Cambridge Judge Business School. His research focuses on issues in social ontology, particularly the nature of technological objects. He is an editor of the *Cambridge Journal of Economics*, and has published on various topics in journals including *Academy of Management Review*, *Economica*, and *the Journal of Economic Methodology*.

Anne-Laure Fayard is Assistant Professor of Management in the Department of Technology Management at the Polytechnic Institute of New York University. Her research interests involve organizational communication, sociomaterial practices, and interdisciplinary collaboration. Her work has been published in several leading management, organization studies, and information systems journals. She holds an MA in Philosophy from La Sorbonne, an M.Phil. in Cognitive Science from Ecole

Polytechnique and a Ph.D. in Cognitive Science from the Ecole des Hautes-Etudes en Sciences Sociales.

Carole Groleau is Associate Professor in the Department of Communication at Université de Montréal where she teaches seminars on information technology in the workplace as well as on qualitative methods. Her research investigates the dynamic relationship between material and social dimensions of organizing processes. She co-authored *The Computerization of Work: A Communication Perspective* (Sage, 2001). She has numerous publications including recent articles in *Organization Science* and *Journal of Organizational Change Management.*

Chris Harty is Lecturer in Socio-Technical Systems and Director of the Health and Care Infrastructure Research and Innovation Centre in the School of Construction Management and Engineering, University of Reading, UK. His research interests are in the socio-technical interactions across IT system development, implementation, and use, especially within complex construction projects. He received his Ph.D. in Sociology from Lancaster University, UK.

Romain Huët is Assistant Professor (Maître de Conférence) at the Department of Communication of the European University of Brittany (Rennes, France). His interests include business ethics and corporate social responsibility. While these issues have gained access to companies' policy agendas, the objective of his research is to analyze and understand how the economic sphere can be organized on these issues through a communicational perspective. His research highlights the conditions of production of new forms of deliberation and invite a reflection on new forms of governance.

Jannis Kallinikos is Professor and Ph.D. Program Director in the Information Systems and Innovation Group, Department of Management at the London School of Economics. His research covers a wide range of topics on the interpenetration of technology with the administrative and institutional arrangements of contemporary societies. Recent books include *The Consequences of Information: Institutional Implications of Technological Change* (Edward Elgar, 2006) and *Governing Through Technology: Information Artefacts and Social Practice* (Palgrave, 2011).

Paul M. Leonardi is the Pentair-Nugent Associate Professor at Northwestern University where he teaches courses on the management of innovation and organizational change in the School of Communication, the McCormick School of Engineering, and the Kellogg School of Management. His research focuses on how companies can design organizational structures and employ advanced information technologies to more effectively create and share knowledge. He is the author of *Car Crashes Without Cars: Lessons about Simulation Technology and Organizational Change from Automotive Design* (MIT Press, 2012).

Bonnie A. Nardi is Professor in the Department of Informatics in the Donald Bren School of Information and Computer Sciences at the University of California, Irvine. As an anthropologist, she has studied the uses of digital technologies in offices, schools, homes, libraries, hospitals, scientific laboratories, and virtual worlds. Her theoretical orientation is activity theory. She is the author of many scientific articles and books.

Her latest books are *My Life as a Night Elf Priest: An Anthropological Account of World of Warcraft* (University of Michigan Press, 2010) and *Ethnography and Virtual Worlds: A Handbook of Method* (co-author, University of Princeton Press, 2012).

Wanda J. Orlikowski is the Alfred P. Sloan Professor of Information Technologies and Organization Studies in the Sloan School of Management at the Massachusetts Institute of Technology. Her research examines the dynamic relationship between organizations and information technologies, with particular emphases on organizing structures, cultural norms, communication genres, and work practices. She is currently exploring the sociomaterial entailments of social media.

Brian T. Pentland is Professor in the Department of Accounting and Information Systems at Michigan State University. His publications have appeared in *Academy of Management Review, Accounting, Organizations and Society, Administrative Science Quarterly, JAIS, Management Science, MIS Quarterly, Organization Science,* and elsewhere. He received his Ph.D. in Management from the Massachusetts Institute of Technology in 1991 and SB in Mechanical Engineering from the Massachusetts Institute of Technology in 1981.

Neil Pollock is Reader in Information Systems at the University of Edinburgh where he teaches and researches on the Sociology of Information and Communication Technologies (ICTs). He has co-authored two books: *Putting the University Online* (with James Cornford) and *Software and Organizations* (with Robin Williams). He is currently working on a third provisionally entitled "The Social Study of the Information Technology Marketplace." His research has appeared in leading journals such as *Information Systems Research, Social Studies of Science, Information and Organization,* and *Science, Technology and Human Values.*

Benoit Raymond is Assistant Professor in the Department of Organizational Information Systems at Université Laval, Canada. He received his Ph.D. in Computer Information Systems from Georgia State University in 2010. His research interests are the organizational and social effects of IT with an emphasis on the materiality of IT artifacts. His dissertation examined how IT artifacts influence the design and performance of organizational routines and he is currently doing research in this area.

Daniel Robey is Emeritus Professor of Information Systems at Georgia State University. He is Editor-in-Chief of *Information and Organization* and is the author of numerous articles in leading journals in the fields of both information systems and organization science. His research includes empirical examinations of the effects of a wide range of technologies on organizations. He received lifetime achievement awards from the Association for Information Systems (2009) and the Academy of Management (2010).

Jochen Runde is Director of the MBA and Reader in Economics at Cambridge Judge Business School, and Professorial Fellow and Director of Studies in Management Studies at Girton College, Cambridge. He is an editor of the *Cambridge Journal of Economics* and has published on various topics in journals including *Academy of Management Review, Journal of Economic Behavior and Organization, Studies in History and Philosophy of Science,* and *the British Journal for the Philosophy of Science.*

Susan V. Scott is Senior Lecturer in the Information Systems and Innovation Group, Department of Management, at the London School of Economics and Political Science. Her research focuses on technology, work, and organization from a management studies perspective. She has published on the implementation of information systems for risk management; electronic trading; strategic organization of post-trade services; organizational reputation risk; enterprise resource planning and best practice. She is currently involved in research projects exploring the sociomateriality of rating and ranking mechanisms.

Harminder Singh is Senior Lecturer in the Faculty of Business and Law at the Auckland University of Technology, New Zealand. His primary area of interest is the tension between change and control in IT-enabled work environments, and how this is governed. He received his doctoral degree from Michigan State University, and for his dissertation, he examined unintended changes in IT portfolios ("drift") using Bourdieu's practice lens and the concept of sociomateriality.

Jennifer Whyte is Professor in Innovation and Design at the School of Construction Management and Engineering, University of Reading. She leads a team researching design in the digital economy. Her research interests are at the interface between engineering and management and include coordination across digital design interfaces; and management practices in the delivery of large complex projects. Her work is published in leading journals, including *Organization Studies, Long Range Planning*, and *Strategic Organization*.

Youngjin Yoo is Professor in Management Information Systems and Irwin L. Gross Research Fellow at the Fox School of Business and Management School of Management at Temple University. His research interests include evolution of digital artifacts, digital innovations, and design. His work has appeared at leading academic journals, including *Information Systems Research, MIS Quarterly, Organization Science, the Communications of the ACM*, and *the Academy of Management Journal.*

I

Setting the Stage

1

The Challenge of Materiality: Origins, Scope, and Prospects

Jannis Kallinikos, Paul M. Leonardi, and Bonnie A. Nardi

Return to the Material

Few people would dispute the claim that new technologies bring changes to the way people communicate, act, and organize their social relations. Technologies such as tablet computers certainly have transformative potential. Users of iPads recognize that their relationship to the office has changed now that they are perpetually connected, and users of social media have found that their private lives are now much more public than they once were. Today, although technologies such as telephones or microwaves are considered quite mundane, it does not take too much thought to envision how our daily use of these tools has changed the ways we maintain social connections, think about our diets, or even conceive of family time.

Ask the random person on the street whether new technologies bring about important social change and you are likely to hear a resounding, "yes." But ask academics who study technology and social practice and the answer is likely to be less definitive. Although most contemporary scholars do not deny the important role that new technologies play in the reconfiguration of social practice, few are ready to admit a direct causal relationship. Instead, they tend to say things like new technologies "enable," "occasion," "afford," "enact," "make possible," "co-construct," or "mutually constitute" the social contexts into which they are introduced. In short, today's scholars from fields as diverse as Communication, Information Systems, Management, Anthropology, and Informatics recognize technologies' transformative potential, but typically stop short of saying that technologies "cause," "shape," "create," or "determine" social change.

Since the dawn of the industrial revolution, scholars and other public commentators have made grand claims about how new technologies have changed or will change the social world. From epochal accounts of riding stirrups creating medieval feudal systems (White, 1962) and the written or printed word altering humanity's collective consciousness (Goody, 1977; Ong, 1982) to systems-level theorizing about the ways new manufacturing technologies would change the way firms organized (Woodward, 1958) and how information systems would inevitably shift a firm's expertise and decision-making processes (Likert, 1961), scholarly work considered technology as a major agent of social change.

From the 1980s onwards, scholars working with more fine-grained levels of analysis began to challenge the assumptions made in this prior research. By charting the actual ways in which people incorporated new technologies into their work, a new set of "social constructivist" studies demonstrated that any effects that new technologies had on the way people worked were mediated by a variety of social processes (for a detailed review of these studies, see Leonardi and Barley, 2008). Consequently, many scholars concluded that the important social changes that occurred in the wake of a new technology's implementation were as likely to be the result of people's choices about how to design and use the technology as attributable to the artifact itself.

As social constructivist thinking has gained in popularity, at least in academic circles, scholars tended to move toward the use of ambiguous terms such as those presented above when talking about technology-induced changes in a way that does not exclude the role of human agency and social choice in shaping technologies' effects. Use of such terminology has led many down a slippery ontological slope. At its extreme, the social constructivist position holds that technologies themselves hardly matter at all in discussions of social change (Grint and Woolgar, 1997). What matters, instead, are the ways in which those technologies are used in the context of work or other social settings. The types of empirical studies that have evolved from such thinking certainly do not trace a particular social change to a particular technology feature. Nor do they describe in any detail what features a technology has or how they are used. Instead, studies taking this rather extreme ontological position document ways in which a new technology becomes embedded in a "web of interpretation," a "practice," or a "cultural meaning system" and how perceptions "enact social practices" or "enable interactive shifts." In short, technologies themselves cease to be important in studies of social change. At best, they become an occasion for studying social relationships in particular contexts of social or institutional life.

Such thinking is problematic for theoretical reasons and empirically untenable in the face of the many studies that document the profound changes resulting from the use of new technologies. Our own experience belies extreme forms of constructivism. Even undergraduate students often point to previous technological eras they lived through, acutely mindful of the differences new technologies sustain.

To steer away from the poles of determinism and radical constructivism, a new trend has emerged in the last five years discussing how "every organizational practice is always bound with [a technology's] materiality" (Orlikowski, 2007: 1436) and arguing that "materiality matters for theories of technology and organizing because the material properties of artifacts are precisely those tangible resources that provide people with the ability to do old things in new ways and to do things they could not do before" (Leonardi and Barley, 2008: 161). In their efforts to reclaim the important role that technologies play in organizational life without resorting to antiquated claims of determinism, organizational researches have followed the broader "material turn" in the social sciences (for a review, see Pinch and Swedberg, 2008: 2) in arguing for a focus on the ways that technology's materiality becomes implicated in the process of organizing and social practice in general.

Although a return to the material sounds like a viable strategy for incorporating the role of materiality into constructivist understandings of the social world, there is, at present, at least one problem with this strategy. The problem is that the term "materiality," and other related terms that scholars have begun to use, such as "material," "material property," "material consequence," "materialize," "materialism," "sociomaterial," and "sociomateriality," are neither well defined nor consistently used (see for further discussion, Leonardi's Chapter 2, in this volume). Moreover, their relationship to concepts in regular use in the social sciences such as "technology," "form," "function," "artifact," and "socio-technical system" is not yet clear.

As Berger et al. (1972) argue, it is difficult to build meaningful theoretical research programs without concepts that are internally consistent and clearly defined in relation to other existing concepts. The goal of this book is not to reach clear definitions about the concept of materiality. Instead, we assembled the chapters included herein so that in conversation with each other, the chapters might begin to surface similarities and differences in the way leading international scholars from the fields of Anthropology, Communication Studies, Economics, Informatics, Information Systems, Philosophy, Science & Technology Studies, and Sociology are thinking about the role of materiality in the organizing process. Before previewing the book's contents, we discuss why "materiality" has been such an elusive term and we offer some observations on what it might mean to take materiality seriously when studying technologies and their effects.

Origins and Problems

The ideas currently conveyed by the concept of materiality may have lived an obscure life for several decades. The origins of the current use of the term (e.g., Hayles, 2002; Barad, 2003; Orlikowski, 2007; Pinch, 2008) can perhaps be traced back to French structuralism and the significance this metatheory has attributed to the material means (phonemes and marks) through which the fugitive character of meaning is captured or molded, expressed, or extracted and conveyed. On this view, ideas are transient, borderless, and evasive. By contrast, the solidity of the material means by which ideas are expressed helps fix and stabilize the transience and evasiveness of ideas, making them the cognitive currency of human communities. One of the key tenets of structuralism that descends from cultural anthropology, linguistics, and philosophy of Claude Lévi-Strauss (1963, 1966) is the derivative status of cultural content. Rather than being at the origin of human action, the primary condition that provides the orientation human communities seem to need, meaningfulness emerges out of basic distinctions, first performed upon the molten character of reality through the materiality of signs (phonemes and marks) and further organized into semantic blocks and narratives by means of a web of primary oppositions or differences (e.g., good/evil, light/dark, earth/haven, mother/father, raw/cooked). Via post-structuralists such as Barthes, Baudrillard, Derrida, Foucault, and Lacan, among others, these ideas have drifted and reached in a massive scale the Anglo-Saxon world, originally in literary theory, linguistics, and semiotics (Bolter, 1991; Jameson, 1991; Hayles, 1999) and later in social theory at large. In social theory, in particular, they have further been disseminated and occasionally popularized by the perspective known as actor-network theory (ANT; Law, 1991; Law and Mol, 1995; Latour, 2005) while their origin has become increasingly obscure.

The strong appeal that the concept of materiality has been exercising over the last few years is indicative of the rich web of associations the concept carries. And yet, what exactly materiality connotes beyond its allure is far from clear. While occasionally recognizable, the original structuralist insights have been lost or weakened and the concept has been vested with all sorts of meanings and implications. At first sight, the current reference to materiality would seem to reecho the original belief in the plenitude and enduring status of material reality. The concept seems to resonate with the power which solidity grants to physical things and the inescapable situatedness of human agents that is among other things tied to the corporeality of the human body (Arendt, 1958; Bowker and Star, 1999). Whatever humans do would seem, in some way or another, to be contingent upon these inexorable conditions and the ways material things mediate and interfere with human affairs. Thus understood, materiality is indicative of both the embodied and embedded nature of human experience, the multiple entanglements of humans with material objects and artifacts, and the various supports these provide to human pursuits.

These appealing ideas are, however, not as unequivocal as they may seem at first glance. For, how is one to distinguish the material from the nonmaterial? Where exactly would the dividing line between material reality and whatever aspects of human living and doing that transcend that reality be drawn? Most of the material things and objects that sustain social life are cultural objects that have been invented and produced by humans to assist with their daily endeavors. Production resonates well with materiality but invention less so. To invent amounts, after all, to imagining a function or use that does not exist yet. Imagination transcends the given or immediate, and in doing so, it construes reality as different from what currently is. From this point of view, the process of inventing is akin to metaphor and thus closely associated with linguistic creativity and the faculty of imagination. Here is how Kenneth Burke captured the issue some years ago (Burke, 1978: 18–19):

> The most practical form of thought that one can think of, the invention of some usable device, has been described as analogical extension, as when one makes a new machine by conceiving of some old process such as the treadle, the shuttle, the wheel, the see-saw, etc. carried over into some sets of facts to which no one had previously felt that it belonged... "Carried over" is itself as etymologically strict translation as you could get of the Greek word "metaphor."

Invention is certainly a material or Promethean practice of "creative destruction" (Schumpeter, 1983/1934) that takes place amidst or against the background of other things, prior inventions, prevailing techniques, and new concerns or experiences. At the same time, it is a practice that is inseparable from imagination and the capacity to think of reality as different from what it is. In a fascinating and, at the same time, perplexing way, invention as social practice ties the plenitude of things to the disembodied status of imagination, and instrumentality and utility to make-believe (Bateson, 1972; March, 1976; Kallinikos, 1998, 2011). And while it might be possible to concur with Latour (1986) on the dependence of thought on material things and instruments, what he calls "thinking with eyes and hands," lumping the two processes together may often end up glossing over important differences that obscure rather than illuminate the intricate ways through which ideas and things turn upon one another (Vygotsky, 1978; Leonardi and Barley, 2008). Imagination is certainly associated with visual supports of various kinds, plotting and measurement systems, and other techniques of "materialization" but these do not make the faculty of imagination and whatever it sustains just a material practice.

Analogous objections could be raised with respect to the evocative yet ambiguous label "material culture." The term carries several undertones (Jameson, 1991; Miller, 2005) but it is often deployed to refer to the omnipresence of material things, that proliferating universe of consumer objects that accompanies contemporary life and is often associated with the embeddedness of capitalism and the hedonistic habits of consumer society. Implicit in the use of the term material culture is the idea of another culture, less or nonmaterial. But which culture? Capitalism is after all the most abstract and artificial (in the sense man-made) social system ever known, and its material culture is accordingly closely tied to its abstractions. Consumption itself is in many respects a semiological practice (Barthes, 1977; Baudrillard, 1988), the very mediation and using-up of impressions by means of material objects. Consumption grows at the confluence of use and status symbolization, physical needs and psychological motivations, reality, and deception, a practice which Thorstein Veblen (1899/2007) identified at the dawn of the consumer society and summarized with his much quoted phrase "conspicuous consumption."

Technology and Materiality: Double Binds

Some of these intellectual puzzles recur in the use of the concept of technology which constitutes a central object of inquiry in this volume. Technology and materiality may seem intimately related to one another. One of the primary associations which tools, technical systems, and artifacts invoke is their solidity, a persisting objectness granted to them by their material or corporeal status. Yet again, upon closer scrutiny, technological objects are not distinguished on the basis of their sheer materiality but rather thanks to the functions they summon or embody, for example, a knife is used to cut, a vessel to contain, a camera to take photographs. Of course, functions are human fabrications grafted upon whatever matter fits or is able to support them, straightforwardly or through elaboration.

Function is so deeply immersed in our ordinary awareness of objects that it is only with an effort, as Searle (1995) notes, that we can transcend the perception of technological objects as functional entities and see, say, a motor vehicle, a bath tub, a screwdriver as a material bundle, a collection, as it were, of molecules. These feelings of wonderment are reinforced by reflection on the nature of software, surely one of the most interesting technological developments in the history of technology. For, ultimately, software is no other thing than a series of logical instructions, possible to execute by a machine. No matter which material supports software needs, reducing its distinctive technological profile to whatever connections it maintains with hardware and human agents amounts to bypassing what is most interesting about the software; that is, its logical and largely immaterial nature. Failing to appreciate the ontological constitution of software and the intimate connections it maintains with logics and mathematics may end up missing the distinctive nature of its far-reaching effects on humans and their pursuits (Borgmann, 1999; Gleick, 2011).

Against the background of these intellectual puzzles and the difficulties they generate, many scholars have, wittingly or unwittingly, turned their attention away from the cardinal issue of how the functional identity of technological objects is or could be tied to social outcomes. Technology and materiality have been claimed to matter only to the degree that they produce effects of one or another kind by providing an occasion or context into which human practice unfolds (several chapters in this volume). What matters here is not the intrinsic functional identity of technological objects (if there is such a thing) but whatever consequences these objects help bring into being (i.e., materialize), with the intentional or unintentional complicity of humans, purposeful engagement, or chance. Conceived in these terms, materiality is not a causal force of social outcomes but a fundamental human condition tied to the material anchorage of human agents and what is often perceived as the inexorable reality that everything that happens cannot but happen in a "here and now" (the leitmotif of situatedness). In some way or another, materiality is in the very end equated with materialization that seems to be considered the passage through which everything that exists ultimately acquires material status.

However, puzzles expelled and not resolved are bound to recur, in some way or another. What effects exactly is such a view of materiality concerned with? How is technology tied to these effects, given the fact that ideas and immaterial conditions may also have consequences, and not infrequently serious ones? Is materiality just another name for realization? Is all that exists in a "here and now" and that comes to pass through the active or latent complicity of humans seen as materialization? Furthermore, are all effects and consequences equal? What technologies or conditions are tied to what effects? And what about unintended and propagating consequences, side effects or, as we claim below, temporally non-contingent yet connected events? How is one to grasp the distant or historically constituted forces that often transcend the confines of particular contexts and operate beyond a "here and now"? This is just a small sample of questions raised by the conception of materiality as materialization and the understanding of technologies solely in terms of the consequences they bring about.

These observations indicate that straightforward and conspicuous as materiality may originally seem, it turns out to be a rather evasive and difficult to pin down concept. Similar to the horizon, an understanding of materiality seems to recede as we approach it. In some way, the issues we have mentioned so far recall major divides in social thought and reprise the inherent tension that is, among other things, associated with whether the focus is on the interactive order (the situated character of human encounters) versus the structural order (the stable conditions that shape and reproduce social practices). Such epistemological and ontological differences recur through the ages and are unlikely to be resolved in any definite way, even though advances can be made in one or another frontier of social inquiry. In this regard, materiality and technology raise new challenges and provide novel avenues and conceptual tools for rethinking these recurrent and persistent questions.

It is interesting in this context to point out a certain drift that the current revival of materiality epitomizes away from the anti-subjectivism, characteristic of scholars of a post-structural bent and the critical attitude many of them have maintained vis-à-vis agency-centric explanations of social life, the making of art objects included (see, e.g., Barthes, 1977; Foucault, 1980, 1988; Hayles, 1999). For, as already indicated, key to the original understanding of materiality has been the idea that the solidity and enduring character of things and, crucially, the different relations they enter with one another (the idea of structure) counterbalance the fugitive character of ideas, and the fragility, limited perspective, and inconclusiveness of human agents, insights that are still discernible in Barad (2003). Rather than being simply constrained by structure, as the typical conventional interpretive understanding wants us to believe, human choice and agency are made originally possible through the very resources that objects and structures dispose.

Activity theory, for example, assumes that human activity is mediated by tools, that is, the relation of a human subject to reality alters through the use of tools (Vygotsky, 1978). Vygotsky proposed a notion of *psychological* and *material* tools (1978). At first glance, this scheme would appear to denote the physicality of the tools themselves, but the notion hinges instead on the "reality" they are changing. Psychological tools enable complex cognitive functions and include, for example, systems for counting, mnemonic devices, algebraic symbol systems, works of art, literary genres, schemas, diagrams, maps, technical drawings, and language (see Cole and Wertsch, 1996). Material tools act on the world and comprise what we normally think of as "tools": knives, hammers, and so on. Critical to this formulation is the capacity of tools to make human choice and agency possible in the first place—not merely to "constrain." For example, Fayard (Chapter 9, this volume) observes that when physical space is mentioned in organizational studies, it is usually as a constraint affecting interaction and communication (e.g., measurable physical distances that separate people) rather than a generative resource (p. 179). Activity theory holds that whether such resources and tools promote more sophisticated cognition or transform the world around us, they are not a kind of external surround like a shell, but the very stuff of who we are, continually producing skills, identities, and culture.

These observations suggest that the mutual interpenetration of the social and the technical—the entanglement of humans with artifacts—can be approached by different routes that put the emphasis either on the interpretive nature of human agency or, alternatively, on those factors that render that agency possible in the first place. In the former case, human agency tends to remain the original explanatory medium for the constitution of social reality, the primary force, as it were, through which material conditions are appropriated to shape the social world. In the latter case, materiality, technology, and agency are not any longer exogenous to one another. Rather, they mingle in an indissoluble bundle of iterative or recursive relations that removes human agency from the center stage, making it just one more force among the dance of forces that express and govern social life.

One way of addressing some of these rather intricate issues is by inserting their understanding within a historical or evolutionary framework. There is, of course, a strong difference in focus between history (culture) and evolution (biology) but common to both is the temporal perspective they invoke and the concomitant assumption that human artifice and material conditions exhibit different constellations across time. Against these considerations, it would seem rather unlikely that materiality and technology have mattered and continue to matter equally across historical periods and contexts (Arthur, 2009; Borgmann, Chapter 17, this volume). Indeed, from such an evolutionary or historical perspective, technology could be seen as an important means through which the human relationship to the world and its materiality is reframed and reordered. Medieval techniques, for instance, epitomize a different relation to matter and the world than contemporary science-driven technologies (Mumford, 1952; Heidegger, 1977). Technological developments construct novel artifacts; discover or invent new materials (e.g., plastics); promote new skills, practices, and habits; and establish processes that alter the parameters by which matter is appropriated, used, and consumed by humans. They also generate effects or consequences of particular kinds (e.g., information search, digital records) that, despite their stochastic, non-determinist nature, are possible to trace back and associate with the cumulative and often interdependent nature of technological processes and systems and their technical lineage (here, search engines and software).

Book Structure

It appears that, while scholars have revived questions of materiality from decades past, they have, in doing so, sometimes construed materiality in ways foreign to, and well beyond, older conceptions of materiality as material culture, machines, physical artifacts, or things. In Part II, "Theorizing Materiality," which follows this introductory chapter, three conceptual chapters retrace, each in their own distinct ways, the origins, scope, and prospects confronting the concept of materiality. In his chapter, entitled, "Materiality, Sociomateriality, and Socio-Technical Systems: What Do These Terms Mean? How Are They Different? Do We Need Them?", Leonardi reviews the history of these various terms and explores ways in which they overlap and depart in meaning from one another in scholars' writings. He concludes with a set of tentative arguments for how scholars might use these terms parsimoniously. For example, materiality might be best viewed as a concept that refers to properties of a technology that transcend space and time, while sociomateriality may be best used to refer to the collective spaces in which people come into contact with the materiality of an artifact and produce various "functions" (to use Kallinikos's language elaborated in this volume). He argues, then, that the concept of a sociomaterial practice is akin to what socio-technical systems theorists referred to as the "technical subsystem" of an organization, or the way that people's tasks shape and are shaped by their use of machines. This technical subsystem is recursively organized alongside the social subsystem of an organization, which is characterized by an abstract set of roles, communication patterns, and so on. The distinctions that Leonardi draws between these terms, and the conceptual apparatus that underlie them, help us to perhaps understand why authors of this book can speak to very distinct processes and outcomes while using such similar sounding words.

The second chapter in Part II, by Faulkner and Runde, takes issue with the perspective of sociomateriality, pioneered by Orlikowski a few years ago, by critically reexamining three fundamental assumptions on which such a perspective is predicated, namely the assumptions of relationality, interpenetration, and agential cuts. Relationality and interpenetration assert the constitutive bonds of entities or processes, agents and things underlying the making of our world, while the notion of agential cuts primarily refers to social agents and the ways they draw boundaries and institute relations, in a world in which humans and things are inherently entangled with one another. Faulkner and Runde question not the truthfulness of these claims, which they find reasonable, but rather their universality and the rather generic or unqualified ways by which they are made. They first note that these are arguments about ontology (not simply methodology or epistemology), that is, arguments making claims about the nature or being of the world. Then, they go on delivering with a nearly uncanny precision a number of arguments that show the basic assumptions of relationality and interpenetration to be specific instances of a much wider range of phenomena in which ontological independence is at least as tenable a position and as widespread a state as that implied by relationality and interpenetration. They similarly question the notion of agential cuts by pointing out the agent-independent and temporarily prior nature of what, following John Searle, they call brute facts (moon, Mount Everest, Tyrannosaurus Rex). An essential and very subtle claim they make in this respect is that categories and classifications drawn out by social agents may have far-reaching social significance but they do not create new entities or things. They additionally criticize the notion of agential cuts for failing to recognize the stable, non-dramatic character of human dealings underlying the ordinary fabric of social life in which the boundaries of entities and established relations are seldom questioned.

The third chapter in Part II, by Kallinikos, "Form, Function, and Matter: Crossing the Border of Materiality," portrays technological objects as variable assemblies of three fundamental constituents: *function*, *form*, and *matter*. Every technological object or set of objects has been designed or thus evolved as to accomplish a purpose or function (e.g., cut, transport, or move, rest or protect, process, contain) that is variously supported by the materials the object is made of and the ways these materials have been given specific form to assist and/or embellish that function. Kallinikos goes on to suggest that technological evolution betrays a trajectory that attests to the increasingly prevailing role *function* has assumed in the design of technological systems and artifacts at the expense of the other two constituents, matter and form. The pivotal role of function in the design and making of technological objects has essentially been assisted by advances in science and the discovery of new matter qualities or attributes that could be technologically extracted from nature or other primary matter, and enriched or manipulated in variety of ways. Kallinikos reads these developments as implying the progressive dematerialization of technological processes and objects and the increasing dissociation or unbundling of functional attributes from their underlying material constitution. Once perhaps a derivative of material attributes (wood, clay, or stone), an abstracted notion of function has progressively provided the primary matrix out of which technological objects and processes emerge. This shift is strongly reflected in contemporary digital technology which is essentially more *logic* (pre-programmed logical instructions) than *techne* (the art of making), and the consequent and expanding infiltration of computer hardware and other physical objects by software.

Part III, "Materiality as Performativity," explores the relationship between the nature of artifacts and techniques, and the ways these are implicated in the making of aspects of the world we tend to confront as given. Pollock's chapter, "Ranking Devices: The Socio-Materiality of Ratings," deals with rankings and the effects these may have on the domains of life they subject to rating. Rankings seldom are passive recordings of a world out there. More often than not, they are constitutive of the world. They do so in the double sense of establishing the dimensions on the basis of which the domains they rank are conceived and perceived, and by influencing the behavior of actors within that domain, as these last seek to accommodate whatever opportunities or threats rankings offer. The idea is well known but Pollock offers a further exploration into how the effects rankings have are associated with the makeup or materiality of the devices by which rankings are achieved. He pays particular attention to visual devices and the way they structure and "economize" perception and attention. In other words, visual devices and graphs have affordances that are intimately tied to the figurative or spatial modes of perception that considerably shape how the domains to which they apply are conceived, as opposed to perceived, and mediated. These ideas are empirically explored in connection with the so-called "magic quadrant," a particular ranking of IT (information technology) vendors, and the social, cognitive, and material process through which the magic quadrant has been constructed over time. A conclusion Pollock reaches is the very construction of markets, here the IT vendors market, through the affordances specific to the devices by which these markets are conceived as they are ranked and the rankings are mediated.

In their chapter, "Great Expectations: The Materiality of Commensurability in Social Media," Scott and Orlikowski examine how contributors and users of the travel website TripAdvisor, along with the algorithms that the site uses to rank hotels and restaurants, together materialize particular value distinctions in the travel sector. Drawing on the concept of commensuration—the process of comparison according to a common metric—the authors examine how the rankings given by TripAdvisor are produced through the distributed reviews and ratings of thousands of user-generated postings, and transformed through filtering and weighting algorithms. Through this process of sociomaterial commensuration, the authors argue that the "truth" about hotel and lodging options given by TripAdvisor is performed through the website's distributed and dynamic materiality, and they consider its influences for the expectations and encounters experienced by guests and hotel owners in the United Kingdom. Although commensurability may be changing the way that travelers and hoteliers experience the travel sector, Scott and Orlikowski suggest that "the processes of materialization that are core to TripAdvisor's rating and ranking mechanisms are neither open nor subject to public scrutiny. In the past, hoteliers received private feedback from their guests on feedback forms, but through TripAdvisor's proprietary algorithms, hoteliers are now digitally exposed" (p. 130). Consequently, when one looks at an entry about a hotel or a restaurant on the social media site, they are experiencing the process of commensuration as they are materialized through TripAdvisor's rating and ranking mechanisms. The consequence of such materialization, of course, is that what one tends to and how one evaluates options for food and lodging are distinct from what they would be if AAA (American Automobile Association) or Michelin provided ratings based on standard, static criteria.

The chapter by Yoo, "Digital Materiality and the Emergence of an Evolutionary Science of the Artificial," finds its point of departure in the claim of an evolving world of artifacts that emerges out of the generative makeup of digital technologies. Such a fundamental condition is the offspring of the distinctive constitution or materiality of digital technologies. While the argument of the distinctive nature of digital technology unfolds along several paths, key to it is the contrast digital architectures offer to traditional product and industry arrangements. Digital architectures are, or can become, modular and, crucially, multilayered, thus establishing a spectrum of possibilities along which the products and services which they produce can be unbounded from specific uses, unbundled, and recombined in variety of ways that cross product- and industry-specific boundaries. Though exhibiting significant variability, the evolution of digital artifacts and services which modular and miltilayered architectures enable could be explained, Yoo suggests, by exploring the lineage of digital artifacts and the similarities they share with one another at a deeper genetic, as it were, level. He accordingly draws on modern biology trying to establish an analytical framework whereby variety at the phenotypic level could be accounted for by a deeper genetic layer whereby a relatively limited number of routines and software components are thus made as to be able to enter into a large number of combinations (observable variety). His ideas of a genetics-based evolution of digital artifacts are illustrated by reference to those digital services known as APIs and Mashups. At the heart of Yoo's explanation is the idea of digital (im)materiality and the spectrum of possibilities that are established as digital artifacts or services get divested from fixed or specific and ultimately physically embedded attributes, thus turning increasingly product or service agnostic. In this respect, Yoo offers the beginnings of a theory that could be drawn upon to explain the lineage of what he calls "artificial reality," that is, the reality of manmade objects and artifacts.

Part IV, "Materiality as Assemblage," explores the shifting imbrications of technologies and social practices. Ekbia and Nardi's chapter, "Inverse Instrumentality: How Technologies Objectify Patients and Players," considers how bodies are materialized or dematerialized as they participate in large computational systems. They analyze two such systems—personal health records and multiplayer video games—drawing attention to the instrumental insertion of physical bodies as cognitive labor in the systems. Unlike software code which can be reproduced at almost no cost as the investment of physical matter is minimal (Faulkner and Runde, 2011), recruiting actual human persons to perform tasks in computational systems requires summoning a body that "responds, becomes skillful and increases its forces" as Foucault (1977: 136) says. For example, in multiplayer games, players write software modifications to extend the gaming software, and in personal health records systems, patients interact with a variety of software programs that provide vital information to medical staff. The trained bodies functioning in these systems of "inverse instrumentality" are distinctly material, deploying their useful brains and dexterous fingers. However, a process of fragmentation may also occur that problematizes simple notions of materialization. In the case of personal health records, the cognitive labor of the patient is cleaved away from his fragile physical body, which is in need of care. To the system, he appears primarily as a set of cognitive functions. His need for a (literal) helping hand or warm touch is muted. With gaming, on the other hand, the player blossoms as she becomes transcendently empowered as a player, gaining in new forms of identity, becoming visible in a community of players as a master of the craft, even someone who can reach out to enable others to find their own new satisfactions and rewards. Over time, in interactions with systems of inverse instrumentality, bodies "improve" as Foucault (ibid.) observes, strengthening the systems. The power of systems as regulative regimes increases (see Kallinikos, 2011), reinforcing either fragmenting or totalizing tendencies.

Fayard argues, in her chapter entitled, "Space Matters, But How? Physical Space, Virtual Space and Place," that physical things—"rooms, walls, buildings"—are important to materiality. Fayard observes that they become "entangled" with "social practices and narratives" to produce *space*. She sees space as not merely a set of restrictive distances and enclosures, but a fundamental resource—locations within which we productively "engage with objects and artifacts," and with one another. Fayard addresses the challenging domain of *virtual space* which is less straightforwardly composed of physical components. She maintains a conceptual distinction between physical and virtual space, but emphasizes the possibility for the emergence of "hybrid spaces" in which interaction and activity flow across physical and virtual spaces. With an artist, she designed an art installation with precisely this flow in mind, ingeniously creating a physical world space in a historic building in Brooklyn, New York, in which words projected from blog posts onto fabric panels formed a "maze" through which participants could wander. The maze nicely conceptualizes more ordinary hybrid spaces such as Internet cafes: by making the virtual "stark," as Fayard says, the activities that generate the creation of space are rendered visible, and the space is realized as an "evocative object" (p. 188) in which we may reflect on time and space. Like Kallinikos (this volume), Fayard points to the form and function of virtual spaces as critical; their status as artifacts depends not on their substance as physical matter but on the precise functions they permit which are fundamentally virtual.

Whyte and Harty's chapter, "Socio-material Practices of Design Coordination Across a Large Construction Project," begins by examining Star and Griesemer's (1989) influential work on boundary objects to understand how interactions between people and the objects they create and use allow them to achieve particular organizational goals. In the authors' case—a study of the design and construction of a new airport terminal, Heathrow Terminal 5, in London—members from multiple occupations collaborate to produce the final building. To enable such collaboration, the teams vowed to coordinate their work digitally via an integrated software system that linked data repositories to computer-aided design software, known as a "single model environment." The authors show how the materiality of this single model environment proved too rigid to enable the desired coordination and how the teams employed various other media that were more malleable and plastic to achieve coordination. The plasticity of these objects had, as the authors argue, important implications for coordination: "Objects that are productive in terms of coordination of ongoing work have a dual epistemic and boundary-spanning role, allowing participants in the project to maintain conventions and legibility of objects, across different locations and times, while allowing flexibility and partiality to enable the development of new ideas and innovation" (p. 201). This analysis refocuses initial concern in Star and Greismer's original work to the materiality of objects themselves and the way that different physical and digital artifacts can have consequences for cross-disciplinary coordination.

Part V, "Materiality as Affordance," takes a different route by conceptualizing the role of technology through the lens of affordance. In their chapter, "Theorizing Information Technology as a Material Artifact in Information Systems Research," Robey, Raymond, and Anderson focus specifically on IT artifacts and question how and under what conditions it makes sense to talk about these artifacts as having "material characteristics." The authors trace the neglect of study of IT artifacts to historical decisions to define and measure technologies, broadly conceived, as abstract contingency variables predicting organizational form. They propose two major challenges to restoring theoretical attention to material artifacts. First, they suggest that novel concepts such as "affordances" must be carefully defined and adapted to a sociomaterial context. Second, they suggest that theoretical relationships among these concepts need to be articulated. They illustrate this second challenge by showing two ways that sociomaterial concepts may be positioned theoretically. The first is to extend existing theories to account for sociomateriality. Examples of adaptive structuration theory to include technology affordances, and organizational routines theory to include material agency, are presented to illustrate this strategy. A second way to position materiality is to extend theories that address substantive issues but neglect IT. The extension of work–life boundary management theory to include technology affordances are presented to illustrate this second strategy.

Like Robey et al., Faraj and Azad are concerned with theorizing the materiality of information systems and, like Robey et al., they believe that adopting an affordance lens is a good way to shed new light on the role technologies play in organizational dynamics. In their chapter, "The Materiality of Technology: An Affordance Perspective," they suggest that IT researchers have largely accepted naming and categorization conventions developed by commercial vendors and the trade press that lump hardware and software into particular product classes and define lists of specific functionalities that warrant such classification. Although such categorization practices might not seem, at first glance, to be problematic, Faraj and Azad argue that they place attention on what a technology is, but not what it allows people to do. To overcome this problem, the authors suggest that a relational theory of affordances—where affordances exist between people and the materiality of IT artifacts—may help to explain when and how materiality becomes intertwined with human agency. As the authors note, "we see potential in the notion of affordances as a relational construct linking the capabilities afforded by technology artifacts to the actors' purposes." To refine the concept of affordance, Faraj and Azad speculate about a number of scope conditions that may help researchers speak more specifically about the role of materiality in organizational life.

In their chapter, "Pencils, Legos, and Guns: A Study of Artifacts Used in Architecture," Groleau and Demers are concerned with confronting the complexity of configurations of artifacts and resources in practice. Many studies highlight a particular technology of interest, or particular practices, but Groleau and Demers observe that sets of technologies and sets of practices mutually inform one another. If we abstract them out, disentangle them, we lose their agentic power. Groleau and Demers use activity theory as a theoretical lens. They note that the specific material enablements of artifacts are rarely considered in activity analyses but that it is critical to focus on them to understand how people accomplish work. They note that the functional attributes of artifacts generate, and are supported by, the "sensitive abilities" of practitioners. Thus, artifacts change us even as we use them in object-oriented activity. At the same time, competing artifacts that enable common practice have distinctive capabilities. People choose them according to desires to reinforce or challenge tradition, as illustrated in the authors' comparative analysis of three different kinds of architectural firms. Artifacts are thus instruments of change or stasis.

Part VI, "Materiality as Consequence," shifts the focus of attention from the nature of artifacts to the consequences which their social involvement engenders. On this view, what is material is what matters or is made to matter in a particular context. Perhaps the most dramatic alteration of conceiving of materiality as "that which matters" is found in Pentland and Singh's chapter, "Materiality: What are the Consequences?," which advocates for this position perhaps most strongly of all. The authors make the simple and provocative claim that "Something is material insofar as it has consequences in a particular context" (p. 292). Drawing on the philosophical perspective of pragmatism, they proceed by analogy, explaining how financial auditors and IT auditors use a variety of heuristics to determine the threshold at which a transaction or a system has consequence for a particular community—in short, that it is material. In this formulation it is the consequence of the transaction, not the transaction itself, that can be said to matter. For this reason, what is seen as material to one party may be immaterial to another. As Pentland and Singh suggest in their discussion of a fictional real estate transaction, "In fact, the *same* transaction might be material for the buyer and immaterial for the seller" (p. 291). The conceptual work done to establish materiality as a valued consequence allows the authors to "reverse figure and ground" and argue that artifacts like information systems do not have materiality, but the actions that those technologies enable do. As the authors conclude, "Materiality is not about artifacts, people, ideas, or any *thing*. Or rather, it's about all of them, but they only become *material* when they influence a particular course of actions or events that we value" (p. 294).

This focus on consequence is echoed in the chapter, "Why Matter Always Matters in (Organizational) Communication," by Cooren, Fairhurst, and Huët who invoke notions of preoccupations/concerns/worries/reasons as central to their theorizing of materiality. They observe that "... we thus need to realize that materiality relates to what is *relevant* or *pertinent* to a given situation, i.e., 'the relation [of something] to the matter at hand' (Webster's Dictionary)" (p. 301). This treatment of materiality is similar to Pentland and Singh's pragmatic approach and Leonardi's observations (2010) of how the term is used in field of law (e.g., "material witnesses" or "material facts" are those pertaining with a high degree of relevance to a case—those facts or witnesses that "matter"). Having made a bold move to broaden materiality so significantly, Cooren et al. argue that discursivity should not be left out of the equation: words matter. Therefore, analysis of *interaction* should be central to analysis of materiality. The authors observe: "[We] should focus on ... the multiple ways by which various forms of reality (more or less material) come to do things and even express themselves in a given interaction" (p. 296). In this framing, the causal relationships assumed in older discussions of materiality (that now perhaps seem vulgarly deterministic) which tied human action directly to the specific forms and functions of artifacts, yield to more open, flexible notions of "doing things" and a variety of human and non-human agents "expressing themselves" in action (p. 296).

Burrell also speaks up for discursivity in her chapter, "The Materiality of Rumor." She examines the agency of rumor—a non-material thing—underscoring its capacity to generate powerful *material effects*. As in Cooren et al.'s discussion, contingent consequence is central. This maneuver radically widens conceptions of materiality beyond conventional associations with artifacts, tying them to broad notions of agency at "the conceptual edges where matter or substance is not so evidently massed as an apparent 'object'" (p. 317). Burrell's argument continues, conveying a dual sense of materiality as, in addition to material effects, physical manifestation is also apposite: "The materiality of rumor specifically is linked to the body and the production of speech through the vocalizing organs, and the functioning of human memory. These are critical in constituting rumor's material aspects" (p. 315). Rumor is thus material in two ways: (*a*) it produces material effects (in Burrell's ethnographic study, a false rumor about an impending earthquake in Ghana generates a good deal of drama as people flee into the streets); and (*b*) rumor is anchored in the physical world of the brain and body. Burrell extends the physical aspect of materiality to the technological substrate of mobile phone service in Ghana that underpinned the rapid dispersal of the rumor.

The final section of this book provides a thoughtful epilogue on the idea of materiality by examining how matter has come to matter more and less at various times throughout human existence. Borgmann, a distinguished philosopher, contributes to this volume his meditation on matter and the ways our beliefs on what matter is and how the world is structured have changed from ancient times to contemporary culture. The ultimate aim of his chapter entitled "Matter Matters: Materiality in Philosophy, Physics, and Technology" is to discern or engrave a path along which the worldview of modern physics and the technological and cultural practices to which it is associated can be brought to bear upon the project of good life. Borgmann finds the current concern with materiality indicative of our troubling perceptions and feelings vis-à-vis matter as immaterial and ephemeral artifacts of all sorts increasingly populate our world, a concern that, no matter how indirectly, is reinforced by the compelling worldview of relativity and quantum theory. Both strands of contemporary physics have done much to dissolve the belief in an ultimate and absolute world, a definitive anchor into which reality can be hooked. However, the way the worldview of physics has been refracted in the perception of lay people and everyday life patterns is less clear. Borgmann discerns parallels between the two. In his own elegant prose: "As Aristotle and Plato had indicated, when we push the question of what things are made of to its deepest level, we arrive at one of two answers—structureless matter or immaterial structure. The solid and substantial world gets dematerialized either way. The cultural counterpart of structureless matter is realized in the cultural space of ever open possibilities" (p. 342). It is within this space of ever-open possibilities that Borgmann retraces the significance of matter. To quote him again: "What centers your life and lends it its identity is (significant) events. An event constitutes and occupies its own place and time. But an event could not matter if it were not material through and through. What matters needs to have depth. It must be grounded without rupture and traceable without loss of meaning. Life is lived out in the interval between such events. That interval is the spine of your identity, and like the space-time interval between events in special relativity, it remains no matter your changing frames of reference" (p. 345).

References

Arendt, H. (1958). *The human condition*. Chicago: Chicago University Press.

Arthur, B. W. (2009). *The nature of technology*. London: Allen Lane.

Barad, K. (2003). Posthumanist performativity. *Signs*, 28(3), 801–31.

Barthes, R. (1977). *Image, music, text*. New York: Hill & Wing.

Bateson, G. (1972). *Steps to an ecology of mind*. New York: Ballantine.

Baudrillard, J. (1988). *Selected writings*. Stanford: Stanford University Press.

Berger, J., Zelditch, M., and Anderson, B. (1972). *Sociological theories in progress*. New York: Houghton Mifflin.

Bolter, J. D. (1991). *The writing space*. Hillsdale: LEA.

Borgmann, A. (1999). *Holding on to reality*. Chicago: The University of Chicago Press.

Bowker, G. and Star, S. L. (1999). *Sorting things out*. Cambridge, MA: The MIT Press.

Burke, K. (1978). Rhetorics, poetics and philosophy. In D. Burks (ed.), *Rhetoric, philosophy and literature*. West Lafayette: Purdue University Press.

Cole, M. and Wertsch, J. (1996). Beyond the individual-social antinomy in discussions of Piaget and Vygotsky. *Human Development*, 39, 250–6.

Faulkner, P. and Runde, J. (2011). The social, the material, and the ontology of non-material objects. Paper presented at the European Group for Organizational Studies (EGOS) Colloquium, Gothenburg.

Foucault, M. (1977). *Discipline and punish*: Trans. Richard Howard. New York: Vintage Books.

—— (1980). *Power/knowledge*. (ed.) Colin Gordon. New York: Pantheon.

—— (1988). Technologies of the self. In L. Martin, H. Gutman, and P. Hutton (eds.), *Technologies of the self*. London: Tavistock.

Gleick, J. (2011). *The information*. London: Fourth Estate.

Goody, J. (1977). *The domestication of the savage mind*. Cambridge: Cambridge University Press.

Grint, K. and Woolgar, S. (1997). *The machine at work*. Oxford: Polity Press.

Hayles, K. (1999). *How we became posthuman*. Chicago: The University of Chicago Press.

—— (2002). *Writing machines*. Cambridge, MA: The MIT Press.

Heidegger, M. (1977). *The question concerning technology and other essays*. New York: Harper.

Jameson, F. (1991). *Postmodernism or the cultural logic of late capitalism*. London: Verso.

Kallinikos, J. (1998). Utilities, toys and make-believe. In R. Chia (ed.), *In the realm of organization*. London: Routledge.

—— (2011). *Governing through technology*. Houndmills: Palgrave.

Latour, B. (1986). Visualization and cognition. In *Knowledge and society* (Vol. 6), pp. 1–40. San Francisco: JAI Press.

——(2005). *Reassembling the social.* Oxford: Oxford University Press.

Law, J. (ed.) (1991). *A sociology of monsters.* London: Routledge.

——Mol, A. (1995). Notes on materiality and sociality. *The Sociological Review*, 43(2), 274–94.

Leonardi, P. M., and Barley, S. R. (2008). Materiality and change. *Information and Organization*, 18, 159–76.

Levi-Strauss, C. (1963). *Structural anthropology.* New York: Basic Books.

——(1966). *The savage mind.* London: Weidenfeld and Nicolson.

Likert, R. (1961). *New patterns of management.* New York: McGraw-Hill.

March, J. G. (1976). The technology of foolishness. In J. G. March and J. P. Olsen (eds.), *Ambiguity and choice in organizations.* Oslo: Universitetsfoerlaget.

Miller, D. (ed.) (2005). *Materiality.* Durham: Duke University Press.

Mumford, L. (1952). *Arts and technics.* New York: Columbia University Press.

Ong, W. J. (1982). *Orality & literacy.* London: Routledge.

Orlikowski, W. J. (2007). Sociomaterial practices. *Organization Studies*, 28(9), 1435–48.

Pinch, T. (2008). Technology and institutions. *Theory & Society*, 37, 461–83.

——Swedberg, R. (2008). Introduction. In T. Pinch and R. Swedberg (eds.), *Living in a material world* (pp. 1–26). Cambridge, MA: MIT Press.

Searle, J. (1995). *The construction of social reality.* London: Penguin.

Schumpeter, J. A. (1983). *The theory of economic development.* New Brunswick, NJ: Transaction Books. Originally published in 1934.

Star, S. L., and Griesemer, J. R. (1989). Institutional ecology, "translations" and boundary objects. *Social Studies of Science*, 19, 387–420.

Ulrich, K. (1995). The role of product architecture in manufacturing firm. *Research Policy*, 24, 419–40.

Veblen, T. (2007). *The theory of the leisure class.* Oxford: Oxford University Press. Originally published in 1899.

Vygotsky, L. (1978). *Mind in society.* Cambridge, MA: Harvard University Press.

White, L. (1962). *Medieval technology and social change.* Oxford: Oxford University Press.

Woodward, J. (1958). *Management and technology.* London: HSM.

II

Theorizing Materiality

2

Materiality, Sociomateriality, and Socio-Technical Systems: What Do These Terms Mean? How Are They Different? Do We Need Them?

Paul M. Leonardi

Many articles about information technology use in organizations published during the past five or so years employ one of the following three terms: "Materiality," "Sociomateriality," or "Socio-Technical Systems." Some critics claim that the use of these terms represents the diffusion of "academic jargon monoxide" and scholars should stick to simply talking about "technology"—a word that is understandable by "normal human beings" (Sutton, 2010).[1] Others argue that a basic term like "technology" is too simplistic because its use creates the illusion that there is some object, device, or artifact out there doing things and it ignores the empirical reality that those objects, devices, and artifacts only come to have meaning and effects when they are enrolled in social practice (Suchman, 2007). Others suggest that using a simple term like "technology" focuses too much attention on particular pieces of hardware or software and, consequently, directs researchers' attention toward the period of adoption as a "special case" instead of recognizing that technologies permeate all aspects of organizational life (Orlikowski, 2007: 1436). And still others argue that studies of technology and organizing have veered too far in the direction of rampant social constructivism and that a way back to a middle ground between the poles of voluntarism and technological determinism is to

[1] This blog post, by Robert Sutton, is characteristic of his iconoclast tone. Sutton is certainly an advocate of and important contributor to studies of technology use in organization. His point about the development of new terms is well advised and was one major impetus in the writing of this chapter.

recognize that technologies have certain material and institutional orders that transcend the particularities of the contexts in which they are used (Kallinikos, 2004). Despite the differences in their arguments, it seems that everyone has a point.

The motivation for this chapter is quite personal. Over the past five years I have attended at least a dozen conferences, workshops, and colloquia that used the terms "materiality," "sociomateriality," or "socio-technical systems" in their titles. Each time some participant at one of these events (often it was me!) asked, "What do we mean when we talk about materiality?" or "How is sociomateriality different from socio-technical systems?" The questioner was critiqued for aiming to exclude some, privilege others, or perhaps worst of all, close off productive debate. These concerns are certainly warranted. But without some definitional clarity, the terms remain jargon—criticized even by scholars who sympathize with this line of inquiry—instead of serving as useful tools for understanding and explaining the symbiotic processes of technological and organizational change. Certainly to outsiders, these terms all look quite similar to one another and appear little more than fancy synonyms for the quotidian word, "technology." This chapter makes a modest attempt at definition by comparison. That is, I explore the history that led to the use of each of these terms in organization studies and I make some tentative arguments about how these terms are similar and different to one another and, ultimately, how we might think about their relationship to one another. I focus specifically on non-physical information technology artifacts in this chapter, but I suspect that many of the arguments will also hold for other physical technological artifacts like hammers and bicycles as well as other non-physical technologies not (information technology artifacts) like language. By no means is this chapter aimed at stamping out debate about what these terms mean. Instead, the goal of this chapter is to begin a movement in the direction of clarity so that scholars can use these terms productively to theorize the complexity of collective endeavors, generally, and organizational dynamics specifically.

Materiality

Since Joan Woodward's (1958) provocative claims about the deterministic relationship between manufacturing processes and organizational structure she uncovered in the 1950s and Charles Perrow's (1967) hospital studies, conducted in the 1960s, out of which he concluded that technologies were independent variables affecting the dependent variable of work organization, researchers have sought to understand what role technologies play in the process of organizing. For many years, organizational scholars who were interested in technological change operationalized "technology" broadly as work processes conducted in conjunction with machines and conducted macro-level research (with organizations as the unit of analysis) into the effects that changing core technology had on an organizations' formal structure (e.g., degree of centralization, span of control, layers of hierarchy, etc.) (for discussion, see Robey, Raymond, and Anderson, this volume). As this generation of research began to offer a diminishing number of interesting insights and die off, a new spate of micro-level research (with the individual or the group as the unit of analysis) began to explore how people in organizations used technologies to accomplish their work. Whereas the first generation of studies looked to make law-like, often deterministic, claims about how particular technological arrangements would or should change formal organizational structures, the studies in this second generation were more comfortable showing how one technology could engender various unexpected shifts in informal organizing processes. In fact, over time, demonstrating emergence and unpredictability seemed to become this second generation's explicit goal.[2]

The zenith of this second generation came when researchers began to argue that technologies did not always bring predictable effects to the informal organization of work, or that one organizational structure best suited a particular type of technology. Instead, it was only once technological artifacts were enmeshed in a web of organizational, occupational, and institutional forces that people interpreted them and variously employed them in the practice of their work. With such recognition, terms like "technology-in-use" (Orlikowski et al., 1995) and "socio-technological ensembles" (Bijker, 1995) began to replace the word "technology" in many discussions about the genesis of organizational change. Taken at its extreme, this constructivist position suggested that technologies themselves mattered very little in the way people worked, but people's interpretations of the technology mattered a lot. The "if a tree falls in the forest and no one is around to hear it, does it still make a sound?" kind of argument that such writing provoked was very interesting on a theoretical front. But from a practical standpoint it proved problematic because the vast majority of studies of technology use in organizations never even described the technology that was under study (Orlikowski and Iacono, 2001; Markus and Silver, 2008).

To combat this problem, some scholars began advocating that researchers should renew their focus on what features a new technology actually had and what those features did or did not allow people who use them to accomplish (Poole and DeSanctis, 1990; Monteiro and Hanseth, 1995; Griffith, 1999). Enter the term *Materiality*.

[2] For a detailed discussion of this point, see Leonardi and Barley (2008).

Orlikowski (2000: 406), for example, wrote about groupware software that the technology embodies "particular symbol and material properties." She provided several examples of the "material properties" of groupware, which included features contained in the menus that were embedded in the program. Volkoff et al. (2007: 843) described the enterprise resource planning software that they studied as having "material aspects" such as algorithms that allowed financial transactions and features that permitted only certain people to authorize accounts and payments. Leonardi (2007: 816) documented use of a help-desk queuing software by IT technicians and argued that its "material features" made possible activities such as assigning jobs or documenting what one did to solve a particular used problem.

The use of the adjective "material" by these authors, and many others like them, seemed carefully chosen to remind readers that there was some aspect of the technology they described that was intrinsic to the technology, not part of the social context in which the technology was used. In other words, when everyone packs up their bags and goes home at the end of the day, those inherent properties of the technology do not go away. Perhaps the slippery language around what exact properties the technologies had or what they were made of came from the fact the these researchers casted their gaze upon software-based digital technologies. If one were to consider a physical technology like a hammer, it would be relatively easy to isolate and describe a set of properties intrinsic to it. For example, one could point to the steel out of which the head was fashioned, the fiberglass that was shaped into the handle, and the rubber that was placed on top of the fiberglass. We could easily say that the materials from which the hammer was made were steel, fiberglass, and rubber. But when one moves from the realm of the physical to the digital, it is much more difficult to isolate the materials out of which a technology is built. Try it! What are the materials out of which a Microsoft Word Document is made? What are the materials that constitute simulation software? What are the materials out of which social media tools are fashioned? Most information technology artifacts like computer programs and various software applications (the kinds of technologies with which I am concerned in this chapter) have no physicality. Such information technological artifacts may be accessible through certain technological artifacts that have physical properties—that are made of identifiable materials (e.g., a computer program is accessible to users through a monitor and keyboard) but the physical properties of the artifacts that serve as "bearers" (Faulkner and Runde, 2011) for the non-physical artifact do not change the composition of that non-physical artifact in any real way.

But, as Kallinikos (this volume) reminds us, matter is not the only thing that identifies a technology. Form is also important. If one were to take the same mass of steel out of which a hammer head is normally made and form it into a long, thin cylindrical shape and reaffix it to the fiberglass handle, the hammer would no longer be useful for driving nails, knocking holes in drywall, or dislodging jammed boards. But it might be useful for poking holes in leather. Thus, matter (or whatever constituent materials out of which a technology is fashioned) and form together constitute those properties of a technological artifact that do not change, by themselves, across differences in time and context. It is this combination of material and form that I call "materiality." To be clear, "materiality" does not refer solely to the materials out of which a technology is created and it is not a synonym with "physicality." Instead, when we say that we are focusing on a technology's materiality, we are referring to the ways that its physical and/or digital materials are arranged into particular forms that endure across differences in place and time. Such a definition suggests that the usefulness of the term "materiality" is that it identifies those constituent features of a technology that are (in theory) available to all users in the same way.

Although the observations above may seem trite, they are important because if a technology did not have a fixed materiality, extreme constructivist theorizing would not be possible. The prototypical constructivist study shows that people in two different organizations use the same new technology differently and, consequently, change (or do not change) their informal organizing in distinct ways (e.g., Barley, 1986; Zack and McKenney, 1995; Robey and Sahay, 1996). The only way that scholars have been able to demonstrate these findings empirically is because the materiality of that technology was the same in both organizations under study.

Given the arguments made above, it makes most sense to use the term "materiality" to refer to those properties of the artifact that do not change from one moment to the next or across differences in location (recognizing that the uses to which they are put can change greatly, as will be discussed in the following section). Faulkner and Runde (2011: 3) refer to this aspect of materiality as "continuance":

> In saying that objects endure, or exist through time, we mean that they are things that are fully present at each and every point in time at which they exist. Objects can therefore be said to be "continuants", in contrast to events or "occurrents" that take place and whose different parts occur at different points in time. The length of time an object typically endures, what we will call its lifespan, depends on the nature of the object under consideration. Thus while an organism such as a housefly might have a lifespan of no more than a few weeks, an artefact such as a hammer or skyscraper might endure for decades or even centuries.

The use of the term in this way seems almost contrary to Orlikowski's well-cited claim (2000) that...

> Technologies are...never fully stabilized or "complete," even though we may choose to treat them as fixed, black boxes for a period of time. By temporarily bracketing the dynamic nature of technology, we assign a "stabilized-for-now" status...to our technological artifacts. This is an analytic and practical convenience only, because technologies continue to evolve, are tinkered with (e.g., by users, designers, regulators, and hackers), modified, improved, damaged, rebuilt, etc. (pp. 411–12)

Orlikowski is undoubtedly correct that technologies, and their uses, continue to evolve over time. To say that materiality refers to the properties of technologies that do not change is not to disagree with her point; it simply changes the time scale. A popular software program like Microsoft Excel evolves over time. Its materiality is far different today (Version Excel 2010) than it was when it was first released for the PC as Version 2.0 in 1987,[3] or when its predecessor, Multiplan, debuted in 1982. Excel's materiality changed when Version 3.0 was released in 1993 and it changed again with Version 2000.[4] Over time, Excel's materiality has evolved. But to discount the five years or so between version changes that its materiality remained quite constant is to adopt a time horizon that exceeds practical utility. Saying that a technology has a materiality is to say that its materiality has indeed stabilized...for now. And it is this stabilization that allows two people working on the same document, drawing, or database to share work with each other.

One argument made by several authors in this book (e.g., Cooren, Fairhurst, and Huët, this volume; Pentland and Singh, this volume) is that even if a technological object is constructed of particular materials, not all of those materials "matter" for all individuals in particular contexts. They argue that certain aspects of technological artifacts are materialized when they have consequence in a particular setting. Extended to the example of the hammer, we might say that the rubber coating on the handle is a material that does not much matter in one's ability to drive a nail into a board in most circumstances. However, if one's hands are wet, the particular material may suddenly matter in that it has consequence for one's efficacy at driving the nail. In the case of advanced information technologies, the argument is less trivial. Consider the following example provided by Leonardi (2010) of the use of the software application Adobe Photoshop:

> ...one can pick from any number of menus and discover a variety of features (e.g., blur, sharpen, pixelate) that can be used at a given time. Some are extremely important to a certain set of users, while others are not. You might imagine that

[3] Excel was released two years earlier, in 1985, for the Macintosh.

[4] There were other small changes in intermediate version. I highlight only the major version changes as promoted by Microsoft.

> the "blur" feature is important for editors of high school yearbooks attempting to hide blemishes, while the "sharpen" feature is important for law enforcement professionals who are attempting to read the license plate numbers on a passing car. Conversely, if an amateur user is trying to touch up a nighttime shot of the Burj Al Arab hotel from his vacation, it is possible that none of these features make a significant difference in his ability to accomplish his goal. Just like a material fact in a case, a piece of software can have certain material features—features that are "more significant" to the user than others. Of course, significance changes across populations of users, and may even change for one user over time. So, researchers should ask, when examining practices of use, which features are "material" (significant) for this user and how those features become significant for the type of work she does, for whom she interacts with, or for maintaining control.

Although it seems inappropriate, at least when speaking of physical and digital artifacts, to define materiality solely as that which matters to users, the point is well taken. Thus, when referring to physical or digital artifacts, specifically, a general definition of materiality would be, "The arrangement of an artifact's physical and/or digital materials into particular forms that endure across differences in place and time and are important to users."

It is important to focus research attention on materiality if we aim to understand social interaction. Take social media tools—like social networking sites, blogs, wikis, micro-blogging platforms—as an example. The materiality of most social media tools enables editability. In other words, a user can edit and re-edit comments and additions to a site before actually clicking "post" or "share." As researchers have shown, editability derives from asynchronicity and spatial distancing (Ramirez et al., 2002; Dennis et al., 2008). The materiality of social media tools also enables persistence of text, images, and sound. That is to say that one's communication is stored in the system where it remains over time and can be accessed later (Erickson and Kellogg, 2000; Binder et al., 2009). But editability and persistence are not germane to social media. Email offers a high degree of editability as does podcasting. Social media differ from technologies like email and persistence because their materiality enables people's posts to be immediately broadcast to a large unknown audience (see, e.g., the Scott and Orlikowski chapter in this volume). In this way, other people have visibility into an individual's actions. This difference in materiality means that people who use social media will have to contend, in some way, with the fact that their posts, comments, and queries are public. Whether or not they realize that their actions are visible to others, this materiality may have direct consequences for organizing. To understand what these consequences are and the conditions under which they are likely, researchers must first recognize that the technology has a materiality that makes certain actions possible and others impossible, or at least more difficult to achieve (see Faraj and Azad, Chapter 12, this volume).

In short, the term materiality seems useful if it can direct attention to the properties intrinsic to technological artifacts and remind researchers that those properties are fixed, at least for some short period of time, and encourage them to explore not only how they become fixed (as researchers in science and technology studies have done so well) but also how their fixedness affects what people deem to be important to their work. Why not simply use the term "technology" instead of materiality? To answer this question, we must turn to a discussion of another term: "sociomateriality."

Sociomateriality

The term "sociomateriality" is, obviously, the fusion of two words: social and materiality. Why should we use this new term? And, why would a term like this exist at all? The simplest answer would be that this term reminds its readers (*a*) that all materiality (as defined in the prior section) is social in that is was created through social processes and it is interpreted and used in social contexts and (*b*) that all social action is possible because of some materiality.

The first point has a long tradition in the sociology of technology and in organization studies. Researchers in the sociology of technology, including the areas of social construction of technology (Pinch and Bijker, 1984), actor-network theory (Latour, 1991; Law, 1992), and large scale systems theory (Hughes, 1987, 1994) have shown, convincingly, that the development of any new technology is the product of contestation and negotiation among groups (Bijker, 1995), redefinition of problems and the alignment of actors' interests by powerful actors (Callon, 1991), and the result of definitions of what it means to say that a particular technological artifact "works" (Pinch, 1996). Actor-network theorists have taken these observations the farthest in their suggestion that the distinction between that which is social and that which is material is a distinction that scholars have invented to demarcate disciplines of study; it is not a distinction that exists in the empirical world (Latour, 2005). Authors such as Mol (2002) and Barad (2003) who are sympathetic to these ideas have argued that the boundaries between the social and the material are not predetermined, but rather are enacted in the practice of one's work.

Organization theorists have also argued for an intertwining of the social with technology's materiality, but they have primarily focused on how a new artifact merges with an organization's social system during adoption and use. Researchers in communication studies such as Fulk and her colleagues (Fulk et al., 1990; Fulk, 1993) and Aydin and Rice (Rice and Aydin, 1991; Aydin and Rice, 1992) pioneered a line of study, suggesting that people's attitudes and beliefs about what new technologies could do and how they would be useful in one's work were influenced by communication processes and social dynamics. Authors such as Orlikowski (1992, 2000) and Poole and DeSanctis (1992, 2004) who adopted a structurational approach toward technology use suggested that people's decisions about how to use the technology were affected by institutional and organizational norms and, once those technologies were used they began to shape the way that future effects of the technology could unfold. Other studies by Barley (1990), Edmondson et al. (2001), and Boudreau and Robey (2005) demonstrated that changes in the use of a technology over time—changes that were negotiated socially—could shift the dynamics of teams, organizations, and occupations. In short, it would be incorrect to say that a technology "caused" a particular change when ample evidence shows that people decide how they will let the technology influence their work.

Scholars straddling the line between these two areas of study—technology development and use—have made the claim that if organizations are as much material as they are social and if technologies are as much social as they are material, then perhaps it makes sense to break down the distinction between the social and the material altogether. For example, in a study of the organization of civil engineering work, Suchman (2000: 316) argued:

> Like an organization, a bridge can be viewed as an arrangement of more and less effectively stabilized material and social relations. Most obviously, of course, the stability of a bridge is a matter of its materiality, based in principles and practices of structural engineering. This material stability is inseparable, however, from the networks of social practice—of design, construction, maintenance and use—that must be put into place and maintained in order to make a bridge-building project possible, and to sustain the resulting artifact over time.

Orlikowski (2007: 1437) has made a similar argument:

> Materiality is integral to organizing... the social and the material are *constitutively entangled* in everyday life. A position of constitutive entanglement does not privilege either humans or technology (in one-way interactions), nor does it link them through a form of mutual reciprocation (in two-way interactions). Instead, the social and the material are considered to be inextricably related—there is no social that is not also material, and no material that is not also social.

Thus, in support of the first point above, the term "sociomaterial" is a bold reminder that when we talk either about technologies or organizations, we do well to remember that social practices shape the materiality of a technology and its effects.

Interest in establishing the second point mentioned above—that all social action is possible because of some materiality—is, perhaps, more political or agenda-setting in nature. Over the last two decades, a number of reviews of papers in the literature on organizational behavior, organizational communication, and organizational theory have concluded that organizational researchers do not spend much time and effort thinking about the role that technologies play in their areas of inquiry (e.g., decision-making, status, strategy making, etc.) (Markus and Robey, 1988; Liker et al., 1999; Orlikowski and Barley, 2001; Rice and Gattiker, 2001). Orlikowski and Scott's (2008) recent review is perhaps the most striking. The authors claim that less than 5 percent of all articles published in top American management and organization studies journals considered the role and influence of technology directly. They argue that part of the reason that technology may be "missing in action" (p. 434) is that most organizational researchers do not consider themselves scholars of technology. Their natural predilection to overlook the role technology plays in the particular organizational processes that capture their interest is further exacerbated by the fact that new technologies change often and the study of them requires that scholars continue to learn about these changes. Orlikowski (2007) has also made the point elsewhere that most existing studies of technology in organizations focus on new technology "implementation." The continued appearance of implementation studies marks technology implementation off as a specific and unique area and people think that if they are going to study technology they need to study implementation.

The term "sociomateriality" has the potential to address these concerns by reminding organization scholars that materiality is present in each and every phenomenon that they consider "social." To be sure, strategies are formed based on the ways people use PowerPoint presentations to share information with one another (Kaplan, 2011); routines are both made possible and performed through the use of checklists and forms (D'Adderio, 2011); and quadrants and algorithms shape perceptions of risk and spur the formulation of institutional categories (see Pollock, Chapter 5, this volume). In short, one need not study new technology implementation to respect the ways that materiality is a constitutive part of all practice that organizational scholars typically call "social."

Scholars who adopt the term "sociomateriality" would likely argue that it is unique from the term "materiality" in that it shifts the unit of analysis from materials and forms to the development or use of materials and forms. In other words, talking about sociomateriality is to recognize and always keep present to mind that materiality acts as a constitutive element of the social world, and vice versa. Thus, whereas materiality might be a property of a technology, sociomateriality represents that enactment of a particular set of activities that meld materiality with institutions, norms, discourses, and all other phenomena we typically define as "social."

Although one could say that a technology has a certain materiality, it would make little sense to talk about a technology's sociomateriality. For this reason, Orlikowski (2007) and others have suggested that what is sociomaterial is not the technology, but the "practice" in which the technology is embedded. Social theorists such as Giddens, Lave, and Bourdieu offer nuanced definitions of what counts as *practice*. For each of these authors, practice is not equivalent to individual activity (e.g., doing something); rather it is a socially shaped arena in which activities are collectively negotiated. In Giddens' (1984) terms, the arena of practice is the medium and outcome of institutional structures that guide individuals' processes of interpretation and evaluation, and hence, their activities. Thus, practice is shared in common by people and its production and perpetuation is a collective accomplishment. For Lave (1988), the arena of practice is a negotiated order in which people's patterns of action are contingent upon specific structural conditions of their own making. Thus, practice is a social space that is shared in common by members of a community. Bourdieu (1977) conceptualizes practice as an arena in which the dialectic of subjective experience and objectified reality is played out. Building on the work of Bourdieu and Lave, Cook and Brown (1999: 388) go so far as to define the term "practice" as "the coordinated activities of individuals and groups in doing their 'real work' as it is informed by a particular organizational or group context."

In this formulation, practice is the space in which the social and the material become constitutively entangled (Orlikowski, 2010). Although most studies up to this point have sufficed to simply show that social and the material are thoroughly intertwined, scholars are just beginning to consider how such intertwinement occurs. Leonardi (2011), for example, has offered one theory about how the social and the material become entangled. This theory suggests that coordinated human agencies (social agency) and the things that the materiality of a technology allow people to do (material agency) become interlocked in sequences that produce the empirical phenomena we call "technologies," on the one hand, and "organizations," on the other.

Human agency is typically defined as the ability to form and realize one's goals (Giddens, 1984; Emirbayer and Mische, 1998). A human agency perspective suggests that people's work is not determined by the technologies they employ. Studies show that even in the face of the most apparently constraining technologies, human agents can exercise their discretion to shape the effects that those technologies have on their work (Boudreau and Robey, 2005). People often enact their human agency in response to technology's material agency.

Material agency is defined as the capacity for nonhuman entities to act absent sustained human intervention. Pickering (1995: 6), for example, observes that the weather "does things" that absent human intervention—it rains, winds blow, and heat and cold fluctuate: "Much of everyday life, I would say, has the character of coping with material agency, agency that comes at us from outside the human realm and that cannot be reduced to anything within that realm." Kaptelinin and Nardi (2006) extend Pickering's discussion of material agency to technological artifacts specifically, arguing that artifacts such as information technologies represent a particular kind of cultural object that produce effects and can realize the intentions of humans (e.g., the people who designed, built, or implemented them), but that they cannot act according to their own biological or cultural needs.

As nonhuman entities, artifacts like information technologies exercise agency through their performativity; in other words, through the things they do that users cannot completely or directly control (see Robey et al., Chapter 11, this volume). For example, a compiler translates text from a source computer language into a target language without input from its user and a finite element solver calculates nodal displacements in a mathematical model and renders the results of this analysis into a three-dimensional animation without human intervention. Although each of these actions is instigated by a human (presumably to address a particular, local need), the technology itself acts (exercises material agency) as humans with goals engage with its materiality.

Coordinated human (social) and material agencies both represent capacities for action, but they differ with respect to intentionality. Pickering (2001) offers a concise and useful empirical definition of human and material agencies that illustrates this difference. For Pickering, social agency is a group's coordinated exercise of forming and realizing its goals. Thus, the practice of forming goals and attempting to realize them is a concrete operationalization of social agency. Material agency, by contrast, is devoid of intention and materiality does not act to realize its own goals because it has none of its own making. In other words, "machine artifacts have no inherent intentionality, independent of their being harnessed to or offering possibilities to humans" (Taylor et al., 2001: 137). Thus, material agency is operationalized as the actions that a technology takes, which humans do not immediately or directly control. Given this important difference with respect to intentionality, even though social and material agencies might be equally important in shaping one's practice, but they do so in qualitatively different ways.

Leonardi (2011) uses the metaphor of imbrication to suggest how social and material agencies become entangled. The word "imbrication" may appear, at first glance, to be more jargon lining the already detritus-filled road to scholarly enlightenment, but its origins are both humble and practical. The verb "imbricate" is derived from names of roof tiles used in ancient Roman and Greek architecture. The tegula and imbrex were interlocking tiles used to waterproof a roof. The tegula was a plain flat tile laid on the roof and the imbrex was a semicylindrical tile laid over the joints between the tegulae. The interlocking pattern of tegulae and imbrices divided the roof into an equal number of channels. Rainwater flowed off the row of ridges made by the imbrices and down over the surfaces of the tegulae, dropping into a gutter. The imagery of tiling suggests that different types of tiles are arranged in an interlocking sequence that produces a visible pattern. A roof could not be composed solely of tegulae nor imbrices—the differences between the tiles in terms of shape, weight, and position prove essential for providing the conditions for interdependence that form a solid structure. Social and material agencies, though both capabilities for action, differ phenomenologically with respect to intention. Thus, like the tegula and the imbrex, they have distinct contours and through their imbrication they come to form an integrated organizational structure.

This perspective takes a complementary yet distinct approach to that offered by authors such as Barad (2003: 818) who claims that "Agencies are not attributes [of either humans or technologies] but ongoing reconfigurations of the world." It argues that the materiality of a technological artifact affords certain uses and actions. Although materiality, itself, transcends variations in space and time, those uses and actions can be different depending upon the context in which the materiality is used. For example, Microsoft Excel has many features that do not change across contexts (materiality). But those features do not automatically calculate modal values in a numerical list (material agency) until some user (with social agency) tells that materiality to do so. Even a simple physical technology like a hammer whose materiality (steel formed into a flat head and hook, fiberglass formed into a semicylinder, and rubber formed into a thin sheet) does not change can have many functions in that the same materiality can support driving nails into wood or holding papers down on a desk so they do not fly away. Whereas materiality refers to properties of the object, material agency refers to the way the object acts when humans provoke it. This distinction between materiality and material agency is akin to the distinction between the arrangement of physical or digital materials into particular forms—what I have called "materiality"—and what Kallinikos (this volume) describes as "function" (what I suggest could alternatively be called "material agency"). What the technology *is* does not change across space and time, but what it *does* can and often changes. Function—or material agency—is a construction that depends, in part, on materiality but also depends on one's perceptions of whether materiality affords her the ability to achieve her goals or places a constraint upon her.

Materiality exists independent of people, but affordances and constraints do not. Because people come to materiality with diverse goals, they perceive a technology as affording distinct possibilities for action. The perceptions of what functions an artifact affords (or constrains) can change across different contexts even though the artifact's materiality does not. Similarly, people may perceive that a technology offers no affordances for action, perceiving instead that it constraints their ability to carry out their goals. In this view, affordances and constraints are constructed in the space between social and material agencies. Peoples' goals are formulated, to an important degree, by their perceptions of what a technology can or cannot do, just as those perceptions are shaped by people's goals. Depending on whether they perceive that a technology affords or constrains their goals, people make choices about how they will imbricate social and material agencies. Thus, while it makes sense to talk about material and social agencies as attributes that are activated in response to one another in the space of practice, it seems empirically inaccurate to say that agencies themselves are "reconfigurations of the world." Social and material agencies are distinct from one another, and it is only once they become imbricated in particular ways that they can then reconfigure technology's materiality and organizations' communication patterns.

To weave the arguments in this section together, I suggest that (*a*) "sociomateriality" is not a property of a technology but the recognition that materiality takes on meaning and has effects as it becomes enmeshed in a variety phenomena (e.g., decision-making, strategy formulation, categorization) that scholars typically define as "social"; (*b*) "Sociomaterial" is an adjective best used to modify the noun "practice" where (*c*) "practice" is understood as the space in which social and material agencies are imbricated with each other and, through their distinct forms of imbrication, produce those empirically observable entities which we call "technologies" and "organizations."

Socio-Technical Systems

Many papers on technology use in organizations published over the last two decades use the term "socio-technical system" (STS) to describe their object of study. In general, when employed in studies of technology development, technologies are often referred to as "socio-technical systems" to bolster the recognition that the technology under design will be implemented and used in a social context that will, to some degree, shape whether and how it is adopted (Bostrom and Heinan, 1977; Benders et al., 2006). In organization studies, authors will sometimes use the term "socio-technical system" to claim that the organization is made up of social systems (hierarchies, communication networks, etc.) and technical systems, which are usually defined as technological artifacts like imaging devices, numerically controlled machine tools, enterprise resources planning systems, and the like (Barley, 1990; Thomas, 1994; Griffith and Dougherty, 2001). Although these contemporary uses of the term "socio-technical systems" rightly point to the interdependencies between people and things, researchers at the Tavistock Institute of

Human Relations who coined the term in the 1950s had a more nuanced definition in mind.

Shortly after World War II, Eric Trist and Ken Bamforth, two researchers at the Tavistock Institute, conducted a series of field studies of the organization of work around a new process of coal-getting in British mines. In an influential paper (Trist and Bamforth, 1951), they charted the response of workers who were migrated from a traditional "hand-got" method of extracting coal from mines to a new semimechanized "longwall" method (sometimes called "conveyor method") for coal-getting. The major difference between these two methods concerned the way that tasks were apportioned among workers. In the traditional hand-got method, small groups of workers labored at individual coalfaces in a large mine. These groups were in charge of their own face. Each member performed a variety of tasks using a number of different tools (pick axes, shovels, etc.) and they substituted for each other frequently. In short, they had a high degree of what Hackman et al. (1975) call "elements of job enrichment"—autonomy, task significance, and task identity.

As Trist and Bamforth observed, the hand-got method eventually gave way to the new longwall method: "With the advent of coal-cutter and mechanical conveyers, the degree of technological complexity of the coal-getting task was raised to a different level. Mechanization made possible the working of a single long face in place of a series of short faces" (p. 9). As the authors argued, new technologies made possible a new system of work in which the coalfaces could now reach lengths of up to 200 yards, meaning that the coal could be extracted much more efficiently than through the old hand-got method. To take advantage of this new semimechanized longwall method, management split work into three different shifts over 24 hours. During the first shift, miners used an electric coal cutter (instead of pick axes) to cut the coal from the seam. During the second shift, it was hand-loaded onto a conveyor (instead of removed manually from the mine in hand-filled tubs) that was placed parallel to the seam. And during the third shift the equipment at the face and the hydraulic jacks that supported the roof and walls of the mine were moved forward. Trist and Bamforth argued that the departmentalization of work into these three shifts represented not only the demise of an intact and interdependent work group but also the loss of team identity, pride, status, and the fractionalization of work into tasks that were boring and repetitive.

Trist and Bamforth documented in great detail (pp. 16–17) the various tasks that each individual conducted with and around the new technological artifacts used in the mines. Interestingly, and contrary to much subsequent interpretation of their study, they operationalized the technical subsystem of the coal mine not simply as the technologies that the miners employed but also those tasks that the miners conducted around the technological artifacts. What makes their analysis so interesting is that Trist and Bamforth identified various ways that miners could organize their tasks around the technologies' materiality. For example, they found that miners who experienced job dissatisfaction after the implementation of the longwall method due to the dissolution of their work teams often created new informal teams on their shifts such that they could change who conducted what tasks and when. These informal groups defied management logic about the kinds of tasks that workers needed to conduct to use the technologies, and the way those tasks were distributed among workers (see pp. 30–5). In other words, they found that there was an indeterminate relationship between tasks and technologies such that a technology's fixed materiality could support multiple task structures depending upon people's desires and goals.

But in addition to discussing this technical subsystem (technology's materiality and the tasks conducted in interact with it), Trist and Bamforth showed how the social subsystem in the mines was changed from its structure during the era of the hand-got method. They document in great detail how communication patterns among miners changed, how status hierarchies became unsettled, and how power relations calcified. These elements comprising the "social subsystem" were entirely abstract in that they were institutionalized ideas about how people could and should relate to one another.

Researchers within the STS tradition drew on this initial, detailed study to suggest that an organization's performance was directly correlated with the degree to which the social and technical subsystems were "jointly optimized" (Emery, 1959)—the demands of one system fit the demands of the other. Rice's (1953, 1958, 1963) work in the weaving sheds in Ahmedabad, India, demonstrated how the social and technical systems could be jointly optimized. He suggested that the social organization of work in the sheds was out of alignment with the demands of the looms used to produce textiles because the workers had organized the social subsystem so that they could work independently, while the technical subsystem demanded that people work interdependently in order to maximize use of the machines. In what has now become the most popular take away of the STS literature, Rice attempted to solve this problem by creating autonomous teams based on interdependent roles. Despite the rhetoric of jointly optimizing both the social and technical aspects of work, Rice's innovation adjusted the social organization of work to fit the demands of the loom technologies, thus privileging the demands of the technical subsystem over those of the social subsystem.

What is interesting about the early work on socio-technical systems theory is that conceptualization of a technical subsystem very much resembles what scholars today call "sociomaterial practice." STS scholars showed, empirically, that the materiality of a new technological artifact could be used in a variety of ways to support various tasks and/or task apportionments. Although they did not have the language to describe it, their findings do suggest that social and material agencies became imbricated in ways that produced various local orientations to work—whether in coalmines or weaving sheds. The technical subsystem, then, was not just comprised of technological artifacts but was instead a sociomaterial practice in which people's goals and the technology's materiality became, to use Orlikowski's term (2007: 1437), "constitutively entangled."

But unlike scholars who study materiality or sociomaterial practices, STS researchers raised the level of analysis to focus on the way that the technical subsystem became integrated into the macro organization of work. The concept of "joint optimization," while it may be criticized for being too normative, was intended to showcase how the abstract properties of a social subsystem could be strengthened or disturbed based on the particular ways in which social and material agencies were imbricated in the technical subsystem. In their formulation, STS researchers seemed to imagine that while the boundaries between materiality and task that characterized the technical subsystem were enacted in practice as opposed to alternating between causalities, the broader relationship between the technical and social subsystems was one of mutual shaping over time. Once a particular set of relations emerged from the technical subsystem, people had to decide if and how they would reconfigure the abstract social subsystem. And, of course, reconfigurations of the social subsystem could then catalyze new cycles of sociomaterial imbrication in the technical subsystem (Cummings and Srivastva, 1977; Pasmore, 1988). Although they seemed to recognize this mutual influence was possible theoretically, in practice they normally advocated that the social subsystem was more influenced by the technical subsystem than the reverse and that it should be modified accordingly.

In summary, the term "socio-technical system" appears distinct from the term "materiality" in that materiality simply refers to the properties of a technology that are used in various ways to support various tasks in the technical subsystem. The notion of a technical subsystem in socio-technical systems theory does not seem very different from the term "sociomaterial practice" because both refer to a space in which work is made possible through the imbrication of social and material agencies. But a "socio-technical system" appears to be distinct form a "sociomaterial practice" in that it refers to the entire organization of work (abstract institutional constructs and patterns of sociomaterial imbrication), as opposed to a group's localized experiences around a particular or various technologies. Thus, an organization might be conceptualized as a "socio-technical system" but not a "sociomaterial practice." Sociomaterial practices (or "technical subsystems," should we choose to use this more antiquated term) influence and are influenced by broader abstract social structures such as roles, statuses, hierarchies, power relations, communication networks, and other similar constructs. Kallinikos (2011) calls such abstract social structures "institutional forces." He suggests that institutions are temporally bound and, consequently, should not be simply seen as a way for researchers to vacillate between micro and macro levels of analysis, but that they are useful for moving from static to dynamic patterns of analysis such that each layer of sociomaterial imbrication becomes more substantial in that it shapes action in a path-dependent manner because of its history of accumulation.

Defining and Interrelating Terms

Early on, I argued that the goal of this chapter was to stimulate debate and discussion about popular terminology used in contemporary explanations of technology and organizing. In this spirit, I have reviewed the historical foundations of the terms "materiality," "sociomateriality," and "socio-technical systems" and I have made some first, undoubtedly contentious, steps to define how these terms relate to one another. In doing so, I have placed certain boundaries around these concepts for the sake of definitional clarity. Below, I summarize the preceding discussions into a rough and entirely tentative glossary of terms:

Materiality: The arrangement of an artifact's physical and/or digital materials into particular forms that endure across differences in place and time and are important to users.

Sociomateriality: Enactment of a particular set of activities that meld materiality with institutions, norms, discourses, and all other phenomena we typically define as "social."

Sociomaterial Practice: The space in which multiple human (social) agencies and material agencies are imbricated (also called a "technical subsystem").

Social Agency: Coordinated human intentionality formed in partial response to perceptions of a technology's material agency.

Material Agency: Ways in which a technology's materiality acts. Material agency is activated as humans approach technology with particular intentions and decide which elements of its materiality to use at a given time.

Socio-Technical System: Recognition of a recursive (not simultaneous) shaping of abstract social constructs and a technical infrastructure that includes technology's materiality and people's localized responses to it.

Figure 2.1 provides an illustration of how these various terms might relate to one another. The shaded boxes at the right side of the figure indicate that people have intentionality and technological artifacts have materiality. As people approach technological artifacts they form particular goals (human agency) and they use certain of the artifact's materiality to accomplish them (material agency). These collective human (social) and material agencies become imbricated in the space of practice. Certain imbrications produce changes in the abstract "social" formulations (e.g., roles, status, etc.) that occupy so much of organization theorists' attention. Alterations in these abstract formulations can shape future patterns of imbrication, which, in turn, can bring changes to an artifact's materiality or a person(s)' intentionality. This mutual shaping of social and technical subsystems (indicated by shaded ovals) is what defines a socio-technical system. We might usefully be reminded that organizations are socio-technical systems.

To be sure, the road to nuanced and empirically grounded understanding of the relationship between technological and organizational change is littered

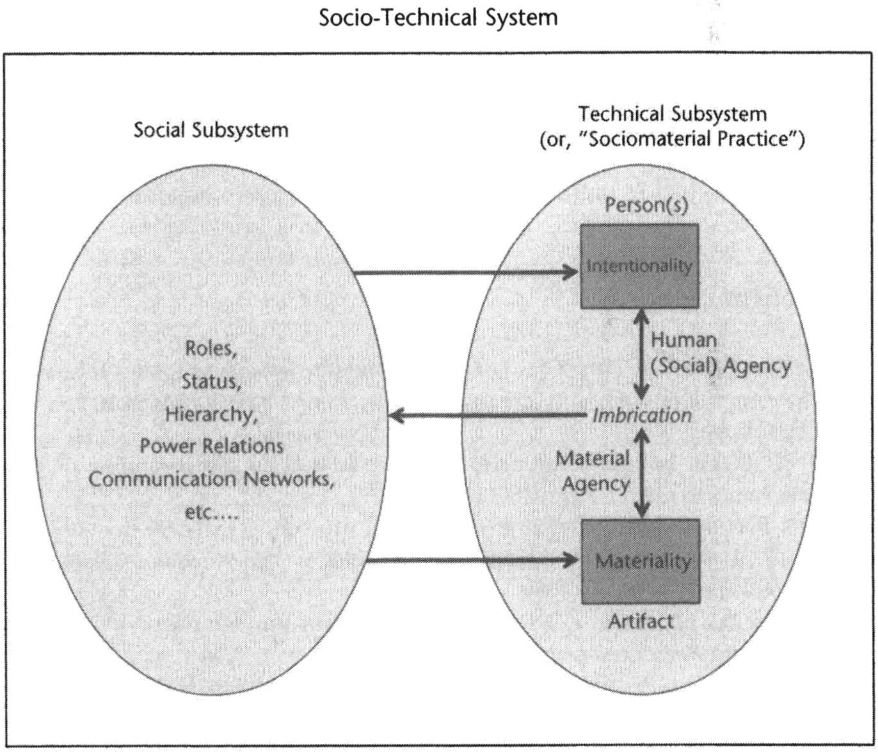

Socio-Technical System

Social Subsystem

Technical Subsystem
(or, "Sociomaterial Practice")

Person(s)

Intentionality

Roles,
Status,
Hierarchy,
Power Relations
Communication Networks,
etc....

Human
(Social) Agency

Imbrication

Material
Agency

Materiality

Artifact

Figure 2.1 Potential relationships between materiality, sociomateriality, and socio-technical systems

with academic jargon. Some of that jargon—terms like "strategic choice," "joint optimization," or "equifinality"—is rarely used today while other jargon—terms like "structuration," "inscription," and "morphogenesis"—is still widely in use. In the past couple of years, students of technology and organizing have added three additional terms to the jargon-lined road: "Materiality," "Sociomateriality," and "Socio-Technical Systems." Sometimes, authors use these terms interchangeably. Sometimes they seem to use them quite distinctly. Sometimes these terms include hyphens. Sometimes they don't. To assure that these terms don't become "academic jargon monoxide" requires some definitional clarity. This chapter has taken an initial step in providing this clarity. All definitions include some ideas and exclude others. Also, all definitions reflect the author's view of the world. This chapter offers these tentative definitions without any aspiration that people will use them, but with only the hope that they will spur debate and seed discussion about what they mean, how they relate to one another, and whether we need them at all.

Acknowledgments

I thank Jannis Kallinikos and Bonnie Nardi for reading through several drafts of this chapter and offering suggestions and critiques that help me refine my arguments. Wanda Orlikowski's work continually inspires me and gives me ideas on which to build and react, so I thank her very much too.

References

Aydin, C. and Rice, R. E. (1992). Bringing social worlds together: Computers as catalysts for new interactions in health care organizations. *Journal of Health and Social Behavior*, 33(2), 168–85.

Barad, K. (2003). Posthumanist performativity: Toward an understanding of how matter comes to matter. *Signs*, 28(3), 801–31.

Barley, S. R. (1986). Technology as an occasion for structuring: Evidence from observations of CT scanners and the social order of radiology departments. *Administrative Science Quarterly*, 31(1), 78–108.

—— (1990). The alignment of technology and structure through roles and networks. *Administrative Science Quarterly*, 35(1), 61–103.

Benders, J., Hoeken, P., Batenburg, R., and Schouteten, R. (2006). First organise, then automate: A modern socio-technical view on erp systems and teamworking. *New Technology Work and Employment*, 21(3), 242–51.

Bijker, W. E. (1995). *Of bicycles, bakelites, and bulbs: Toward a theory of sociotechnical change*. Cambridge, MA: The MIT Press.

Binder, J., Howes, A., and Sutcliffe, A. (2009). The problem of conflicting social spheres: Effects of network structure on experienced tension in social network sites. In Proceedings of the 27th International Conference on Human Factors in Computing Systems, Boston, MA, April 4–9.

Bostrom, R. P. and Heinan, J. S. (1977). MIS problems and failures: A socio-technical perspective. *MIS Quarterly*, 1(4), 11.

Boudreau, M.-C. and Robey, D. (2005). Enacting integrated information technology: A human agency perspective. *Organization Science*, 16(1), 3–18.

Bourdieu, P. (1977). *Outline of a theory of practice*. Cambridge: Cambridge University Press.

Callon, M. (1991). Techno-economic networks and irreversibility. In J. Law (ed.), *A sociology of monsters: Essays on power, technology and domination* (pp. 132–61). London: Routledge.

Cook, S. D. N. and Brown, J. S. (1999). Bridging epistemologies: The generative dance between organizational knowledge and organizational knowing. *Organization Science*, 10(4), 381–400.

Cummings, T. G. and Srivastva, S. (1977). *Management of work: A socio-technical systems approach*. San Diego, CA: University Associates.

D'Adderio, L. (2011). Artifacts at the centre of routines: Performing the material turn in routines theory. *Journal of Institutional Economics*, 7(2), 197–230.

Dennis, A. R., Fuller, R. M., and Valacich, J. S. (2008). Media, tasks, and communication processes: A theory of media synchronicity. *MIS Quarterly*, 32(3), 575–600.

Edmondson, A. C., Bohmer, R. M., and Pisano, G. P. (2001). Disrupted routines: Team learning and new technology implementation in hospitals. *Administrative Science Quarterly*, 46(4), 685–716.

Emery, F. (1959). *Characteristics of sociotechnical systems*. London: Tavistock Institute.

Emirbayer, M. and Mische, A. (1998). What is agency? *American Journal of Surgery*, 103(4), 962–1023.

Erickson, T. and Kellogg, W. (2000). Social translucence: An approach to designing systems that support social processes. *ACM Transactions on Computer-Human Interaction*, 7, 59–83.

Faulkner, P. and Runde, J. (2011). The social, the material, and the ontology of non-material technological objects. Paper presented at the European Group for Organizational Studies (EGOS) Colloquium, Gothenburg.

Fulk, J. (1993). Social construction of communication technology. *Academy of Management Journal*, 36(5), 921–51.

—— Schmitz, J., and Steinfield, C. (1990). A social influence model of technology use. In J. Fulk and C. Steinfield (eds.), *Organizations and communication technology* (pp. 117–40). Newbury Park, CA: Sage.

Giddens, A. (1984). *The constitution of society*. Berkeley, CA.: University of California Press.

Griffith, T. L. (1999). Technology features as triggers for sensemaking. *Academy of Management Review*, 24(3), 472–88.

Griffith, T. L. and Dougherty, D. J. (2001). Beyond socio-technical systems: Introduction to the special issue. *Journal of Engineering and Technology Management*, 18, 207–18.

Hackman, J. R., Oldham, G., Janson, R., and Purdy, K. (1975). A new strategy for job enrichment. *California Management Review*, 17(4), 57–71.

Hughes, T. P. (1987). The evolution of large technological systems. In W. E. Bijker, T. P. Hughes, and T. Pinch (eds.), *The social construction of technological systems: New directions in the sociology and history of technology* (pp. 51–82). Cambridge, MA: The MIT Press.

—— (1994). Technological momentum. In M. R. Smith and L. Marx (Eds.), *Does technology drive history? The dilemma of technological determinism* (pp. 101–13). Cambridge, MA: MIT Press.

Kallinikos, J. (2004). Farewell to constructivism: Technology and context-embedded action. In C. Avgerou, C. Ciborra, and F. Land (eds.), *The social study of information and communication technology: Innovation, actors and contexts* (pp. 140–61). Oxford: Oxford University Press.

—— (2011). *Governing through technology: Information artifacts and social practice*. Basingstroke: Palgrave Macmillan.

Kaplan, S. (2011). Strategy and powerpoint: An inquiry into the epistemic culture and machinery of strategy making. *Organization Science*, 22, 320–46.

Kaptelinin, V. and Nardi, B. A. (2006). *Acting with technology: Activity theory and interaction design*. Cambridge, MA: MIT Press.

Latour, B. (1991). Technology is society made durable. In J. Law (ed.), *A sociology of monsters: Essays on power, technology and domination* (pp. 103–31). London: Routledge.

—— (2005). *Reassembling the social: An introduction to actor-network theory*. Oxford: Oxford University Press.

Lave, J. (1988). *Cognition in practice*. Cambridge: Cambridge University Press.

Law, J. (1992). Notes on the theory of the actor-network: Ordering, strategy, and heterogeneity. *Systems Practice*, 5(4), 379–93.

Leonardi, P. M. (2007). Activating the informational capabilities of information technology for organizational change. *Organization Science*, 18(5), 813–31.

—— (2010). Digital materiality? How artifacts without matter, matter. *First Monday*, 15 (6), Available from: http://www.uic.edu/htbin/cgiwrap/bin/ojs/index.php/fm/article/viewArticle/3036/2567.

—— (2011). When flexible routines meet flexible technologies: Affordance, constraint, and the imbrication of human and material agencies. *MIS Quarterly*, 35(1), 147–67.

—— Barley, S. R. (2008). Materiality and change: Challenges to building better theory about technology and organizing. *Information and Organization*, 18(3), 159–76.

Liker, J. K., Haddad, C. J., and Karlin, J. (1999). Perspectives on technology and work organization. *Annual Review of Sociology*, 25, 575–96.

Markus, M. L. and Robey, D. (1988). Information technology and organizational change: Causal structure in theory and research. *Management Science*, 34(5), 583–98.

—— Silver, M. S. (2008). A foundation for the study of it effects: A new look at DeSanctis and Poole's concepts of structural features and spirit. *Journal of the Association for Information Systems*, 9(10/11), 609–32.

Mol, A. (2002). *The body multiple: Ontology in medical practice*. Durham, NC: Duke University Press.
Monteiro, E. and Hanseth, O. (1995). Social shaping of information infrastructure: On being specific about the technology. In W. J. Orlikowski, G. Walsham, M. R. Jones, and J. I. DeGross (eds.), *Information technology and changes in organization work*. London: Chapman and Hall.
Orlikowski, W. J. (1992). The duality of technology: Rethinking the concept of technology in organizations. *Organization Science*, 3(3), 398–427.
—— (2000). Using technology and constituting structures: A practice lens for studying technology in organizations. *Organization Science*, 11(4), 404–28.
—— (2007). Sociomaterial practices: Exploring technology at work. *Organization Studies*, 28(9), 1435–48.
—— (2010). The sociomateriality of organisational life: Considering technology in management research. *Cambridge Journal of Economics*, 34, 125–41.
—— Barley, S. R. (2001). Technology and institutions: What information systems research and organization studies can learn from each other. *MIS Quarterly*, 25, 145–65.
—— Iacono, C. S. (2001). Research commentary: Desperately seeking the "it" in IT research—a call to theorizing the IT artifact. *Information Systems Research*, 12(2), 121–34.
—— Scott, S. V. (2008). Sociomateriality: Challenging the separation of technology, work and organization. *The Academy of Management Annals*, 2(1), 433–74.
—— Yates, J., Okamura, K., and Fujimoto, M. (1995). Shaping electronic communication: The metastructuring of technology in the context of use. *Organization Science*, 6(4), 423–44.
Pasmore, W. (1988). *Designing effective organizations: The sociotechnical perspective*. New York: John Wiley and Sons.
Perrow, C. (1967). A framework for the comparative analysis of organizations. *American Sociological Review*, 32, 194–208.
Pickering, A. (1995). *The mangle of practice: Time, agency, and science*. Chicago: University of Chicago Press.
—— (2001). Practice and posthumanism: Social theory and a history of agency. In T. R. Schatzki, K. Knorr-Cetina, and E. von Savigny (eds.), *The practice turn in contemporary theory* (pp. 163–74). London: Routledge.
Pinch, T. J. (1996). The social construction of technology: A review. In R. Fox (ed.), *Technological change: Methods and themes in the history of technology* (pp. 17–35). Amsterdam: Harwood.
—— Bijker, W. E. (1984). The social construction of facts and artifacts: Or how the sociology of science and the sociology of technology might benefit each other. *Social Studies of Science*, 14, 399–441.
Poole, M. S. and DeSanctis, G. (1990). Understanding the use of group decision support systems: The theory of adaptive structuration. In J. Fulk and C. Steinfield (eds.), *Organizations and communication technology* (pp. 173–93). Newbury Park, CA: Sage.
—— —— (1992). Microlevel structuration in computer-supported group decision making. *Human Communication Research*, 19(1), 5–49.

Poole, M. S. and DeSanctis, G. (2004). Structuration theory in information systems research: Methods and controversies. In M. E. Whitman and A. B. Woszczynski (eds.), *Handbook of information systems research* (pp. 206–49). Hershey, PA: Idea Group.

Ramirez, A., Walther, J. B., Burgoon, J. K., and Sunnafrank, M. (2002). Information-seeking strategies, uncertainty, and computer-mediated communication. *Human Communication Research*, 28(2), 213–28.

Rice, A. K. (1953). Productivity and social organization in an Indian weaving shed: An examination of some aspects of the socio-technical system of an experimental automatic loom shed. *Human Relations*, 6(4), 297–329.

—— (1958). *Productivity and social organization: The Ahmedabad experiment*. London: Tavistock.

—— (1963). *The enterprise and its environment*. London: Tavistock Institute.

—— Aydin, C. (1991). Attitudes toward new organizational technology: Network proximity as a mechanism for social information processing. *Administrative Science Quarterly*, 36(2), 219–44.

—— Gattiker, U. (2001). New media and organizational structuring. In F. M. Jablin and L. L. Putnam (eds.), *The new handbook of organizational communication: Advances in theory, research, and methods* (pp. 544–81). Thousand Oaks, CA: Sage.

Robey, D. and Sahay, S. (1996). Transforming work through information technology: A comparative case study of geographic information systems in county government. *Information Systems Research*, 7(1), 93–110.

Suchman, L. (2000). Organizing alignment: A case of bridge-building. *Organization*, 7(2), 311–27.

—— (2007). *Human–machine reconfigurations: Plans and situated actions*. Cambridge: Cambridge University Press.

Sutton, R. I. (2010). Sociomateriality: More academic jargon monoxide [web log comment]. http://bobsutton.Typepad.Com/my_weblog/2010/10/sociomateriality-more-academic-jargon-monoxide.Html

Taylor, J. R., Groleau, C., Heaton, L., and Van Every, E. (2001). *The computerization of work: A communication perspective*. Thousand Oaks, CA: Sage.

Thomas, R. J. (1994). *What machines can't do: Politics and technology in the industrial enterprise*. Berkeley, CA: University of California Press.

Trist, E. L. and Bamforth, K. W. (1951). Some social and psychological consequences of the longwall method of coal-getting. *Human Relations*, 41(1), 3–38.

Volkoff, O., Strong, D. M., and Elmes, M. B. (2007). Technological embeddedness and organizational change. *Organization Science*, 18(5), 832–48.

Woodward, J. (1958). *Management and technology*. London: HMSO.

Zack, M. H. and McKenney, J. L. (1995). Social context and interaction in ongoing computer-supported management groups. *Organization Science*, 6(4), 394–422.

3

On Sociomateriality

Philip Faulkner and Jochen Runde

The contributions to this volume illustrate well the growing interest in the topic of materiality in the social sciences, and the variety of different approaches that exist to studying materiality or that take materiality seriously. One of the most interesting recent developments in this rapidly expanding field is the emerging "sociomateriality" perspective championed by Orlikowski (2007, 2010) and Orlikowski and Scott (2008) in the area of technology in organizational theory. Of particular significance is that this perspective represents an attempt to introduce an ontological dimension into a literature dominated by social constructivism, a tradition that has hitherto shown little interest in questions of existence, modes of being, and so forth. We regard this as an exciting departure, one that brings to the surface many of the themes and questions that lie at the heart of contemporary debates involving the notion of materiality, not least those discussed elsewhere in this volume.

Our aim in this chapter is accordingly to identify and examine some aspects of the social ontology outlined by Scott and Orlikowski.[1] We recognize that all this might be seen as placing an undue weight on what is still quite a small body of work, and which both its authors freely admit is at an early stage. However, it seems to us that providing a commentary on the three papers concerned is consistent with the spirit in which they were written, avowedly to throw out some ideas and see how they were taken up (Orlikowski and Scott, 2008: 456). Further, and as will become evident below, we are sympathetic with the broad thrust of much that they contain. While we will be

[1] In their Chapter 6 of the present volume, Scott and Orlikowski use the sociomateriality perspective to study commensuration in the context of online rating and ranking mechanisms. Since our primary focus is on the (ontological) foundations of the sociomateriality perspective, we will concentrate on the three articles highlighted in the text rather than on this more applied contribution.

critical in places, therefore, it is our intention to add momentum to the ideas put forward by Orlikowski and Scott, rather than to put a brake on them.

We begin with a little background on the sociomateriality perspective and by setting out what we regard as its three core themes. These themes are then examined in turn in the sections that follow. We close with a brief conclusion.

Background

Orlikowski (2007, 2010) and Orlikowski and Scott (2008) portray sociomateriality as a "theoretical perspective" or "genre of research." However, while these phrases sound quite methodological and some of what Orlikowski and Scott have to say does have epistemological overtones, the distinguishing feature of sociomateriality is its intended ontological orientation. That is, all three papers focus on arguments concerning the nature of the world rather than on those concerning how research should be conducted. We are fully supportive of this. All researchers bear ontological commitments, whether or not they take the time to spell them out, and these commitments inevitably exercise an important influence on the choice of research questions, research methods, and research results. It is therefore useful to get them into the open.

Following authors such as Barad (2003, 2007), Introna (2007) and Suchman (2007), Orlikowski and Scott set up their account of sociomateriality by contrasting it with what they call an "ontology of separateness." By an ontology of separateness they mean a commitment to the idea that the world is made up of atomistic entities that, even if they should come to interact or combine at some stage, exist and are given in their nature prior to any such interactions or combinations. They associate this view with Cartesian dualism, which they regard as limiting precisely because it treats the social and the material as separate phenomena (Orlikowski, 2007: 1437; 2010: 134).

The picture that Orlikowski and Scott offer as an alternative to this kind of atomism/dualism is of a highly relational and constitutively entangled world. In their own words:

> ... the social and the material are *constitutively entangled* in everyday life. A position of constitutive entanglement does not privilege either humans or technology (in one-way interactions), nor does it link them through a form of mutual reciprocation (in two-way interactions). Instead, the social and the material are considered to be inextricably related—there is no social that is not also material, and no material that is not also social. (Orlikowski, 2007: 1437)

> ... entities (whether humans or technologies) have no inherent properties, but acquire form, attributes and capabilities through their interpenetration. This is a

> relational ontology that presumes the social and the material are inherently inseparable. (Orlikowski and Scott, 2008: 455–6)

> Scholars here have been working within a *relational ontology*, which rejects the notion that the world is composed of individuals and objects with separately attributable properties that 'exist in and of themselves'... Such an ontology privileges neither humans nor technologies... nor does it treat them as separate and distinct realities. (Orlikowski, 2010: 134)

We will unpack various aspects of these passages below. But before we go any further, it is necessary to say something about the terms "material" and "technology." As will be apparent from the above quotations, Orlikowski and Scott tend to use these two terms interchangeably. We are not going to follow in this regard (on the distinction between these two terms, see also Kallinikos and Leonardi, Chapters 4 and 2, respectively). By "material" we mean that the things described as such have a physical mode of being, namely possess spatial attributes—a unique location, shape, volume, and mass.[2] "Technology" refers to a wide variety of things, but like Orlikowski and Scott for the most part, we will concentrate on technology as represented by technological objects.[3] For the purposes of this chapter, we take technological objects to be any artifact or naturally occurring object that has one or more agentive function assigned to it.[4] An agentive function is the use one or more people assign to an object (of a certain kind) in pursuit of their practical interests (Searle, 1995). While agentive functions are typically closely related to the physical or logical structure of the object concerned, they are never intrinsic to it. Agentive functions are always relative to the individuals or communities who assign them.

Orlikowski and Scott do not make these distinctions in the papers under review, and it is not clear if they would accept them. But they do have two immediate implications. The first is that, since only some parts of the material world are technological objects by our criterion, the contents of the material world at any point in time stretch well beyond the world of material technological objects. The second implication is that our definition of technological objects admits both material and nonmaterial varieties. By nonmaterial

[2] In using the word "material" to refer to the physicality of entities, we therefore adopt a conception of materiality similar to Kallinikos (Chapter 4, this volume), in contrast to the accounts advanced by others such as Leonardi (Chapter 2) and Pentland and Singh (Chapter 14).

[3] We take "objects" to be entities that endure and, save for those that are so basic so as not to be composed of constituent parts, are structured. By "endure" we mean that the relevant entity is fully present at each and every point in time at which it exists. Objects may therefore be said to be "continuants," in contrast to events or other kinds of "occurrent" whose different parts occur at different points in time. By "structured" we mean that the relevant entity is composed of distinct parts that are organized or arranged in some way.

[4] Our criterion for something to be considered an item of technology is therefore quite broad, with objects such as notes and coins, sniffer dogs, and policemen considered to be technology within the communities whose members assign a use to those things. In this regard, our account differs from those of others, such as Kallinikos (2011), who adopt a narrower view of technology.

technological objects we mean those that do not have a physical mode of being, such as computer programs, search algorithms, technical standards and protocols, and so on (Faulkner and Runde, 2011*a*). The inclusion of nonmaterial technological objects implies that the set of technological objects in existence at any point in time is far larger than the set of material technological objects.

We can now return to Orlikowski and Scott and the passages quoted above. On our reading, two key themes emerge from them:

Theme 1: *Relationality*
Entities, whether human or technological, are in some way constituted by the relationships in which they stand to each other.

Theme 2: *Interpenetration*
The human and the technological are fused or interpenetrate.

We will explore the relationship between these two themes in detail below, but it will already be apparent that they beg an important question: if there are indeed "no independently existing entities with inherent characteristics" (Orlikowski, 2007: 1438), on what basis are we able to distinguish between things—"bodies, clothes, food, devices, tools" (Orlikowski, 2007: 1438), and so on—be this for analytical or indeed even simple practical purposes? The answer to this question leads to a third key theme:

Theme 3: *Agential Cuts*
The lines between things ("agential cuts") are drawn and enacted by "agencies of observation" and are "performed" rather than being intrinsic properties of those things.[5]

We regard these three themes as lying at the heart of the sociomateriality perspective. We will now state each of them in more detail, suggest where their limits might lie and, in places, propose where they might usefully be clarified and/or amended.

Relationality

The first of the three themes we associate with the sociomateriality perspective is that the social world is one in which humans and items of technology are in some way constituted by the relationships in which they stand to one another. We saw this idea in the passages quoted above. Here are two more examples:

[5] Performativity refers to that by which peoples' intentional attitudes toward something, as expressed in their practices, utterances, and so on, contribute to the constitution of that thing. In the language of Searle (1995), which we prefer, performed phenomena are "ontologically subjective" but may nevertheless be the subject of "epistemologically objective" statements.

> Humans are constituted through relations of materiality—bodies, clothes, food, devices, tools, which, in turn, are produced through human practices. The distinction of humans and artifacts, on this view, is analytical only; these entities relationally entail or enact each other in practice. (Orlikowski, 2007: 1438)

> ... people and things only exist in relation to each other... (Orlikowski and Scott, 2008: 455)

We should say immediately that, even as sometime representatives of the discipline that is most guilty of treating the social world as if it were atomistically constituted (i.e., economics), we too regard the social world as highly relational. The problem that arises is that this claim alone does not advance matters very much, since it is far too general to have much bite. There are after all innumerable ways in which things may be related, including by geographical location, ownership, gender, color, and so on, and surely many ways in which any one thing can be thought of as being related to any other.

However, and although they do not explicitly restrict themselves in this way, Orlikowski and Scott appear to have a particular species of relation in mind, namely *constitutive* relations. Constitutive relations are ones in which the relation in question contributes to making what one or more of the relata are. Let us follow convention and call all constitutive relations "internal relations" (Moore, 1919–20) and the residual set of all nonconstitutive relations "external relations." If we restrict ourselves to binary relations, there exists an internal relation between any pair of relata X and Y if:

> X would not be what it is but for the existence of Y, or
> Y would not be what it is but for the existence of X, or
> X would not be what it is but for the existence of Y and Y would not be what it is but for the existence of X.

Because they go only one way, the first two of these relations are less demanding than the third. River fishing, for example, is internally related to the category of rivers, since without rivers there would be no such thing as river fishing. But rivers can exist without river fishing. The internal relation is unidirectional in this case, as only one of the relata presupposes the other. Two-way internal relations are nevertheless quite common in the social world, including human-to-human relations (e.g., tenant to landlord, husband to wife), human-to-nonhuman relations (e.g., painter to paintbrush, footballer to football) and nonhuman-to-nonhuman relations (e.g., nut to bolt, computer to computer software). In each of these cases, the relata presuppose each other and the internal relation is a reciprocal one. Note that we regard all of these relations as social relations, that is, including those between nonhumans, and that we therefore regard the world of technological objects as part of the social world.

We fully accept the ubiquity of internal relations of the kind we have described, and therefore agree with Orlikowski and Scott that there are all kinds of things in the world that relationally entail each other. As we see it, such relations constitute instances of social structure that are drawn on and reproduced in human activities in the manner described by authors such as Giddens (1984), Bhaskar (1989), and Archer (1995). We are therefore quite happy with statements to the effect that material means are "constitutive of both activities and identities" (Orlikowski and Scott, 2008: 455). Reverting to our earlier example, the social position of footballer presupposes the social position occupied by the technological object we call footballs and vice versa.[6] The two positions are internally related, and this relation is drawn on and reproduced ("performed") in and through the practices of footballers (as well as those of third parties such as referees, coaches, fans, and so on). Further, as the function assigned to the objects we call footballs is community-relative and so not intrinsic to those objects,[7] the same is true of the identity of footballs that flows from their being implicated in the practices of footballers.

On the subject of relationality, then, we can go a long way with Orlikowski and Scott. There are, however, a number of caveats we would add that bear upon the sociomateriality perspective. The first is that, however many bits of the world may be related to other bits in all kinds of ways, it is probably not the case that everything is internally related to everything else (or if this is indeed what is being claimed—and there is a small group of philosophers including Hegel who are sometimes interpreted as doing so—then some argument needs to be provided in support of this claim). For example, while postmen and dogs often have an intimate and typically fraught relationship, it is probably not the case that this relationship is constitutive of either postmen or dogs. That is to say, at least as we see it, having interacted with a dog is not a necessary condition for someone to be a postman, just as having interacted with a postman is not a necessary condition for a four-legged creature that barks to be a dog. And if so, then all relations between postmen and dogs are external rather than internal relations. The same goes for the relations between many other things.

Second, and again contrary to what sometimes appears to be suggested on the sociomateriality perspective, it is quite possible to talk about entities being related without having to assume that those relata are preconstituted. That is to say, references to relata are not necessarily a sign of a commitment to an

[6] We have argued elsewhere that technological objects occupy identity-providing social positions as much as humans do (Faulkner and Runde, 2009, 2011*a*, 2011*b*).

[7] It is an interesting question whether all relations are observer-relative, and our intuition is that most are similar to functions in this respect. However, there are some species of relation that may not be observer-relative, such as apparently invariant relations between numbers that are often thought of as being "discovered" rather than "created" by humans.

"ontology of separateness." Thus the social positions of husband and wife are the relata in the relation between the social positions of husband and wife, even though the relation between them is an internal one and thus partly constitutive of the relata. Nevertheless, there is also a sense in which many social relations (and therefore their relata) are preconstituted, namely that they are often already in existence before people or objects draw on and help reproduce them in their activities. The social relation between husband and wife, for example, is one that had already been in existence for generations before the world's current population of husbands and wives came into being.

Third, when making claims about the existence of internal relations between things, it is important to be clear exactly what the relata involved are and whether or not all of the relata are constituted by that relationship. Our own examples, for the most part, involve relations between social positions, for instance between the social positions occupied by technological objects and the social positions occupied by human beings in their capacity as users of those objects. In this case, the internal relation between these two kinds of position flows in both directions, with each being part constituted by that relation.

Finally, while the identity of technological objects is inseparable from the social position they occupy, the particular *objects* that occupy social positions, once they have been created, are to some extent separable from those positions. We know this to be true because the same objects may move into a new position and acquire a new identity as a result (e.g., the gramophone turntable moving from its traditional function as a playback device to a new function as a musical instrument, first among the DJ and turntablist community in hip hop music, and then in other musical genres (Faulkner and Runde, 2009)); may already be in existence before they are put into a particular position (e.g., the tree used as a parasol or the stray reed used as a drinking straw); and may continue to exist as an object even when their social position and attendant identity have been forgotten (e.g., things that are obviously human artifacts sometimes found in museums, but the function and technical identity of which have been lost in time). Indeed, it is also possible that there exist social positions for objects that have yet to come into existence, such as the space shuttle that only existed in the minds and drawings of aeronautical engineers before the object itself was eventually constructed.

Interpenetration

The second of our three themes is the idea that the social and the technological are "fused" or "interpenetrate":

> ... the focus is on agencies that have so thoroughly saturated each other that previously taken-for-granted boundaries are dissolved. Our analytical gaze is drawn away from discrete entities of people and technology, or ensembles "of equipment, techniques, applications, and people" ... to composite and shifting assemblages ...
>
> ... entities (whether humans or technologies) have no inherent properties, but acquire form, attributes and capabilities through their interpenetration. This is a relational ontology that presumes the social and the material are inherently inseparable. As Barad (2003: 816) argues, this is a constitutive entanglement that does not presume independent or even interdependent entities with distinct and inherent characteristics. The portmanteau "sociomateriality" (no hyphen) attempts to signal this ontological fusion. Any distinction of humans and technologies is analytical only, and done with the recognition that these entities necessarily entail each other in practice. (Orlikowski and Scott, 2008: 455–6)

Careful reading of these and other passages that contain observations to the effect that humans and technologies "saturate each other" and acquire distinctive properties only in virtue of their "interpenetration" suggests that Orlikowski and Scott regard this ontological fusion or "constitutive entanglement" of entities as coextensive with what we have called internal relationality. We have separated the two because, as we will explain below, talk of composite entities, ontological fusion, saturation, interpenetration, and so on, suggests something rather stronger than mere internal relations. We will call this something extra the "interpenetration thesis."

The interpenetration thesis owes much to the feminist philosopher Karen Barad's notion of "entanglement," an ontological variant of the epistemic version originally introduced by Pickering (1993), which she states most succinctly as follows: "To be entangled is not simply to be intertwined with another, as in the joining of separate entities, but to lack an independent, self-contained existence" (Barad, 2007: ix).[8] In Orlikowski's own words:

> ... the notion of constitutive entanglement presumes that there are no independently existing entities with inherent characteristics (Barad, 2003: 816). Humans are constituted through relations of materiality—bodies, clothes, food, devices, tools, which, in turn, are produced through human practices. The distinction of humans and artifacts on this view, is analytical only; these entities relationally entail or enact each other in practice. (Orlikowski, 2007: 1438)

What then is to be made of general claims of this kind that humans and technological objects ("artifacts") interpenetrate to the extent that "there are

[8] The idea of entanglement is closely related to that of "intra-action" (as distinct from "interaction") in Barad's writings. As interpreted by Barad, whereas interactants exist prior to and therefore independently of their interaction, intra-actants only come into being via their intra-action. Barad associates the concept of interaction with a commitment to the atomism or "individualist" metaphysics that Orlikowski and Scott are themselves opposing.

no independently existing entities with inherent characteristics"? As we spelled out in the previous section, we accept that the social positions that humans and technological objects occupy are often internally related. And we have argued elsewhere that these social positions inform the identity of both the humans and technological objects that occupy them, as well as the practices in which those objects are implicated (Faulkner and Runde, 2011*a*, 2011*b*). Further, as far as technological objects themselves are concerned, they are almost always "socially shaped" in the sense of being the product of the activities of designers, users, trial-and-error learning, and so on, as well as being bearers of meaning in virtue of the positions they occupy (Pinch and Bijker, 1987; Pinch, 2008, 2010; Bijker, 2010).

The interpenetration thesis goes some distance beyond all this, however, bordering on a world picture in which machine operators meld physically with the machines they operate, computer programmers with the computer programs they write, and architects with the drawings they produce. Now it is possible that this image is intended metaphorically on the lines of Haraway's famous cyborg manifesto (1991: ch. 8). And of course we do not deny that there are real cases of technological objects that penetrate—and perhaps even merge with—the human body, such as artificial hearts and hips, bullets, and so on. Similarly, there is even a sense in which humans might be regarded as penetrating technological objects, such as when they enter other bodies such as a submarine, a bus, or an elevator. But if ontology is supposed to be about what exists, then it is perhaps one discipline in which one should be chary about positing metaphorical cases of interpenetration. And it seems to us here, first of all, that cases of the human body being penetrated by technology are the exception rather than the rule, and second that even the exceptions mentioned above involve very little in the way of an *inter*penetration of entities or entities "fusing" or "saturating" each other. While the needle of a syringe may penetrate the body, does the body penetrate the needle at the same time? We suggest not. The same goes for human bodies penetrating technological objects. In most cases, we submit, things like submarines, buses, or elevators do not get to penetrate the human body, or at least if they do, then with pretty dire consequences for the human bodies concerned.

In short, then, we would argue that most technological objects qua objects are separate from their users, in the sense that they have structures that, once in existence, are often quite stable and do not depend on relations with particular users (see also Leonardi, Chapter 2, this volume). We know this partly because we know that they are usually designed and built to be relatively robust across users and the types of environment in which they might be expected to be used, but also because we know from experience that they can often be relocated across space, time, and users without discernible effect on their structure and capacities. The physical object that touches the ground at Edwards Airforce Base in California is the same object that took off from Cape Canaveral two weeks earlier for a space shuttle mission (and was used by a different crew on an earlier mission), just as the physical object that we inherit from our grandfather is the same object that was once worn as a pocket watch by our great grandfather. Note, again, that this is not to say that the function, identity, and the social position of these objects are intrinsic to them, any more than the cultural connotations and political symbolism they may bear, the power relations they may reflect, and so on. But it is to say that, as objects, once they have come into existence, they are continuants that typically continue to exist even when not implicated in particular human activities (Kallinikos, 2011: ch. 2).

On the basis of these essentially empirical considerations, we feel that the interpenetration thesis must be rejected as a general claim about the nature of social reality. It is true that in some places Orlikowski and Scott appear to be thinking of human practices as opposed to human bodies fusing with or interpenetrating technological objects. They also talk about "composite and shifting assemblages" that could presumably penetrate, and be penetrated by, humans and technological objects. But this too seems quite metaphorical and not a little vague. Human practices are surely influenced and shaped by technology, just as technology is the product of human practices and shaped by them. But again, we find it hard to see in what sense there is an ontological fusion going on here. Human practices are activities, processes, etc., which are surely different from the objects that are implicated in those practices. There is perhaps more mileage for the interpenetration thesis in the assemblage idea, and we have suggested that computer networks may be regarded as things that involve true interpenetration (Faulkner and Runde, 2011*a*). But this hardly forms the basis for interpenetration being a generalized feature of the social world. And if so, we suggest that the sociomateriality perspective would be all the stronger if it took a more guarded position on the interpenetration thesis. Crucially, and as we have pointed out, doing so would continue to leave considerable scope for claims about internal relationality in the social world.

Agential Cuts

The third of our three themes flows from the first two: if the contents of the world are not separated by intrinsic boundaries as Orlikowski and Scott would have it, on what basis do we distinguish between things, be this in our capacity as researchers or simply as people going about our everyday affairs? The answer they give to this question, again following Barad (2003), is that the boundaries between, and possibly even the properties of, things are the products of "agencies of observation" and "performed" in "specific material-discursive practices." That is to say, they argue that, in the absence of intrinsic boundaries between things—observer-independent "joints" at which the world can be unambiguously "cut"—the boundaries we live by must be "agential cuts." As Orlikowski (2010: 135–6) puts it, "[M]aterial-discursive practices... enact specific local resolutions to ontological questions of the nature of phenomena." The question, then, is what exactly agencies of observation and agential cuts are, and, importantly, what is being assumed about the humans involved in making them.

According to Barad, agencies of observation are situated entities comprising humans and nonhumans (artifacts) standing in a particular relationship. The idea is an aspect of her development of Niels Bohr's reflections on the wave–particle dualism and double-slit experiments of quantum physics, where the relevant agencies of observation were an amalgamation of humans (experimental scientists) and nonhumans (interference devices, detection screens, etc.). The notion of agential cuts, as we have already hinted, seems to be a play on the old idea that nature has joints at which it can be "cut." This idea is in turn closely linked to a commitment to natural kinds, namely that there exist categories that do not depend on human observers and, which, on some accounts, is one of the goals of science to discover and document. Orlikowski and Scott would presumably follow Barad in rejecting the existence of natural kinds. Nevertheless, and like Barad, they seem to regard the drawing of boundaries (making agential cuts) as intimately connected with creating categories of things. We will for the most part follow them in this regard.

The question that arises at this point is how much these ideas apply outside of laboratory situations in the natural sciences, and in particular to what extent they might inform analysis of the social realm. This is where questions about what agencies of observation might be, and specifically the identity of the humans implicated in them, become important. Do social scientists make agential cuts in the course of carrying out research into the social domain? And to what extent do ordinary individuals going about their day-to-day business get to make agential cuts, or is this only the privilege of scientific researchers? Can isolated individuals even make agential cuts, or is this only something they can do as members of communities? And do humans make agential cuts all the time? That is to say, are people constantly creating their worlds as they are negotiating them—enacting "specific local resolutions to ontological questions of the nature of phenomena"—or do they also proceed on the basis of agential cuts performed by prior generations of agencies of observation and which they take as more or less given?

Orlikowski and Scott do not address these questions directly, although it is possible to tease out their position on some of them from their examples: in general, that they take it that ordinary people do make agential cuts as part of their day-to-day activities in the kind of dynamic, technology-rich environments in which Orlikowski and Scott are interested. We go along with this insofar as we accept the general point that it is materially and discursively situated humans who draw boundaries between things and create categories, and that these boundaries and categories are drawn on and performed in human practices. For example, and using our language, the social positions of different items of technology, and on the basis of which they are classified in one way rather than others, involve community-relative assignments of function. If these assignments of function bear wider currency in the community concerned, then they will guide practice and reinforce the social position in that community. Thus occupants of the social position of footballer "perform" the social position and associated identity of the objects we call footballs in their practices, just as the social position and identity of the objects we call footballs are integral to the social positions and associated practices of footballer, football referee, football fan, and so on.

Further, we accept the view that boundaries and categories are sometimes fluid, and that what is ostensibly the same object may sometimes be categorized in different ways and that existing categories may change. For example, the compound sildenafil citrate was initially regarded as a potential remedy for hypertension and angina, only to become most widely regarded as an effective counter to erectile dysfunction under the label of Viagra. And in a somewhat different vein, and as we have already observed, the identity of the object that we know as the gramophone turntable is one that has undergone a radical transformation in certain communities, from being regarded purely as a playback device to being regarded as a musical instrument.

However, examples of this kind notwithstanding, we would argue that many if not most of the boundaries and categories we live by in our day-to-day lives are generally quite stable, at least relative to our life histories, and that the same is true of most of the objects classified within them (Kallinikos, 2011). Otherwise, it would be hard to account for the apparent stability of social institutions and our artifactual worlds. Further, once particular boundaries and categories have become generally accepted, we would argue that whether or not something falls into one category or another depends not on observers but on how the world is. That is to say, once we have accepted a category—for example, once it is generally, if perhaps only implicitly, agreed that bicycles are two-wheeled, rider-propelled devices that include a frame, saddle, and handlebars, and whose function is a means of transport—then whether or not the object in front of us is indeed a bicycle and not a porcupine is not something that depends on us as observers, but on the extent to which it has the features just listed.

As before, however, Orlikowski and Scott seem to have something stronger than what we have allowed so far in mind, namely that, following Barad, objects are at least partly constituted by the activities of agencies of observation. Again, by "partly constituted" we take it they mean things being manifested ontologically, that is, that their coming into being and the objects they become depend in part on the activities of agencies of observation. An immediate question that arises is whether the view just described applies to all phenomena or only to those that depend on human intentionality. Orlikowski and Scott are silent on this question, so let us see what is involved here.

Let us call things whose existence does not depend on human intentionality "brute" objects. Examples of brute objects include those we call the moon, Mount Everest, and Tyrannosaurus Rex, all of which we believe there is considerable evidence to suggest were in existence long before the emergence of humankind and which are therefore not dependent on any agency of observation in which humans are implicated. We are not entirely clear whether Orlikowki and Scott would accept the existence of such "ontologically prior" objects, or whether they would regard them as something that only acquire shape, separation, and possibly even being through the explorations of agencies of observation. But going the latter route does have at least two uncomfortable consequences. The first is that it leads to the conclusion that there is no such thing as brute objects—no distinct objects with inherent boundaries and properties such as the ones that we humans have labeled the moon, Mount Everest, and Tyrannosaurus Rex—prior to there being agencies of observation to provide those labels. The second is that it leads to a problem of regress: if it is indeed the case that objects are only constituted via the activities of agencies of observation, this then raises the question of who/what are the agencies of observation at the next level that contribute to constituting those at the first level, and then who/what are the agencies of observation that contribute to constituting the agencies of observation at the second level, and so on.

Note again, however, that there is no conflict between being a realist about brute objects and the same brute object being categorized in different or changing ways. For example, while penguins were traditionally regarded as distant from all other living birds by Linnaeus (the distinguished Linnaean ornithologist and avian paleontologist Alexander Wetmore putting them in a superorder by themselves, with all other non-ratite birds in a different superorder), under the more recent phylogenetic species concept they are classed as being in the same superfamily as divers (loons), tubenoses, and frigate birds.[9] This would be the case of a new classification but, crucially, would not imply a change in the feathered (and fluffy when young) object that is now classified in a new way. That is to say, the brute object would remain exactly what it was

[9] We are grateful to Mike Rands for this example.

before it was put into a new category, although we do of course accept the possibility that a particular (re)classification may have a causal impact by feeding back on what was formerly a brute object (e.g., if some species is classified in a way that leads to human-led changes in its environment that affect its survival prospects).

So much for brute objects that do not depend on human intentionality. What about those that do? It is clear that the phenomena in which Orlikowski and Scott are particularly interested are of this second kind, namely technology and the relationship between humans and technological objects. The question that arises is whose intentionality is at issue here, whether it is:

1. of ordinary people whose day-to-day activities involve in some way the technology under consideration; or
2. of social science researchers who are investigating that technology and its use.

We have already said that we accept the idea that the functions, identities, and, more generally, the social positions of technological objects are performed by ordinary people going about their day-to-day affairs. Further, we accept that the distinctions between humans and nonhumans, as well as between different kinds of nonhumans, are indeed agential cuts in the sense of being made by human observers. But we have also suggested that it is the world that fills our categories, and that once a category has been accepted it is not an arbitrary thing whether or not some object falls into a particular category. In addition to this, we would suggest that the language of "agencies of observations" is probably not appropriate for ordinary people going about their day-to-day affairs. Of course, we acknowledge that human beings display reflexivity, that boundaries and categories do sometimes change, and that the world periodically throws up things that call for new boundaries and categories. But for the most part it seems to us that, while people may occasionally make new agential cuts and redefine and reshape parts of their world, many of the boundaries and categories they live by are ones they were born and socialized into and typically do not question. That is to say, they are part of what Searle (1995) calls the Background, part of our taken-for-granted world that forms a largely subliminal backdrop of peoples' more straightforwardly conscious deliberations and doings.

The tem "agency of observation" is probably more apposite when applied to social scientists, many of whom regard the phenomena they are investigating as something that exists independently of their investigations. Of course, researchers who investigate social phenomena are in a different position from that of natural scientists, insofar as they are able to exploit their own participation in the social world as a source of information about whatever it is they are studying, its meaning, and so on (the techniques of *Verstehen*). But that should not be taken to imply that they are constituting, in the sense defined earlier, the object of their investigations. So in our own study of the appropriation of the gramophone turntable as a musical instrument (Faulkner and Runde, 2009), we took the DJ/turntablist phenomenon to be something real and existing independently of our investigation, not least because we were writing well after most of what we were describing had run its course. We were indeed able to enhance our understanding of that phenomenon by interacting with a number of DJs, and we accept that it is just possible that these interactions might have been sufficient to affect their practices in some way. But even then it seems to us that the idea that we may thereby have been constituting DJ-ing/turntablism through our investigation would be making exaggerated claims about the causal power of our research activities.

There are, nevertheless, at least two ways in which researchers might affect the object of their investigations. First, they may do so at the time of carrying out research, when their activities during fieldwork, data collection, and so on, may have an impact on the object under investigation. For example, the actions of a team of economists appointed by the government to assess the strength of the balance sheets of major banks might well affect those balance sheets significantly if it should begin to emerge that their findings will be different from what the markets had been anticipating. Second, the knowledge researchers generate may subsequently feed back on the object of study by influencing the humans implicated in that object, for example, as described in Donald MacKenzie's (2006) account of how theoretical work on financial markets became part of what it examines. The former case is probably closer to Bohr and quantum physics, where the relevant agencies of observation interact with the object of study in a way that alters that object. The second case is unlikely to occur in the natural sciences of course, since the entities under investigation cannot read, reflect, and change in response to what scientists are saying about them. But even where this does occur in the social realm, researchers are not constituting the object under investigation as much as influencing its subsequent form (e.g., the capital asset pricing model altering subsequent behavior of finance professionals). It is then not so much that the investigation contributed to the creation of the object under investigation, as that object subsequently changing to some degree as a consequence of that investigation.

Conclusion

We said at the beginning of this chapter that our intention was to address the topic of materiality in the form of the emerging sociomateriality perspective, critically examining the three themes we regard as lying at the core of this approach with the aim of contributing to its continued development. We close with a reminder of these themes and a quick summary of our response to each of them.

Theme 1: *Relationality*
Entities, whether human or technological, are constituted by the relationships in which they stand to each other.

We agree that the social world is characterized by all kinds of social relations and that many of these are the constitutive variety of internal relations emphasized by Orlikowski and Scott. However, we would argue that it is not the case that everything is internally related to everything else. Further, we suggest that technological objects qua objects exist in a way quite apart from their social positions and technical identities, in the sense that, once they have been produced, they form part of the external world (and in most cases would continue to do so even if all human observers were suddenly to disappear).

Theme 2: *Interpenetration*
The human and the technological are fused or interpenetrate.

We argued that the interpenetration thesis should be rejected, at least as a general claim about the nature of social reality, and we did so on a largely empirical basis. The kind of exceptions such as artificial hearts, syringe needles, submarines, and so on notwithstanding, most technological objects do not penetrate human bodies (or vice versa), far less interpenetrate one another. But none of this is to deny that technological objects are shaped by the activities of humans, that technological objects in turn shape human activities, or that their functions and identities are dependent on the social relations in which they are embedded.

Theme 3: *Agential Cuts*
The lines between things ("agential cuts") are drawn and enacted by agencies of observation and are performed rather than being intrinsic properties of those things.

We argued that there are many complex issues involved here, among them what an agential cut entails, who the humans implicated in agencies of observation are, and which objects are considered within the theory's domain. Our own position is that the boundaries and categories we live by are indeed the constructions of humans. But that is emphatically not the same thing as saying that the things between which lines are drawn and/or that are being categorized are constituted by those constructions. Indeed, on the argument we have been developing, it is possible to be a realist about not only brute objects but also social phenomena, in the sense of taking the view that whatever we contemplate as observers exists independently of those contemplations. In the case of social phenomena, to use the terminology of Searle (1995), it is possible to make epistemologically objective statements about phenomena that are ontologically subjective. While it is true that there are cases in which the activities of human observers may affect whatever it is being observed and perhaps performed, that is not incompatible with realism about the phenomena concerned.

Acknowledgments

The authors are grateful to the editors for their comments on an earlier draft, and to the Cambridge Social Ontology Group for acting as a sounding board for many of the ideas reflected in this chapter. Jochen Runde would like to thank his MBA Programme colleagues Karen Siegfried and Jennifer Hersch for their support in helping to create the time to work on it.

References

Archer, M. (1995). *Realist social theory: The morphogenetic approach*. Cambridge: Cambridge University Press.

Barad, K. (2003). Posthumanist performativity: Toward an understanding of how matter comes to matter. *Journal of Women in Culture and Society*, 28(3), 801–31.

—— (2007). *Meeting the universe halfway: Quantum physics and the entanglement of matter and meaning*. Durham, NC: Duke University Press.

Bhaskar, R. (1989). *Reclaiming reality: Critical introduction to contemporary philosophy*. London: Verso Books.

Bijker, W. E. (2010). How is technology made?—That is the question! *Cambridge Journal of Economics*, 34(1), 63–76.

Faulkner, P. and Runde, J. (2009). On the identity of technological objects and user innovations in function. *Academy of Management Review*, 34(3), 442–62.

—— —— (2011*a*). The social, the material and the ontology of non-material technological objects. Unpublished.

—— —— (2011*b*). Creating space for technology, unpublished.

Giddens, A. (1984). *The constitution of society*. Cambridge: Polity Press.

Haraway, D. (1991). *Simians, cyborgs and women: The reinvention of nature* (pp. 149–81). New York: Routledge.

Introna, L. D. (2007). Towards a post-human intra-actional account of sociomaterial agency (and morality). Paper prepared for the Moral Agency and Technical Artefacts Workshop. The Hague: Netherlands Institute for Advanced Study.

Kallinikos, J. (2011). *Governing through technology: Information artefacts and social practice*. Basingstoke: Palgrave Macmillan.

MacKenzie, D. (2006). *An engine, not a camera: How financial models shape markets*. Cambridge, MA: MIT Press.

Moore, G. E. (1919–20). External and internal relations. *Proceedings of the Aristotelian Society*, 20, 40–62.

Orlikowski, W. (2007). Sociomaterial practices: Exploring technology at work. *Organization Studies*, 28(9), 1435–48.

——(2010). The sociomateriality of organizational life: Considering technology in management research. *Cambridge Journal of Economics*, 34(1), 125–41.

——Scott, S. (2008). Sociomateriality: Challenging the separation of technology, work and organization. *Academy of Management Annual*, 2, 433–74.

Pickering, A. (1993). The mangle of practice: Agency and emergence in the sociology of science. *American Journal of Sociology*, 99(3), 559–89.

Pinch, T. (2008). Technology and institutions: Living in a material world. *Theory and Society*, 37(5), 461–83.

——(2010). On making infrastructure visible: Putting the non-humans to rights. *Cambridge Journal of Economics*, 34(1), 77–89.

——Bijker, W. E. (1987). The social construction of facts and artifacts: Or how the sociology of science and the sociology of technology might benefit each other. In W. E. Bijker, T. P. Hughes, and T. Pinch (eds.), *The social construction of technological systems*. Cambridge, MA: MIT Press.

Searle, J. R. (1995). *The construction of social reality*. Middlesex: Allen Lane The Penguin Press.

Suchman, L. A. (2007). *Human–machine reconfigurations: Plans and situated actions*. Cambridge: Cambridge University Press.

4

Form, Function, and Matter: Crossing the Border of Materiality

Jannis Kallinikos

> In a hallway I saw a sign with an arrow pointing the way, and I was struck by the thought that the inoffensive symbol had once been a thing of iron, an inexorable, mortal projectile that had penetrated the flesh of men and lions and clouded the sun of Thermopylae and bequeathed Harald Sigurdson, for all time, six feet of English earth.
>
> Jorge Luis Borges, Mutations from the collection *The Maker*

Introduction

In this chapter, I examine whether and to what degree the physical or material substratum of technologies and technological objects affect the shaping of their functionalities, and the social relationships that develop around their conception and use (Pinch, 2008). Straightforward as such an objective may seem, actually achieving it is elusive. Even simple objects such as a chair or a knife entail much more than the materials of which they are made. Do the functions objects fulfill in social settings stem from their material makeup? A chair, for instance, is so shaped as to serve sitting rather than cutting and yet invoking the function of sitting is a purposeful (meaningful) act. To the degree that function is a modality of meaning, a particular way through which beliefs or intentions are expressed with respect to technological objects, it could be seen as predominantly nonmaterial (Searle, 1995; Faulkner and Runde, 2011).[1]

[1] Such a position does not assert that meaning is unrelated to contextual or material conditions and even less that it is deprived from material consequences. But it does assert that the ontology of meaning is predominantly immaterial. In other words, there is difference between ideas and things and something important is lost when this difference is glossed over.

How are then matter and function related to one another? These are tricky questions and they do not admit easy answers.

These difficulties, though, should come as no surprise. The reexamination of the role material conditions play in social life reprises major divides in Western thought such as those between mind versus matter, representation versus action, or rationality versus experience, to name but a few. The current interest in materiality and the material conditions of social action is, however, of more recent descent. Though variously taking side for matter, action, and experience (Suchman, 2007), such an interest reflects a growing awareness (see, e.g., Miller, 2005; Orlikowski, 2007; Leonardi and Barley, 2008) that the ordinary distinction between social and material relations may often obscure more than it reveals. Approaching the relevant issues through the lens of technology provides, I contend, an opportunity for shedding new light into these old and vexed questions. There would seem scarcely better focus on which to reassert the crucial role material conditions play in social life (Pinch, 2008). Whatever shifts to the meaning of technology that software-based systems and artifacts have brought about, the dominant associations technology invokes still evolve around the object world. Material embodiment is essential to technology. The functions objects embody either by deliberate design or haphazard use are variously conditioned by their material constitution. Cotton, for instance, cannot be used to construct hammers or screwdrivers, while wood, glass, or iron are not appropriate materials for clothing.

Thus, conditioning the deployment and shaping of technological objects or processes, the *materialness of matter* enters into human affairs. The constitution of matter would seem to circumscribe the field of possible objects or functions that can be accommodated by that matter. It is also implicated in the profile of skills and social practices that are associated with the mastery and use of materials. Traditionally, craftsmanship has developed around the mastery of specific materials such as wood, stone, clay, gold and iron, and textiles (Mumford, 1934, 1952; Sennett, 1994, 2008). Object agency, craftsmanship, and materiality are thus closely related to one another. Such a relation undoubtedly varies between different objects or fields and has been shifting in history through the evolution of crafts.

Function and matter are therefore closely related to one another, and matter has traditionally been used, elaborated, and formed to serve particular functions. Granted the ambiguity and diversity of meanings tied to materiality (Kallinikos, Leonardi, and Nardi, Chapter 1, this volume; Leonardi, Chapter 2, this volume), the observations put forth so far should be taken to indicate that I take the concept of materiality to literally signify the *material or physical constitution of technological objects* (or lack of it) and the implications (social and technical) such a constitution has for the design, making, and use of such objects. No other signification is attributed to matter (e.g., what matters, context embeddedness) if this does not stem from, or is closely related to, the material constitution of technologies.

A corollary of this is that technologies and technological objects are understood to grow at the confluence of form, function, and matter and thus entail much more than their material constitution. By the same token, technologies and technological objects are distinguished from standardized social behavior crystallized in routines, programs, and standard operating procedures. Despite far-reaching standardization and various material supports, routines, programs, and standard operating procedures lack the elaborate function structure of technological objects mapped to an equally elaborate structure of physical components (Ulrich, 1995). These defining attributes set technology apart from routines, programs, and standard operating procedures (see, e.g., Kallinikos, 2006, 2011*a*). A *clock*, for instance, is a technology, a *schedule* a program. They are, of course, closely related to one another, and programs, routines, and standard operating procedures are often supported by technologies (Sismondo, 1993; D'Adderio, 2011) but this does not justify lumping them together.

In the section that follows this introduction, I seek therefore to outline the basis on which function, form, and matter as defining attributes of technology could be analytically disentangled from one another. I then put forward the key argument of the chapter that is predicated on a reading of cultural and technological evolution that accords modern technology and technological design (function and form), a growing emancipation from the materials with which they are entangled. Counterintuitive as it may seem, such a claim makes sense against the background of the progressive dissociation of function, form, and matter. This dissociation has been driven by the decomposition of matter afforded by scientific advances and the growing entanglement of science with technology. I take the development of computation as a powerful and distinctive manifestation of the analytic predilection of technology that industrial techniques once inaugurated (Mitchell, 1996, 2005*b*; Borgmann, 1999). I therefore dedicate a section to discussing the distinctiveness of computation and the ways by which it relates to matter and form. In the final section of the chapter, I summarize my claims reflecting upon the explanatory potential of the concept of materiality and the implications my ideas have for empirical research.

Object Architecture

Deployed in a broad and associative fashion, materiality may seem as a straightforward condition that recalls the tangible, physical constitution of things and the inescapable material anchorage of life. However, upon closer scrutiny, the term turns out to be more slippery than this. Most of the things that populate our everyday personal or working space are cultural objects, that is, things, tools, or utensils constructed by humans to serve specific purposes. Such a condition sets human-made artifacts apart from objects and forms found in nature. It is precisely the imprinting of purpose and use upon matter that makes them cultural objects. This unavoidably raises the question concerning the materiality of cultural objects. To what extent is a simple and straightforward object as a table material? A table is made of some matter; yet what we call table is the designation of a certain *function* that has been perceived and imposed upon whatever matter can be made to serve that function, that is, wood, iron, synthetic material, stone (Searle, 1995). By function, then, I predominantly mean the purpose or purposes an object, or a set of objects, fulfills.

In the context of modern life, function is more often than not the outcome of deliberate design, that is, the activity of shaping matter to accomplish an end. I use the term design in this broader sense to designate not simply the process of form-giving as an epiphenomenon but to include as well the knowledge-based processes through which problem-solving arrangements are conceived, invented, and materialized (Simon, 1969; Ulrich, 1995; Flusser, 1999). Thus viewed, design is a cultural practice with strong artisan roots but also an activity closely associated with the faculty of imagination and the formal or experience-based knowledge on the basis of which it is exercised. Regardless of how deeply rooted in practice design may be, the invention and embodiment of function and form upon matter are heavily conditioned by knowledge, thinking, and imagination. This holds particularly true for those technological fields in which scientific knowledge plays a significant role (Ulrich, 1995).

Thanks to the spectacular development of science and technology in modern times, the uses of matter have expanded remarkably through the improvement of the conversion (matter to object) process and the discovery of new elements or matter qualities. Nuclear power furnishes a dramatic, perhaps, example showing how the disentanglement of matter (the identification of uranium and plutonium) can be used to support new processes (fission) out of which energy is produced. But there are plenty of humbler examples from petrochemicals (oil refinement), paper (extracting cellulose fibers out of wood), the evolution of computer processors afforded by new materials or matter qualities, to food crops. These examples reveal that the identification and discovery of matter qualities brought about by advances in science and technology affords penetrating the interior of matter, decomposing its compact nature, and promoting *selective use and (re)combination of matter qualities* along a broad spectrum of matter (Arthur, 2009). Humans, says Paul Valéry, in his imaginative and eloquent text on architecture *Eupalinos or the Architect*, extract from matter just some qualities (smoothness, hardness, resilience, plasticity) that fit their purpose and, within this context, remain as a rule indifferent to the rest (Valéry, 1950). Extraction is a physically conditioned state but one mediated by the implicit or explicit conceptual operations of abstraction: that is, the analytic ability to disentangle a quality or set of qualities from the bundled status by which they usually occur in physical states.

A case could therefore be made for the fact that, while important, the material substratum of technological objects does not suffice to define their instrumental identity. This last is contingent on form and function that are the deliberate outcomes of shaping or using matter to serve particular purposes. *Form,* in particular, provides the mold to which matter enters and is, as we know since Aristotle, the *causa formalis* (the receptacle, design, *eidos*) of a particular object or artifact, distinguished and to some degree juxtaposed to the *causa materialis* (matter, *hyle*). Technological artifacts combine then form/design and matter in different proportions or patterns. A lever (function) is more matter than design, while a software program perhaps exemplifies the opposite.

Form relates to function, which in some respects is the quintessence of technology. As indicated, function is the purpose which an object or artifact is made to serve, what can be accomplished by it, the *causa finalis,* in Aristotle's terms. While form may serve aesthetic sensibilities, within the overall performativity context technology establishes, it is closely tied to and, to some degree, derives from function, an idea that has succinctly been captured by the modernist motto of architecture *form follows function.* In the design of simple artifacts (tools and utensils), form is, as it were, phenotypic, betraying the function that it serves and with which it is closely associated. A knife must be sharp, the surface of a table flat, a vessel concave. The close relationship between function and form, however, becomes often blurred in complex technological systems that may thus give the impression of amorphous assemblages but persists, certainly, in such complex artifacts as aircrafts, motor vehicles, ships, or skyscrapers.

Form and function admit various types of matter; for example, a table may be of wood, metal, or other material. A particular matter, in turn, can assume different forms and serve different functions; for example, wood can be formed to table, chair, or door. These observations suggest that a certain freedom exists as to how function, form, and matter can be brought to bear upon one another. In the context of contemporary design, function assumes a guiding role. Technological objects often result from the conception of a function or set of functions that are subsequently mapped to physical components. While I have so far spoken of function predominantly in the singular, the fact is that technological objects result from the conception of more than one function, often organized in a layered system or function structure

(Ulrich, 1995; Yoo et al., 2010; Yoo, Chapter 7, this volume). Even a simple object such as a desk is composed by several functional elements (e.g., surface, legs, drawers), each of which is designed to perform one or more functions.

Once conceived, functions are mapped to physical components and brought to bear upon one another via the links or interfaces that connect functions and their physical components. That mapping entails a considerable degree of freedom with respect to the choice of matter and form of the physical components and, crucially, the way they should be connected to one another. The sum of these activities (functional elements, function structure, physical components, and links), which Ulrich (1995) refers to as product architecture, decides whether an object would acquire *en bloc* makeup (integral architecture) or be made of decomposable subassemblies (modular architecture). Ultimately, links and interfaces are functions as well, indicating the pivotal role function assumes in the conception, design, and making of technological objects. Rather than simply providing the overall purpose an object serves (Aristotle's *causa finalis*), functions in principle cascade throughout the entire spectrum of elements and physical components that make up an object and provide the framework within which matter is made to matter.[2] Any talk of material agency (Leonardi, this volume) has therefore to contemplate this intricate network of functional relationships which technological objects embody.

Such a state of affairs is closely associated with the aforementioned decomposition or unbundling of matter and the technological extraction of matter qualities or attributes. As with all analytic reasoning, scientific knowledge pierces deep into the compact and opaque nature of matter and identifies qualities and attributes possible to isolate and extract through technological processes (Borgmann, 1999). This opens up a space of deliberation, whereby a matter quality detached from its original, bundled physical state can be further enhanced, enriched, or combined with other matter qualities to create new qualities and, by extension, new technological objects, states, or processes. Much of technological evolution can be read along these lines. From this standpoint, technology emerges as a *combinatorial methodology* for molding (decomposing and recomposing) matter and reality: an "open language," as Arthur (2009: 45) suggests, entailing a vast and, certainly, miscellaneous repertory of principles and techniques out of which matter can be brought to accommodate new functions and uses. Once the site of

[2] I am not hereby claiming that the design of functions and the making of technological objects is a rational process, in the sense of entailing a linear and unambiguous progression from the conception of function to object-making. In some cases, it may indeed be; in others it may involve various trade-offs and negotiations between interest groups or trial-and-error and experiential learning. In yet others, it may even result from haphazard or serendipitous events or outcomes (Bijker et al., 1987; Margolin, 2002). The point I make is that, however this happens, it will always involve the consideration, confrontation, or negotiation of the role function and functional elements are assumed to play in the making of technological objects.

craftsmanship (technique) and "a means of production," technology "is becoming a chemistry" (Arthur, 2009: 45), a field in which matter is made to matter according to the principles that rule that chemistry.

Technological evolution has consequently transformed the game of technology from a system of fixed processes and ends to a *generative matrix* out of which new functions and forms are constantly produced. Placed in such a context, matter, as an important resource of technological ingenuity, becomes a space of possibilities, supporting a variety of human purposes as these are manifested in the functions and forms which *mater combinatoria*[3] is able to sustain. The disentanglement of the bundled ways by which form, function, and matter are shaped into technical objects and the pivotal role function assumes in this process are therefore essential prerequisites to accounting for the distinctive nature of such objects and assessing the implications of technological dynamics.

These ideas suggest that function, form, and matter are key constitutive elements of technological objects and processes. Different technologies exemplify varying strategies (designs) through which function, form, and matter are made to bear upon one another. Such empirical variability makes it difficult to extract a common pattern that applies across technological domains. At the same time, experience indicates that technological development disrupts established strategies and conventions and establishes new relationships between form, function, and matter. This is dramatically exemplified by major technological shifts such as the one signified by the transition from agriculture to industrial production but also technological changes of smaller scale within particular domains such as, for instance, those of building and architecture (Mitchell, 2005*a*, 2005*b*; Terzidis, 2005) or software development (Ekbia and Nardi, Chapter 8, and this volume; Yoo, 2010 Chapter 7, this volume). In a sense, technological development coincides with the invention of new functions and forms and/or the invention of a new material (dis)order (Valéry, 1950; Hughes, 2004) in which existing functions and forms are differently supported by matter. It is in this respect that technological evolution loosens the ties connecting matter and material conditions with the functions and uses which technological objects and processes are designed to serve.

Technology and Materiality

Placed against this backdrop, a case could be made for the fact that, in the context of modern science and technology, the technological domestication

[3] I juxtapose *mater combinatoria* to what, since Leibniz, is known as *ars combinatoria*, see http://en.wikipedia.org/wiki/Gottfried_Leibniz accessed March 10, 2012.

of matter achieves massive dimensions. In this process, technological projects are increasingly being driven by the vision of design rather than the exigencies of matter encountered in natural states, characteristic of less advanced technological stages. Technologically advanced societies epitomize a different relationship to materiality whereupon *hyle* (matter) gives way to *eidos* (function/form) and craftsmanship to techno-science.

These claims should not, however, be taken to imply that technologically advanced societies are less dependent on matter. In many respects, industrial societies manifest a growing yet qualitatively different dependence upon matter (resources). The more matter qualities are discovered, the broader the dependence on matter becomes, yet that dependence is of different kind. A tower or sailing ship exemplifies a different dependence on matter than adobe houses or dugout canoes, domestic heating systems other than the oil stove or the traditional fireplace (Borgmann, 1984; Arthur, 2009). The same holds true for the hydroelectric power plant compared to the traditional water or windmill (Mumford, 1934; Heidegger, 1977), or, to mention more recent examples, an analog camera and a paper document compared to a digital camera and a computer file, respectively. It is that difference that I am at pains to bring into focus. As technological sophistication grows, dependence on matter becomes increasingly contingent on particular matter *qualities* that are possible to extract, exchange, and combine across a wider spectrum of matter. A tower building can be made of stone, glass, or concrete of various kinds possible to combine in large variety of ways. Construction of adobe houses is critically contingent on the availability of the right type of clay. Thus, comments Flusser (1999: 28) on the issue in his characteristically oracular yet evocative style:

> [N}ow what we have is a flood of forms pouring out of our theoretical perspective and our technical equipment and this flood we fill with material so as to 'materialize' the forms. In the past, it was a matter of giving formal order to the apparent world of material, but now it is a question of making a world appear that is largely encoded in figures, a world of forms that are multiplying uncontrollably. In the past, it was a matter of formalizing a world taken for granted, but now is a matter of realizing the forms designed to produce alternative worlds.

Some of these ideas are given a colorful illustration in Lévi-Strauss' imaginative juxtaposition of the archetypical figures of the *scientist/engineer*, the *bricoleur*, and the *magician*, representing modern, medieval, and pre-technological times, respectively. The scientist/engineer, the emblem of the technological sophistication of modern times, proceeds with his tasks in a deductive/analytic fashion, whereby the problems confronted are, despite one or another deviation, approached by and large through the lenses of what is already formally known, controlled, and practiced. Confronting a task, the scientist/

engineer draws on an extensive body of cumulative knowledge and available techniques that provide him with the analytic and practical resources for defining problems and addressing them accordingly. In Lévi-Strauss' elliptic yet vivid language, the scientist/engineer tends to produce *events out of structures*, that is, the entanglements of formal knowledge, practices, and technologies that define the state of the art in the relevant field at a given time. The contingent and particular (the event) or what some of the authors in this volume will gauge in terms of materialization and consequence (see Kallinikos, Leonardi, and Nardi, Chapter 1, this volume) is always subordinated to the enduring and general (structure).

By contrast, the bricoleur (a handyman, jack-of-all trades and professions) exemplifies an approach to problem-solving in which each problem occasion is negotiated *in situ*. What is in each case addressed is not strictly predefined, even though a general objective (e.g., unlock a door, fix a broken machine) may provide the context for the undertaking. Through playful exploration (rather than deductive/analytic reasoning) of what is possible, the bricoleur adjusts past solutions, reframes, or manipulates technical memories of past events and projects to make them bear upon or even redefine the problem at hand. This is accomplished by relying on a miscellaneous toolbox of materials and quasi-tools whose functional capabilities are tried on the spot. Rather than having been designed in advance with specific purposes in mind, the bricoleur's instrumental resources are experiential leftovers that summarize his technical life journey. They have accrued over the years, as the outcome of the bricoleur's haphazard confrontation with a variety of projects. In their unpolished, underspecialized nature, the bricoleur's instrumental resources exemplify the interplay of function and matter and the shifting invasion of one by the other that his game exemplifies. "A particular cube of oak could be a wedge to make up the inadequate length of a plank of pine or could be a pedestal... in one case it will serve as an extension, in the other as material" (Lévi-Strauss, 1996: 18–19).

The bricoleur's method thus exemplifies a procedural trajectory that stands opposite to that of the engineer, ascending, as it were, from the contingent/particular to the general/enduring, building up *structures* (recurrent solutions) *out of the events* that he encounters in his diverse undertakings and technical confrontations. The bricoleur, Lévi-Strauss claims, may not accomplish his goals but derives pleasure from the challenge and poetry of his undertaking. At the same time, the very poetry and capacity for contingent responses that derive from the bricoleur's nature of instrumental resources place severe constraints on what can be accomplished by them. The constraints derive both from the finite number of the elements that make up the bricoleur's instrumental set and the limited extendibility which the fixed and stuck materialization of each of these elements affords. Function is tightly tied to matter and to the forms that his tools have acquired as the outcome of experiential use rather than deliberate design. Each project is an occasion in which new tools or materials can be added, while older ones may break down, be modified, and perhaps abandoned. In other words, function, form, and matter occur in bundled ways that considerably limit the space of possible uses in which the bricoleur's toolkit can be brought to bear upon. While allowing for contingent exploration, the limited and underspecialized nature of the elements of the bricoleur's instrumental set forcefully constrain the number of possible combinations they can enter with one another.

By contrast, the abstract nature of the knowledge of the scientist/engineer and the standardized character of the techniques that rule his practice, Lévi-Strauss claims, makes them in principle, if not in practice, open toward a much wider spectrum of possible opportunities or combinations. Like currency or standardized information, abstract entities, Lévi-Strauss claims, can enter into a much wider range of relations and combinations with one another, a condition that Yoo et al. (2010) and Yoo (Chapter 7) closely associate with the agnostic or non-product-specific functions or components that they claim are characteristic of many contemporary technological ecosystems (see also Varian, 2010). In such systems, particular elements, while heavily specialized, are still divested from single functions and particular materializations (hence agnostic) and given the possibility of entering into a much broader spectrum of combinatorial relations (Kallinikos, 2011*a*), as it often happens these days in the case of the so-called apps tied to one or more platforms. These claims, I feel, resonate well with the view of modern technology outlined earlier in this chapter as open language or chemistry (Arthur, 2009), a generative matrix, as it were, resulting from the analytic and practical disentanglement of matter and the many combinations and functions this enables.[4] As soon as the bonds tying form, function, and matter are loosened, a new space of action and deliberation opens in which technological functions are less constrained by the material substratum of technological objects or processes and can break through specific functional boundaries usually associated with specific products, industries, or platforms (Varian, 2010).

There may be good reasons to question Lévi-Strauss' idealized description.[5] The advance of science and engineering managed to keep bricolage at bay but never entirely extinguished it from the technological culture of modernity (March, 1976; Barley and Orr, 1997; Ciborra, 2003). An interesting and timely

[4] The scientist/engineer and the bricoleur are further distinguished from the all-embracing determinism of magic (and religion) and the pretension of the magician to produce outcomes through imaginary and noncontingent connections between causes and effects, means and ends, for example, healing by praying or talking, rain by dancing.

[5] See for instance Derrida (1978: ch. 10) for a thoughtful commentary on and "deconstruction" of Lévi-Strauss's distinction between the bricoleur and the scientist/engineer.

analogy to make is perhaps between the portrait of bricoleur painted by Lévi-Strauss and Zittrain's *procrastination principle*. According to Zittrain (2008), both the personal computer and the Internet owe much of their innovative and generative potential to the fundamental condition that they remain intentionally "unfinished" technologies, hence the principle of procrastination. Similarly, perhaps, to the bricoleur's toolkit, the procrastination principle does not close up, through specialization, the future use of computing technologies and the Internet. Rather than being defined in advance through specialization and fine-tuning, the components and procedures that populate the digital ecosystem can be brought to a variety of combinations by experimenting individuals. The PC, the Internet, and much of software innovation are the outcomes of deviant and experimenting routes pursued by amateurs who have an intrinsic predilection for bricolage.

Yet, the experimentation of these amateurs and hackers differs essentially from the technical poetry of bricolage, as this is portrayed by Lévi-Strauss. For, the experimentation opened up by the unfinished nature of computing technologies is still mediated by a highly developed and standardized technological landscape without which its innovative outcomes would be severely curtailed. Indeed, the combinatorial prospects of the "unfinished" technologies, functional elements, and physical components characteristic of the internet ecosystem are the outcome of the far-reaching standardization of that ecosystem along a number of dimensions, including components, protocols and standards, and programming techniques (Varian, 2010; Yoo et al., 2010). The fuller appreciation of Zittrain's argument requires the understanding of the conditions that define the generative capacities of computing technologies. And these are closely associated with a series of critical separations, among which figures prominently the disjuncture of hardware from software and function from matter. Let us turn to these issues.

Computation: Architectures and Principles

The flexibility and functional capacity of computing technologies are widely understood as being related with the abstract or logical nature of software, and its loose coupling to the underlying physical machine or hardwired infrastructure. Being a set of logical instructions written in a specialized code, software stands at the epicenter of the issues that concern us here. To some degree, it is a *technology without matter*, a conceptual scheme or frame in which a number of cognitive relations and procedures are laid out and ordered in ways that make possible their machine execution.

The transition to a technology instrumented as set of logical relations represents a long evolutionary trajectory whose origins can perhaps be traced back to systems of writing, counting, and calculation as methods and techniques of ordering reality, mastering production, and aiding economic exchanges. But it is instructive for the issues that concern us here to recall the initial or proto-environment of computing, as this is represented by the efforts of Charles Babbage to construct the first mechanical arrangement able to perform computational operations. On the one hand, the voluminous nature of Babbage's machine forcefully asserted its materiality and the indissoluble bond tying abstraction to matter. On the other hand, the machine differed conspicuously from all other known species of machines. Despite its imposing material presence, it neither processed material inputs nor produced material outputs. Thus describes Gleick (2011: 80–1) that innovation:

> Like the looms, forges, naileries, and glassworks he studied in his travels across northern England, Babbage's machine was designed to manufacture vast quantities of a certain commodity. The commodity was numbers. *The engine opened a channel from the corporeal world of matter to a word of pure abstraction.*[6] The engine consumed no raw materials—input and output being weightless—but needed considerable force to turn the gears. All that wheel-work would fill a room and weigh several tons. Producing numbers, as Babbage conceived it, required a degree of mechanical complexity at the very limit of available technology. Pins were easy compared with numbers.

The contrast between the weightless inputs and outputs of Babbage's machine and its imposing material presence is interesting and instructive. On the one hand, the often invisible yet massive infrastructures on which modern computation relies could be seen as the analog of Babbage's voluminous machine. On the other hand, the original separation of the two orders (matter and abstraction), which the machine pioneered, has progressed along the lines I have been at pains to describe in this chapter. However the developments since Babbage's original attempt to build a bridge between ideas and things could be construed, epitomize a number of separations among which figure the dissociation of *information processing* from *communication* and the construction of software systems able to operate independently from the semantics and pragmatics of communication (Shannon and Weaver, 1949). Such a context-free ideal of information processing is an instance of the unbundling of information processing (conceived as a function or set of functions) from the dense context of communication within which information processing is traditionally embedded. Once conceived as separable function, information processing could be given technological embodiment through programming and ultimately, via appropriate machine architectures, by the separation of *software* from *hardware*. This last separation permits the development and

[6] My own stress.

functioning of software without immediate interference from the underlying hardware infrastructure. The two systems stand loosely coupled to one another. The higher level technological operations the software embodies, including their regular update and development, can be pursued relatively independently from the prevailing hardware constraints. This is key to understanding how the paradigm of computation epitomizes a new relationship between function, form, and matter.

It would be relevant in this context to briefly comment on the exchange mechanism or channel through which matter and function, logical or cognitive relations, and hardwiring solutions are brought to bear upon one another in the context of computation. The mechanism is no other than binary coding, which, as Leibniz already perceived (see, e.g., Borgmann, 1999), is or can be rendered the fundamental and elementary operation of cognition. For, cognition and communication could be conceived as coinciding with the making, packaging, and transfer of distinctions derived or performed upon a primary (including emotions) reality (Bateson, 1972, 1979; Gleick, 2011).

Each binary alternation can be made to signal or carry a difference (distinction) and several differences (bits) assembled together in a pattern (bytes and beyond) could be used to describe reality (convey meaning) or, prescriptively, lay out the bits and pieces that make up a task or procedure. A consequence of this is that the compact (continuous, analog) nature of reality can be reduced to elementary binary alternations and scaled up again as assemblies of differences to cultural and perceptual units of apprehension and cognitive behavior that allow interaction with human agents (Bateson, 1972; Borgmann, 1999; Kallinikos, 2009). On the other end, the machine (the hardware) can just support the processing of binary alternations without needing to get involved in higher level operations of semantics and communication. For that reason, all software needs, at the bottom, to secure the link to the hardware and be numerical (Shannon and Weaver, 1949). Recall that, in the digital world, pictures, images, and forms that have the habitual appearance of reality (simulations or representations) are assembled by means of numerical computations and manipulated and dissolved accordingly (Manovich, 2001; Kallinikos, 2006, 2009). Digital pictures and images are just patterns of pixels or equations,[7] ultimately numbers (digits).

The relative independence of the critical layer of advanced technological functions that the software embodies from the underlying hardware infrastructure has a series of implications that cascade across the social relations that develop around the use or consumption of software-based objects and processes. At the human/machine interface, users confront cognitive inputs/

[7] Pixels and equations correspond to the *raster* and *vector* software engineering techniques of image composition and manipulation, respectively.

outputs and templates organized in procedures and operations that must be apprehended and acted upon accordingly. This makes imperative to perceive, fathom, and relate to whatever cognitive output the use of software produces. The use of software and the cognitive content it mediates have so far appealed primarily to apprehension and addressed by and large perception and cognition rather than action or bodily skills (Simon, 1969; Zuboff, 1988; Flusser, 2003).

This has been a development with far-reaching social implications. Whatever humans are able to accomplish with software presupposes user understanding of the functionality of the software and its cognitive output (or content). In information-rich environments, work, Zuboff (1988) observed more than two decades ago, becomes literally reading (see also Hayles, 2009), a condition that has been aggravated but also obscured from the bewildering array of inventions that have since her astute observations marked the development of computing, media, and the Internet. The trend, in particular, away from desktop computing that is often predicted to be imminent (e.g., ubiquitous computing) is expected to dissolve the gap separating design from use, information processing from behavior (Greenfield, 2006), reconnecting technology and human drives (Norman, 1999). The separation of form, function, and matter that is at the heart of the technological developments associated with computing suggests, however, that there are good reasons to doubt the implications of these predictions.[8] The growing mass of data characteristic of the internet ecosystem suggests that reading and comprehending software output and its structure are still vital requirements for relating to it and whatever it affords. The diffusion of software in organizational and social life continues to put cognition and cognitive mastery of the content the software helps produce at a premium (Zuboff, 1988). This has, no doubt, been a slow social change preceded and moderated by the culture of writing and printing but a deep going change, nonetheless, toward a culture of the virtual or nonmaterial (Hayles, 2009; Gleick, 2011; Kallinikos, 2011*a*).

Let it be clear that the software admits as inputs sign tokens (data or information items) and equally produces its output again exclusively in this cognitive medium. The software makes the semiotic medium of the *sign* the fundamental, universal, and pervasive "stuff" of social life. This is a radical shift away from the original confrontation of technology with matter and matter qualities and an important outcome of the dissociations (information processing from communication, hardware from software) described above.

[8] Greenfield himself cites Intel Research's Elisabeth Goodman who claims that "(t)he promise of computing technology dissolving into behavior, invisibly permeating the natural world around us cannot be reached... technology is... that which by definition is separate from the natural" (Greenfield, 2006: 28).

A serious implication is the profusion of sign tokens of every kind and the impressive expansion of a complex and, crucially, steadily accruing techno-cognitive net spun by data items and the cognitive patterns (knowledge or information) underlying them (Gantz et al., 2008; Edwards et al., 2009; Hayles, 2009). This constantly accruing net of information and knowledge, of which the Internet is the most conspicuous manifestation, becomes increasingly an important vehicle through which context-free models of cognition and communication penetrate social practice and mediate situated encounters (Borgmann, 1992, 1999; Lévy, 1997; Kallinikos, 2006).

These observations make it relevant and perhaps fruitful to conceive of *digital content* as yet another layer distinct and above the habitual layers of hardware and software (Weinberger, 2007). Individual machines and the Internet as the most encompassing computing environment exemplify the separation or, perhaps more correctly, loose coupling of digital content from the software through which that content is produced and acted upon and the range of physical devices (PCs, handsets, laptops) and hardware infrastructures supporting software. The centrality and growing importance that digital content assumes in the digital ecosystem is indicative of the layered and modular architecture of the digital ecosystem (Weinberger, 2007; Zittrain, 2008; Yoo et al., 2010; Yoo, Chapter 7, this volume) and marks an important development of technology toward the ether of the virtual and the immateriality of cognition and communication, supported underneath by increasing selectivity in the use of matter.

The relative independence of the layers of the digital ecosystem may, however, obscure the steadily shifting boundaries between hardware and software. A very interesting development in this respect is what is often referred to as hardware virtualization (Carr, 2008). At some basic level, hardware virtualization is nothing other than the deployment of software to redistribute memory and processing capacity and, in cases like cloud computing, to lease the use of software in novel and granular ways. In this respect, it bespeaks the steering role software assumes in the digital ecosystem. And yet, in some other more profound way that recalls Borges's "iron mortal projectile" that became a direction sign (the passage quoted at the outset of the chapter), hardware virtualization betrays the expansiveness of software and its capacity to penetrate, extend, and, ultimately, domesticate hardware. In this sense, the developments associated with hardware virtualization and cloud computing are indicative of the trends that I have been trying to depict in this chapter, concerning the predominance of function over matter and form, and the changing parameters of social action that result from this (see also Nardi and Ekbia, this volume).

Aren't these claims derived from an approach to computation predicated on *representational* models of computation (Greenfield, 2006; Yoo, 2010)? What about *material hermeneutics* and the reentry of things in the space of everyday living afforded by embedded, *experiential* computing and the so-called Internet of things (Ihde, 1990, 2011; Greenfield, 2006)? The developments that are associated with experiential computing (Yoo, 2010) are complex enough to warrant their own lengthy treatment. However, the ideas put forth in this chapter do carry some important implications with respect to how the involvement of computing technologies in the space of everyday living should be understood. What I have primarily sought to do in this chapter is to analytically disentangle the *shifting articulations* of *function*, *form*, and *matter* that underlie technological evolution. If I am right, then experiential computing is bound to interfere with the primary processes of perception (Kallinikos, 2011*b*) and reality framing, the original *circumspection* that Heidegger saw as the basis on which things are apprehended and absorbed into everyday living (Dreyfus, 1991). The voice that embedded computational functions lend to things is after all premeditated, designed, and embodied even though these functions are often activated on the fly. Experiential computing is computing nevertheless, a significant evolutionary step in the modes of computation that pulls agents and things together, instead of separating them as traditional modes of computation have done, an undeniably significant development (Greenfield, 2006; Kaptelinin and Nardi, 2006; Yoo, 2010). At the same time, experiential computing allows technology to tread upon aspects of human living that have traditionally escaped technological mediations of this sort (Hayles, 2005).

Postscript

In this chapter, I have claimed that the bonds tying the invention and making of technological objects and patterns to matter have increasingly become loosened over the course of technological evolution. Technological sophistication measures the degree to which technological objects or processes result from considerations (function and form) that transcend the qualities of their material substratum (Flachbart and Weibel, 2005; Faulkner and Runde, 2011). In this regard, technological operations could ultimately be seen as decontextualized conceptual arrangements (templates or matrices) on the basis of which reality is ordered to objects or patterns. Such arrangements can materialize as when a piece of music represented as lines of code is burned to a disk or other material but they do not need to do so (Faulkner and Runde, 2011). The software, then, most clearly epitomizes this immaterial/incorporeal ontology of technology and the gradual and progressive dissociation of function from matter and, of lately, form.

I do not associate any moral value to this and I should not be understood as saying that the trends I seek to describe are good (or bad). If I am allowed to make a confession, then I am convinced that at least as much is lost as it is gained (Nardi and Kallinikos, 2010; Nardi and Kow, 2010). Also, powerful political and economic interests undergird the developments I describe about which I have been virtually silent (Margolin, 2002; Feenberg, 2005), while the ways these developments are refracted across the social fabric vary rather substantially from one societal context to another. I can console myself in front of these limitations by the fact that this chapter was meant to serve a different purpose. I have predominantly sought to deconstruct the notion of materiality, hoping to get away from what I often perceive as a frivolous use of the term and a superficial understanding of the depths of collective historical experience from which our relations to the material world ascend (Borgmann, this volume). Despite the differences separating my approach to technology from those of Pinch (2008), Orlikowski (2007), and Suchman (2007), I deeply regret that in the course of just few years the original promise of the term "materiality" as outlined in their deeply engaging research has been shattered and the term has grown to an inflating buzzword, an empty vessel able to contain an amazingly large and ambiguous web of significations.

But I feel less certain today than the moment I started few years ago, which is, I assume, what a good intellectual journey always affords. I will still hold, albeit with less firm a conviction, that *the progressive dissociation of function, form, and matter from one another is the most remarkable attribute of technological evolution* whose implications are poorly understood and, with few exceptions (Zuboff, 1988; Hayles, 2005, 2009; Kallinikos, 2009, 2011*a*), seldom investigated to a sufficient degree. Such dissociation has shifted and continues to do so the parameters (some may prefer to call them affordances or mediating conditions) of social action as habitual forms of associating to people and things have several times been disrupted or refigured (Borgmann, this volume). The appreciation of these shifts seems to me an essential prerequisite to any reflection and research on the role material relations play in technological and, by extension, social processes.

Short of appreciating the relative independence of function, form, and matter, characteristic of technological sophistication, the involvement of technology in human affairs runs the risk of being misconstrued. Important technological change disrupts the prevailing equilibrium between abstract operations and real things, intangible templates, and their materialization. Empirical research on technology and technological change therefore needs to recognize the wider effects of that disruption and assimilate it to the study of particular technologies and practices (Feenberg, 2005). It is vital to appreciate the ways technological functions become implicated in the pursuits of social agents with or without their awareness, consent, or preference

(Zuboff, 1988; Kallinikos, 2011*a*). By the same token, it is important to historically understand how human agency is framed and constructed by important technological developments. Before they contemplate history, humans, wrote the renowned British historian Carr (1986/1961), are the products of history and a similar claim could be advanced with respect to technology and technological evolution. This recursive relationship between technology and agency has to be contemplated and its crooked history disclosed.

Function, understood that way, is of course as much a technological as a social or communal accomplishment.[9] What is recognized as technological function takes place against the background of established beliefs and social practices and cannot be studied apart from them (Pinch, 2008; Orlikowski, 2007; Suchman, 2007). But function is equally crafted upon the material or logical embodiments of technological objects and systems. Software and the painstaking computations it embodies are not trivial things and they certainly cannot be wished away. Crucially, the enactment of function requires the formation of skills and capabilities distributed throughout communities and carried out by entrenched educational and vocational institutions that transcend situated practice. It is along these paths, I feel, that materiality and social theory could be linked (Leonardi, 2010).

Acknowledgments

Many people have helped me improve this chapter. I would, above all, like to thank my coeditors in this volume Bonnie Nardi and Paul Leonardi. I am also very much indebted to Daniela Piana, Elena Esposito, Giovan Francesco Lanzara, Jochen Runde, and Mireille Hildebrandt, who provided me with valuable comments over the rather extended time I have been writing this text.

References

Arthur, B. W. (2009). *The nature of technology*. London: Allen Lane.

Barley, S. R. and Orr, J. E. (eds.) (1977). *Between craft and science: Technical work in U.S. settings*. New York: Cornell University Press.

Bateson, G. (1972). *Steps to an ecology of mind*. New York: Ballantine.

——(1979). *Mind and nature*. New York: Dutton.

Bijker, W. E., Hughes, T. P., and Pinch, T. (eds.) (1987). *The social construction of technological systems*. Cambridge, MA: MIT Press.

Borges, J.-L. (1998). *Collected fictions*. Trans. A. Hurley. London: Penguin.

[9] I am thankful to Jochen Runde for raising my awareness on this.

Borgmann, A. (1984). *Technology and contemporary life*. Chicago: University of Chicago Press.

——(1992). *Crossing the postmodern divide*. Chicago: University of Chicago Press.

——(1999). *Holding on to reality: The nature of information at the turn of the millennium*. Chicago: University of Chicago Press.

Bowker, G. (2005). *Memory practices in the sciences*. Cambridge, MA: MIT Press.

Ciborra, C. (2003). *The labyrinths of information*. Oxford: Oxford University Press.

Carr, E. H. (1986) (originally published in 1961). *What is history?* Harmondsworth: Penguin, revised edition by R. W. Davies.

Carr, N. (2008). *The big switch: Rewiring the world, from Edison to Google*. New York: Norton.

D'Adderio, L. (2011). Artifacts at the centre of routines: Performing the material turn in routines theory. *Journal of Institutional Economics*, 7(2), 197–230.

Derrida, J. (1978). *Writing and difference*. London: Routledge.

Dreyfus, H. L. (1991). *Being-in-the-world: A commentary on Heidegger's being and time*. Cambridge, MA: MIT press.

Edwards, P. N., Bowker, G. C., Jackson, S. C., and Williams, R. (2009). Introduction: An agenda for infrastructure studies. *Journal of the Association for Information Systems*, 10(5), 365–74.

Faulkner, P. and Runde, J. (2011). The social, the material, and the ontology of non-material objects. Paper presented at the European Group for Organizational Studies (EGOS) Colloquium, Gothenburg, 2011.

Flachbart, G. and Weibel, P. (eds.) (2005). *Disappearing architecture*. Basel: Birkhauser.

Feenberg, A. (2005). Critical theory of technology: An overview. *Tailoring Biotechnologies*, 1(1), 47–64.

Flusser, V. (1999). *The shape of things: A philosophy of design*. London: Reaktion Books.

——(2003) (originally published in German, 1987). *Die Schrift: hat Schreiben Zukunft?* Athens: Potamos (in Greek).

Gantz, J. F., Chute, C. et al. (2008). The diverse and exploding digital universe: A forecast of worldwide information growth through 2011. IDC White Paper, sponsored by EMC. www.emc.com/digital_universe

Gleick, J. (2011). *The information: A history, a theory, a flood*. London: Fourth Estate.

Greenfield, A. (2006). *Everyware: The dawning age of ubiquitous computing*. Berkeley: New Riders.

Hayles, C. (2005). Computing the human. *Theory, Culture and Society*, 22(1), 131–51.

——(2009). RFID: Human agency and meaning in information-intensive environments. *Theory, Culture and Society*, 26(2/3), 47–72.

Heidegger, M. (1977). *The question concerning technology and other essays*. New York: Harper.

Hughes, T. P. (2004). *Human-built world: How to think about technology and culture*. Chicago: University of Chicago Press.

Ihde, D. (1990). *Technology and the lifeworld: From garden to earth*. Bloomington, IN: Indiana University Press.

Ihde, D. (2011). Smart? Amsterdam urinals and autonomic computing. In M. Hildebrandt and A. Rouvroy (eds.), *Law, human agency and autonomic computing: The philosophy of law meets the philosophy of technology*. London: Routledge.

Kallinikos, J. (2006). *The consequences of information: Institutional implications of technological change*. Cheltenham: Elgar.

——(2009). On the computational rendition of reality: Artefacts and human agency. *Organization*, 16(2), 183–202.

——(2011*a*). *Governing through technology: Information artefacts and social practice*. Houndmills: Palgrave.

——(2011*b*). Technology and accountability: Autonomic computing and human agency. In M. Hildebrandt and A. Rouvroy (Eds.), *Law, human agency and autonomic computing: The philosophy of law meets the philosophy of technology*. London: Routledge.

Kaptelinin, V. and Nardi, B. (2006). *Acting with technology*. Cambridge, MA: MIT Press.

Kittler, F. (1997). *Literature, media, information systems*. Amsterdam: OPA.

Leonardi, P. M. (2010). Digital materiality? How artifacts without matter, matter. *First Monday*, 15 (June), 6–7.

——Barley, S. R. (2008). Materiality and change: Challenges of building better theory about technology and organizing. *Information and Organization*, 18(3), 159–76.

Lévi-Strauss, C. (1966). *The savage mind*. London: Weidenfeld and Nicolson.

Lévy, P. (1997). *Collective intelligence: Mankind's collective world in cyberspace*. Cambridge, MA: Perseus Books.

Manovich, L. (2001). *The language of new media*. Cambridge, MA: MIT Press.

March, J. G. (1976). The technology of foolishness. In J. G. March and J. P. Olsen (eds.), *Ambiguity and choice in organizations*. Oslo: Universitetsfoerlaget.

Margolin, V. (2002). *The politics of the artificial: Essays on design and design studies*. Chicago: University of Chicago Press.

Miller, D. (Ed.) (2005). *Materiality*. Durham: Duke University Press.

Mitchell, W. J. (1996). *City of bits: Space, place, and the Infobahn*. Cambridge, MA: MIT Press.

——(2005*a*). After the revolution: Instruments of displacement. In G. Flachbart and P. Weibel (Eds.), *Disappearing architecture*. Basel: Birkhauser.

——(2005*b*). Constructing an authentic architecture of the digital era. In G. Flachbart and P. Weibel (Eds.), *Disappearing architecture*. Basel: Birkhauser.

Mumford, L. (1934). *Technics and civilization*. San Diego: HBJ.

——(1952). *Arts and technics*. New York: Columbia University Press.

Nardi, B. and Kallinikos, J. (2010). Technology, agency and community: The case of modding in the world of warcraft. In J. Holmstrom, M. Wiberg, and A. Lund (eds.), *Industrial informatics design, use and innovation: Perspectives and services*. New York: IGI Global.

——Kow, Y. M. (2010). Digital imaginaries: How we know what (we think) we know about Chinese gold farming. *First Monday*, 15, 6–7 (June).

Norman, D. A. (1999). *The invisible computer*. Cambridge, MA: MIT Press.

Orlikowski, W. J. (2007). Sociomaterial practices: Exploring technology at work. *Organization Studies*, 28(9), 1435–48.

Pinch, T. (2008). Technology and institutions: Living in a material world. *Theory & Society*, 37, 461–83.
Searle, J. (1995). *The construction of social reality*. London: Penguin.
Simon, H. A. (1969). *The sciences of the artificial*. Cambridge, MA: MIT Press.
Sennett, R. (1994). *Flesh and stone: The body and the city in western civilization*. London: Faber and Faber.
—— (2008). *The craftsman*. London: Penguin.
Shannon, C. and Weaver, W. (1949). *The mathematical theory of communication*. Urbana: University of Illinois Press.
Sismondo, S. (1993). Some social constructions. *Social Studies of Science*, 23(3), 515–53.
Suchman, L. (2007). *Human-machine configurations, second edition of plans and situated actions*. Cambridge: Cambridge University Press.
Terzidis, C. (2005). *Algorithmic architecture*. London: Elsevier.
Ulrich, K. (1995). The role of product architecture in manufacturing firm. *Research Policy*, 24, 419–40.
Valéry, P. (1950) (originally published in French in 1921). *Selected writings of Paul Valéry*. New York: New Directions.
Varian, H. A. (2010). Computer mediated transactions. *American Economic Review*, 100 (May), 1–10.
Weinberger, D. (2007). *Everything is miscellaneous: The power of the new digital disorder*. New York: Times Books.
Yoo, Y. (2010). Computing in everyday life: A call for research on experiential computing. *MIS Quarterly*, 34(2), 213–331.
—— Henfridsson, O., and Lyytinen, K. (2010). The new organizing logic of digital innovation: An agenda for information systems research. *Information Systems Research*, 21(4), 724–35.
Zittrain, J. (2008). *The future of the Internet—And how to stop it*. New Haven: Yale University Press.
Zuboff, S. (1988). *In the age of the smart machine: The future of work and power*. New York: Basic Books.

III

Materiality as Performativity

5

Ranking Devices: The Socio-Materiality of Ratings

Neil Pollock

Introduction

Today, it appears that there are rankings to rate the quality and value of most things: of art (Becker, 1982), books (David and Pinch, 2008), theater (Shrum, 1996), restaurants (Blank, 2007), films (Hsu, 2006), music (Karpik, 2010), the performance of various public services such as hospitals, schools, and universities (Strathern, 2000; Free et al., 2009), the efficiency of the latest consumer products (Aldridge, 1994), the reputation (Schultz et al., 2001) and competence (Pollock and Williams, 2009*a*) of companies. There are even ratings listing the "best places" to live and work (Kornberger and Carter, 2010), the "top holiday destinations" (Scott and Orlikowski, Chapter 6), and so on.

Despite their simple and often contested nature, there is much evidence to suggest that ratings are constitutive of domains (Aldridge, 1994; Blank, 2007; Karpik, 2010). Speaking about one of the most well known rankings, the *Red Michelin Restaurant Guide*, for instance, Karpik (2010: 77) has conceptualized this suggestively as a "paper engine." To label something an "engine" is to imply (to use the popular Actor Network Theory assertion) that it is not simply passive but "active" in the world. As MacKenzie (2006: 12) writes: "an active force transforming its environment, not a camera passively recording it."

In terms of how rankings are seen to act, this is predominately described in terms of the way they encourage processes of "reactivity" among people (Espeland and Sauder, 2007). Rankings change how people make "sense of" and "interpret" situations, which can encourage them to adapt their behavior in some way, often toward that suggested by the measure (ibid.). This begs the question as to whether there are further agential aspects of rankings beyond simply how they create processes of reactivity. In what other ways, besides these predominately "social" accounts, are markets constituted by rankings?

In this chapter, we argue for a greater focus on the socio-material agency of rankings (see also Scott and Orlikowski, this volume). This is in particular on the various material devices and equipment that surrounds and contains rankings and which has a bearing on their constitutive capacity. Such is the importance of materiality that we suggest that there can be no rankings without the *devices of ranking*. The very idea of the "100 top restaurants," "10 leading law schools," or "20 best cities to work and live," for instance, could not exist without the device of *the list* (Goody, 1977).

The general aim of this chapter is to show how rankings are shot through with various forms of material equipment; that it is only through working with devices that ranking organizations can produce and communicate assessments. Our more specific aim is to demonstrate, through an empirical study, that these objects do more than simply facilitate communication of a judgment calculated prior to its incorporation in a material form. Rather, these equipment/devices come to shape the kind of review produced. They do so because they offer various "affordances" (Gibson, 1977) that ranking organizations are encouraged to take into account. This means that ranking devices *afford* a particular kind of ranking. This can have a number of intended and unintended consequences for those subject to these measures. In particular, the chapter will focus on how *realizing* these affordances require the ranking organization to make changes to the domain. The setting, in other words, is changed to fit the ranking rather than the other way around—a process we attempt to capture with the word "affordizing."[1]

The empirical focus of the chapter is an influential ranking from within the information technology (IT) domain. This is a device called the "magic quadrant," which rates IT vendors according to their "vision" and "competence" in realizing that vision (see Figure 5.1). The magic quadrant is a particularly intriguing rating to study because it is a *visual* device. Rankings communicated in a figuration not only look different to those found in a "list" or "table" but contain and are surrounded by specific affordances as well as "graphic resources" (Lynch, 1988) and "conventions" (Espeland and Stevens, 2008).

Our work sits in the productive space that exists between Science and Technology Studies (STS), Information Systems (IS) research, and a number of allied disciplines focusing attention on processes of economization and materiality (Caliskan and Callon, 2010), as well as issues of "graphic

[1] "Affordizing" is a concatenation of two words: affordance and realization. It is not enough to say that a device has affordances in and off itself without also focusing on how individuals have an awareness of these capacities and, importantly, the ability to realize them (see Chapter 11 by Robey et al. in this volume for a detailed discussion of the relational view of affordance adopted here).

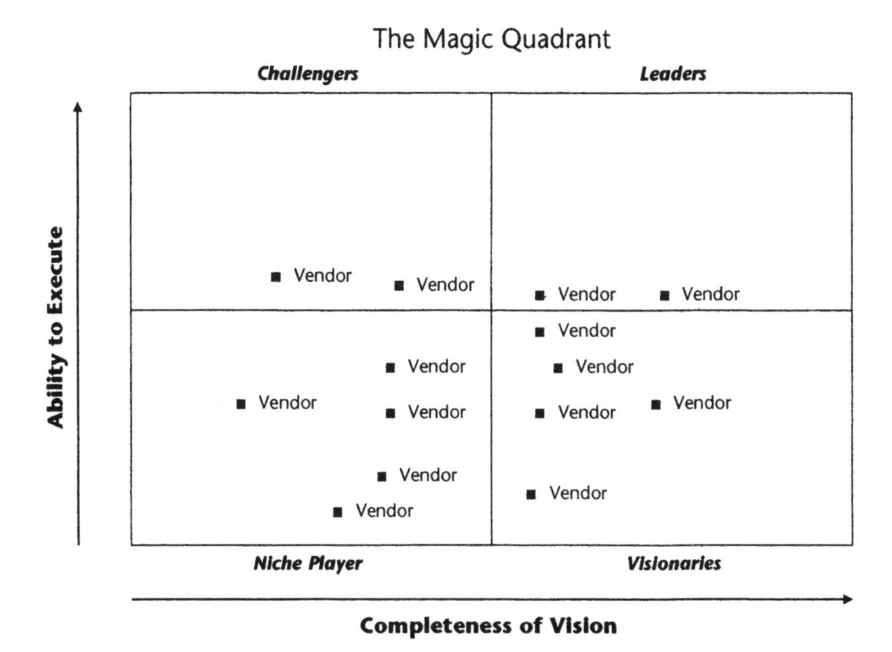

Figure 5.1 The magic quadrant

visualization" (Lynch, 1985, 1988; Latour, 1986; Myers, 1988; Quattrone, 2009). The study draws on observations and interviews conducted over a period of several years on one of the most influential ranking organizations in the IT area. Given there have been few, if any, detailed studies of the internal practices of ranking organizations thus far, this chapter potentially contributes to a better general understanding of the construction of a successful ranking as well as its socio-material constitution.

Rankings are Engines Within the Economy...

An increasing number of scholars argue that rankings do more than simply describe a setting but that they also *intervene* within a situation (Espeland and Sauder, 2007; Karpik, 2010; Kornberger and Carter, 2010). While there appears, on one hand, a growing consensus developing about the constitutive aspect of rankings, there has been, on the other, a lack of clarity or detail around exactly *how* rankings shape settings. Espeland and Sauder's (2007) discussion of university law schools, where they suggest that rankings are engines because they encourage "mechanisms of reactivity," is one exception. We review it here both because the notion of "reactivity" is systematically described and also because it might possibly be strengthened through attending to the issue of materiality.

Mechanisms of reactivity appear to be predominately about the reconfiguring of interpretations: "...ranking are reactive," they write, "because they change how people make sense of situations." Thus, rankings "offer a generalized account for interpreting behavior and justifying decisions within law schools, and help organize the 'stock of knowledge' that participants routinely use" (p. 11). It follows that, because new interpretations of a situation are introduced, people then conform behaviors according to this understanding. To evidence this, Espeland and Sauder cite the words of a respondent who notes how "[r]ankings are always in the back of everybody's head. With every issue that comes up, we have to ask, 'How is this impacting our ranking?'" (p. 11). Their thesis is that, as a result, rankings can become *self-fulfilling*:

> One type of self-fulfilling prophecy created by rankings involves the precise distinctions rankings create. Although the raw scores used to construct [Law School] rankings are tightly bunched, listing schools by rank magnifies these statistically insignificant differences in ways that produce real consequences for schools, since their position affects the perceptions and actions of outside audiences. (p. 12)

This leads them to suggest that "[r]ankings are a powerful engine for producing and reproducing hierarchy since they encourage the meticulous tracking of small differences among schools, which can become larger differences over time" (p. 20). However, while changes in interpretations brought about by rankings are clearly important, to suggest that a ranking is only about a change in "perception" gives the impression that it is an entirely "social" phenomenon. Moreover, a ranking that exists simply in the heads of people suggests it has a rather ephemeral status (MacKenzie, 2006). Rankings are *also* incorporated in material devices. The *list*, for instance, is highlighted as significant in Espeland and Sauder's work but not something brought out in the analysis. Paraphrasing their words, the list magnifies small differences that produce real consequences. This would suggest that there are further agential characteristics to rankings than simply how they encourage processes of reactivity. We develop this insight in the remainder of this chapter.

... But Ranking Devices Matter Also

STS have recently become interested in the study of markets (Callon, 1998, 2007; MacKenzie, 2006), particularly from the point of view of the market "devices" and "equipment" that facilitate their operation and functioning (Callon et al., 2007; MacKenzie, 2009). There is much in the emerging discussion that we find useful for a discussion of rankings, particularly the emphasis given to the *material* and *distributed* character of agency. Briefly, in contrast to a notion like "reactivity" that is centrally concerned with capturing the agency of people (Espeland and Sauder, 2007: 7), this foregrounds the (now familiar Actor Network Theory) point that actions are seldom performed by individuals alone but that people are "propped up" and "equipped" through the mobilization of various kinds of material artifacts. Importantly, in enrolling the latter, it is seldom the case that these material objects simply increase the efficiency of activities that an actor would have carried out anyway. Instead, they prescribe or permit certain behaviors (Akrich and Latour, 1992). The market devices have certain *affordances* that the actor is prompted to accept, such that these can reshape their actions and practices (Callon and Muniesa, 2005).

Rankings are a particularly ripe ground for identifying various kinds of distributed material agency. Once we enquire into the devices and equipment surrounding ratings, for instance, we find there are a number of formats available, each replete with particular affordances and limitations, and accompanied by different kinds of "furniture" and "resources." The visual ranking device described below, for instance, has different capacities and constraints to those found within "lists" or "tables." To give proper consideration to what exactly these differences are, we find it necessary to turn to scholarship that has considered graphical representations.

Figurations

Latour famously argued that "he who visualises badly loses the encounter" (1986: 13). The "scientific graph," for Latour, was one factor that gave science its influence over other forms of knowledge production. Graphs were influential because they could render complicated and often invisible phenomena into readable curves. Thus, they played a central role in both producing and communicating scientific knowledge. This was because the graph is a highly mobile form of knowledge (it formed the basis of Latour's concept of the "immutable mobile") but also because it "simplified" the kind of knowledge that was represented. Quattrone (2009: 109), in his discussion of the history of the book, develops a similar point and argues that it is because graphs are "partial" and "simplified" that they can become "performative":

> Graphical representations... are always so partial and simplified that they essentially contain very little; they have little truth in them; for, if it ever existed, it has been lost in the process of diagrammatic representation which has sacrificed details and context for the sake of clarity. This is the only way in which they can effectively communicate and engage the user in a performative exercise.

Espeland and Stevens (2008: 423), in their review of the field of Communication Studies, suggest that graphs can do this because they are produced according to "aesthetic ideals." This includes the fact that graphs should have particular characteristics: "... people who make pictures with numbers typically prize representations whose primary information is easily legible (clarity), and which contains only those elements necessary and sufficient for the communication of this primary information (parsimony)." This is because those who construct and read graphs as part of their professional activities want them to be "not only errorless but also compelling, elegant, and even beautiful" (Espeland and Stevens, 2008: 422).

Here, we also highlight the views of Lynch who, by contrast, argues that visual displays in science do more than simplify because they also *add* features not found in the original knowledge. According to Lynch, even the simplest graphs *add* rather than reduce information. An example can be found in an earlier paper where he discusses a common but little discussed graphic resource, the "device of the dot" (Lynch, 1985: 43). Analyzing a field manual that describes the anatomy of a lizard, Lynch makes the following point:

> Note that each observation of a marked individual is rendered equivalent to all others through the use of the device of the "dot." The only material difference between one dot and another on the chart is its locale. Locales are reckoned in terms of the grid of stakes, and all other circumstantial features of observation "drop out."

Dots are "additive" rather than simply "reductive" (we get this terminology from Ingold's discussion of another type of notation, "the line" [2007]). The "dot" renders the things on a graph equivalent and encourages the reader to compare different bits of data. What we find useful is Lynch's suggestion that these graphic resources go onto *merge with* and come to *embody* the thing being mapped. He writes: "... one theme which applies to many, if not all, graphs is that of how the commonplace resources of graphic representation come to embody the substantive features of the specimen or relationship under analysis" (Lynch, 1988: 226), and in turn: "... efforts are made to shape specimen materials so that their visible characteristics become congruent with graphic lines, spaces, and dimensions" (p. 227).

To summarize, below we want to show that rankings do not just take away but also *add* to a situation. In particular, they add these commonplace "resources of graphic representation" (Lynch, 1988), which, in our case, are also "dots." The contribution of our chapter will be to demonstrate how the placing of these dots is shaped by three things: these are, the affordances of the graph itself; the "conventions" of graphic representation (Henderson, 1999), which are similar to the "aesthetic ideals" that Espeland and Stevens (2008) identify as guiding the drawing up and presentation of graphs; all of which is further constrained by an organizational apparatus that is developing within the ranking organization.

The Magic Quadrant

The ranking device discussed here is produced by the industry analyst firm Gartner Inc. (hereafter Gartner). Gartner is just one of a number of such consultancy organizations operating in the IT area but it is widely recognized as the largest and most influential. While it does not have a monopoly over the sector, industry commentators suggest it has something close (Hopkins, 2007). We flag two particular developments that indicate this influence. Gartner has been highly active in identifying and classifying new technological trends. For instance, it coined and subsequently went onto shape the enterprise resource planning (ERP) terminology, as well as a number of later technologies (Pollock and Williams, 2009*b*). The designation of a new technological field of activity is far from trivial. If successful, it can—as with ERP and subsequent technologies—draw boundaries around a set of artifacts and their suppliers, thereby creating a space in which some sorting and ranking may be possible (ibid.). The magic quadrant has been particularly relevant in this latter respect. It has paved the way for a comparative analysis of the relative advantages of particular offerings for an adopting organization. Although difficult to measure, it is said to have some effects on procurement choices (if not dictating choices at least dictating shaping the decision about *which* vendors make it onto the table) (Hopkins, 2007). Many acknowledge also how it influences the strategies of vendors (ibid.) Some are said to assess offerings, their promotion, and enhancement in relation to the features of broadly comparable products found within their particular magic quadrant (ibid.).[2]

[2] We have been studying the magic quadrant for several years now. During that time, we have visited their European headquarters in London, interviewed a number of analysts, observed their interactions and dealing with their clients, as well as attended several of their conferences and events. Because the bulk of analysts interviewed work in the area of customer relationship management (CRM), our discussions in this chapter focus primarily on this technology. In particular, the material presented below stems primarily from two sources: interviews conducted with the Gartner CRM team and attendance at the annual Gartner "CRM Summit." In terms of the latter, these turned out to be a particularly fertile ground for studying rankings. Since the meetings were run a little like academic seminars, it was relatively easy to engage analysts in conversations about their work or to simply hang around and listen while others quizzed them about particular vendor placements. All of the formal interviews were taped and fully or partially transcribed. During our participation in Gartner conferences, we took extensive notes. The collection of data at these venues was facilitated by the fact that Gartner video-record all sessions and make these available to participants after the event. This feature meant that we could return to relisten to sessions while back in our university office.

Opening the Black Box of a Ranking

Graphical rankings devices (like the magic quadrant and its main rival, the "Forrester wave") are interesting because much activity occurs around the placing and moving of dots. The dot has become the obligatory point of passage upon which actors focus—and some have gone as far as to develop strategies and plan for modes of interaction with industry analysts to move these dots (see Pollock and William, 2009*a*). If the dot is the focus, then this begs the question as to just how these are placed in the first instance.

Standardizing the Dots

The placing of dots remains the most contentious point surrounding rankings such as the magic quadrant. We had the opportunity during our research to quiz a number of Gartner analysts about this: "The accusation we were always given," responded one to our question "was that we threw darts at the chart" (interview, Gartner analyst). This particular analyst then went on to refute this accusation, through spelling out how exactly they compiled their assessments. In particular, the question people often asked was whether the magic quadrants were drawn by hand or by some other means. In fact, they are plotted by a spreadsheet populated with numbers from a "standardized scoring mechanism." The scoring mechanism is made up of "evaluation criteria" that have been divided along the two axes of the magic quadrant: "completeness of vision" and "ability to execute." These then break down to reveal a number of further criteria, which are given a weighting within the spreadsheet as "high," "standard," "low," or "no rating." If a "no rating" is applied, this means that this particular factor will not be counted. Furthermore, the analyst describes how there is a standardized process across the organization:

> ...individual analysts have to follow the same procedure, and we have to document that, and you have to have an audit trail of how it was created, and usually you have to have scoring sheets to demonstrate how you got to that point but on the actual spreadsheet that creates the quadrant there is a scoring, a whole scoring system which is standardized across the whole company. (Interview, Gartner analyst)

What was interesting about the standardized process was that it differed significantly from how graphs were plotted just a few years ago. He describes how:

> ...they were more comprehensive in those days but they weren't consistent. So the way I would have my criteria would be nothing like my colleague sitting

> next to me. We weight in a very different way and the dots are arrived at very differently. And the vendors didn't like that. The vendors didn't like being top right in one and bottom left in another and not knowing why. Often that was because they were trying to negotiate about how they were treated. (Interview, Gartner analyst)

Magic quadrants were "more comprehensive" because vendors could be scored according to criteria that analysts themselves felt were important at the time. However, this meant that the process of plotting the dots differed across magic quadrants—and this caused problems for Gartner's relationship with vendors. He goes onto describe how certain aspects of the process had "improved" through standardization but also hints that not all these were leading to improvements in the "value" of the magic quadrant:

> So I would argue that the value of the magic quadrants ten years ago was actually better, even though they were less accurate in some ways... there were bigger movements on magic quadrants from year to year. But the point being made was that analysts were changing the weightings much more dramatically to reflect what the customers were telling them. Now we reflect the customers much more less well, because we have to go through a lot more steps to reflect what the customers are asking. So, it is an interesting trade-off really. Who is the value for? (Interview, Gartner analyst)

He goes onto describe how there is growing pressure on the ranking organization to "standardize" and make "public" their assessment measures for fear of litigation: "...I think a lot of this is to protect the company against vendors prosecuting them. So it is defensive rather than beneficial to the customer. As a result, I think it is less beneficial to the customer but maybe I am the only one who has that view. I suspect more have. Anyway, it is a lot more rigorous than it used to be" (interview, Gartner analyst).

Because the tool is seemingly "more rigorous," this has brought about some interesting changes in the placing of dots. One impression we had from interviewing this particular analyst was that he was describing the decrease in freedom and control that he and his colleagues felt over the construction of magic quadrants. The ranking was being circumscribed by a new material and organizational reality. We press this point further through looking at some specific constraints.

New Organizational Machinery

In this section, we show how the placing of dots is dependent on a complex organizational apparatus. We focus on two aspects: technology and increased forms of bureaucracy.

The Spreadsheet

The spreadsheet is central to our discussion because it vividly captures the point that technologies can both enable and constrain (often both at the same time). The spreadsheet represented the primary means by which analysts recorded, calculated, and plotted graphs. It literally, to paraphrase Latour (1986), was able to draw a number of things together (Law, 2002). Yet an interviewee noted how the spreadsheet also created a particular difficulty for analysts. Once graphs had been drawn, it was difficult, if not impossible, to move an individual dot. This was because, as one analyst noted, within the technology the dots were all connected together: "...you just can't put the dots where you want. The dots are all related to each other. So if you move one score up it impacts all the dots on the chart" (interview, Gartner analyst). To move one particular vendor up, down, or across once the graph had been generated would create further widespread movements on the graph. Even the smallest change could affect the position of all the vendors and this was problematic because it would attract the attention of colleagues elsewhere in the organization (see below).

Therefore, the spreadsheet had become a further factor constraining the ability of analysts to place dots as they saw fit. However, one of the *few* things they were permitted to do once a graph was plotted was not to move an individual dot but to:

> ...move the box around a bit. So, in other words, if all the dots are clustered in the centre you can reset the axes to get the box more spread out so they look more attractive. Otherwise, you would have a scale where all the dots are clustered around the centre or clustered around one spot. The idea there is just to make them spread out so you can actually read who compares to who. So, there is a little bit of flexibility on the edges, but frankly, you can't really rig it anymore. (Interview, Gartner analyst)

The analysts had the freedom, this "little bit of flexibility," to adjust the overall scale within the spreadsheet. If the dots were clustered in one spot, it was possible to adjust the box to create distance between the various players. That is, to enhance or develop a greater distinction between dots than the spreadsheet had initially revealed. This seemingly was to attempt to make the rankings more "attractive" (we return to this point below).

Review Cycle and Committees

One of the reasons why it might difficult (in their words) to "rig" the dots was because the magic quadrant was also subject to an internal "review cycle":

> ... a typical review on any piece of research, magic quadrants are just slightly more rigorous, you have to have two peer reviews, then one mandatory review, which is somebody who is an agenda manager for a research area... Then it goes to a team manager who sits above multiple agendas who has a sort of more general overview. And then it should be able to go. (Interview, Gartner analyst)

Added to this, it could also be that the dots of larger vendors were also watched over by a "lead analyst": "Those people have to review where the dot is and what the wording of the text is" (interview, Gartner analyst). One final part of the review process was that graphs were also sent out to the vendors prior to publication who, in turn, were free to comment (though it was acknowledged that a comment from a vendor rarely had any effect on the placing of a dot).

Another committee process relevant to our concerns was the one to approve the creation of new magic quadrants. This was notable because, in the past, developing a new device had apparently been straightforward:

> ... there is a procedure now, before you could just do it, 10 years ago you could just create one if you wanted to, you just had to negotiate with the boss, but now you have to go to a committee. There is a senior research committee that has to approve all new proposals for magic quadrants. So you have to justify there is a market, it's big enough, it's growing at this rate, there's lot of market clients, here's the enquiry volume coming from the customers, 'OK then, you've got a magic quadrant'. (Interview, Gartner analyst)

Alongside this fact that creating these devices was increasingly bureaucratic, there appeared to be a number of quite pragmatic reasons as to why setting up a new magic quadrant was difficult. We consider this issue in the next section.

Aesthetic Conventions

There are a set of socio-material affordances and constraints surrounding the production of rankings. What we want to do now is describe how these include "aesthetics conventions" in relation both to what and when things should be represented on a graph.

It was reported to be difficult to set up a magic quadrant both at the *outset* and at the more *mature* stages of a technological lifecycle. There could be difficulties in the initial stages of a new technological phenomenon because there could simply be *too many* vendors: "When there is a 100 [vendors], that's not very good for us... because then it is not mature enough for us to actually say, so what we are doing is watching that very carefully, and going, I will give you an example, Social Media Monitoring devices. There is tonnes of them at the moment" (interview, Gartner analyst). When asked to explain *why* too many vendors were a problem for creating magic quadrants, our respondent replies: "... graphically, you can't, you can we've done it, you can have a

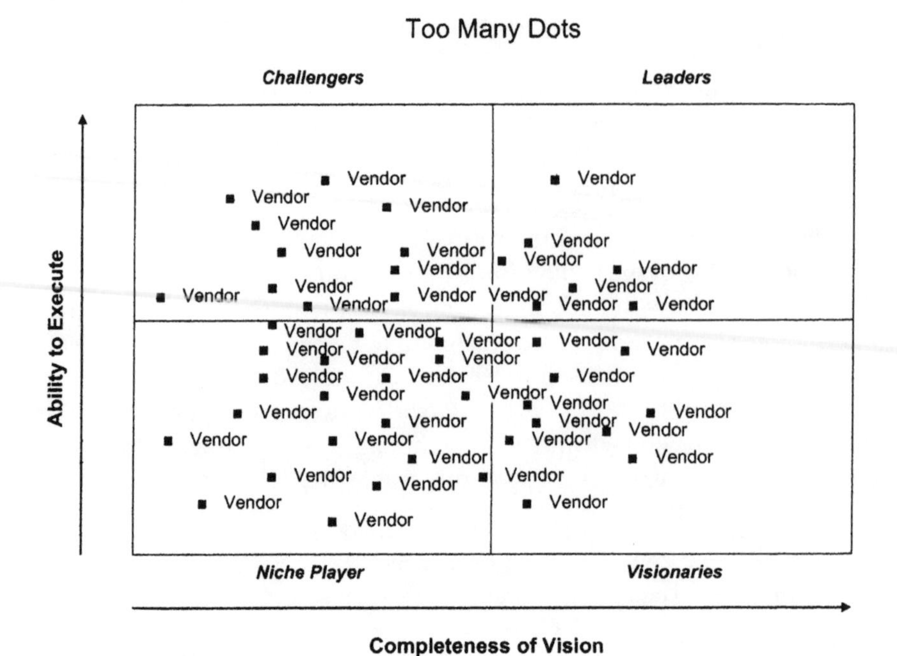

Figure 5.2 Too many dots

100 dots on the chart but it is unreadable. It is just garbage. It is just a bunch of dots" (interview, Gartner analyst). Seemingly, if all players producing a particular new technology were to be included on a graph, then this would mean that the ranking would be too cluttered. There would be too many dots and vendor names and this could present problems for those attempting to read and make sense of rankings (see Figure 5.2).

The analyst then goes onto describe how equally, too few vendors was also a problem: "And likewise when there is 3 dots on it, it is meaningless. What's the point of having a magic quadrant with 3 dots?" (interview, Gartner analyst). Apparently, too few dots means that there was very little being described in terms of how the market was developing (see Figure 5.3). He gives a recent example:

> ...we used to do things like operating systems. Operating systems, ironically, now are actually increasing in number again. But when Microsoft started dominating operating systems on desktop or desktop applications it was pointless having 4 dots on a chart... But the ones that I have seen that have gone, have basically just dwindled to a point where through mergers and acquisitions they are down to less than 8 vendors, and the colleagues all turn around and go 'what was the point in that?'. The clients don't read them anymore, they are not so interesting. The only people who read them then are clients who want to justify what they are

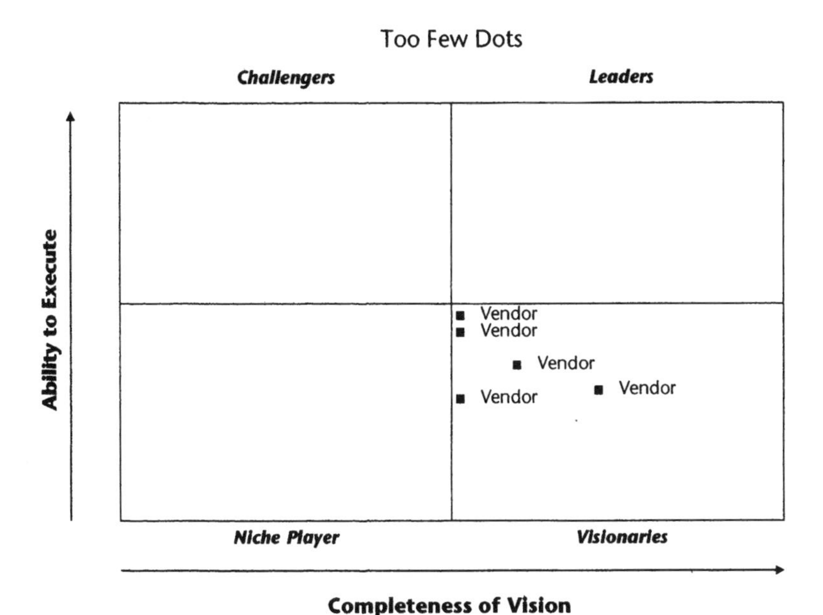

Figure 5.3 Too few dots

> already doing—it is an insurance policy kind of thing. But their value is very, very low. The dots hardly move. And nobody is very interested. (Interview, Gartner analyst)

In contrast to situations where there are too many/too few vendors to be mapped, the analyst described how he had come to realize that there were an *ideal* number of dots that could be pictured:

> So, I would argue that magic quadrants are almost like, if you imagine a market always going theoretically going a 100 down to 10, to 5 vendors or something as it consolidates and the barriers to entry get put up by the incumbent. Gartner's magic quadrant is the beautiful picture when you have gone down to about 20, 25 to 15, or 10, and then once you go below that it ceases to be useful. And before that it is not particularly useful. (Interview, Gartner analyst)

The ideal is somewhere between ten and twenty-five dots. This is what he identifies as the "beautiful picture." It is where the graph is neither too crowded nor too empty. It is a beautiful picture because it keeps Gartner in the "game," to use the words of this respondent: "So, while it is in that sort of state between about 25 down to maybe 10 vendors, there is a choice, there's a multiple different dimensions to it, and different ways of evaluating, how you write each vendor up. There is complexity in it, and therefore there is a game for us to play" (interview, Gartner analyst).

Capturing the Beautiful Picture

One of the problems about beautiful pictures was that they were temporally bound. This was because at times the number of players in an emerging market was changing so fast that Gartner was not able to *capture* it. Sometimes Gartner was simply too slow to respond to it, or, by the time it had reacted, the beautiful picture had long gone. To illustrate this point we include the comments of an analyst talking about the case of "Web Analytics":

> Sometimes they move through so fast that... Gartner's magic quadrant never quite... hits it. And a good example of that would be Web Analytics where... it was 68 vendors about 4 years ago and now there is about 20 or so. But there is only 3 big ones who control a vast majority of the market, followed by Google which is free and then there's a couple of specialists. So really to have a magic quadrant with about 5 or 6 on, there is not much point anymore. So it went from 68 to 6 in about 3 years and so there was little window there where Gartner could have managed to get a snapshot of the market when there was 20 in, but then it was gone. (Interview, Gartner analyst)

In the case of Web Analytics, there were initially too many vendors and then later too few for them to "get a snapshot." This meant it was unable to produce a magic quadrant for this particular technology/market. We postulate Gartner is unable to capture beautiful pictures not only because markets are changing fast but also because it finds it increasingly difficult to mobilize its large (and growing) organizational machinery in a timely fashion.

Creating the Beautiful Picture

What we also found was that Gartner was not completely *passive* in searching for the beautiful picture. If the required picture was not there, then they would help *create* it. From fieldwork we noted how it had (at least) two strategies for doing this. The first was related to the standardized evaluation criteria described above. When there were too many vendors to be included in a magic quadrant, an analyst would continually set and reset these criteria in order to reduce the field. One analyst describes this by talking through the example of Social CRM (customer relationship management): "... I am looking at Social CRM at the moment and we've identified 92 vendors in the last three days. Can't put 92 dots on a chart! So, it is pretty clear that we will set some high criteria to cut people out. And that is what the big debate will be about is how you set those criteria" (interview, Gartner analyst).

The second strategy analysts deploy is to divide markets up to get the required picture. An interviewee describes how: "I mean, so clearly there is a kind of optimal number of dots on a chart which Gartner kind of ends up almost dividing markets up in order to get that number of dots on a chart, which is readable, which is about 15 to 25" (interview, Gartner analyst). Let us unpack the implications of what is being described here. The analysts will set the bounds of the market to ensure that they arrive at what they think is an "optimal number" of vendors in the magic quadrant. Because there are too many vendors in an area—and since the market in its entirety cannot be captured on a magic quadrant—analysts will literally divide markets up. This means they will create new technology groupings (new kinds of distinctions between technologies). Moreover, the easiest way to do this is seemingly to introduce new nomenclatures (Pollock and Williams, 2010). During the period of our research, we observed how Gartner introduced several different terminologies within the category of "Social Software."

Intervening to Create New Technological Fields

Social Software is a relatively new area where there is still a great deal of activity and interest. Gartner describes Social Software as the area where it fields most questions from clients and prospective purchasers. One key issue is that Social Software is something of an "umbrella term" (also described as "Social CRM" or "Social Media"). The problem, according to one analyst, is that large numbers of vendors are rebadging their products as "social" in some way. We attended a Gartner CRM summit in London, for instance, where an analyst makes this point to the audience:

> Social CRM is a huge topic. There has been tonnes of calls about it. I am tracking currently about 90 vendors who have some area of Social CRM. Some vendors are calling themselves that and they are *not*. Some people *are* that. Some people *don't know* that they have it when they have it. So there is a lot of movement going on as people try to make sense of just Social Media in the first place, and that is a hard nut to crack: 'What is Social Media?' (Conference presentation, Gartner analyst)

In the last couple of days alone, Gartner had identified nearly a hundred new players claiming to offer some kind of Social Software. There appears within the market a need for some form of clarity. Gartner's response therefore has been to break this market down into further subsegments. It has therefore defined Social Software as containing: "Social CRM," "Social Software in the Workplace," and "Externally Facing Social Software" (EFSS). In other words, from out of one category, and partly because of the difficulty of representing all the possible vendors in the Social Software magic quadrant, three new submarkets have been crafted. Interestingly, creating these new kinds of technological categories is not a straightforward process, as we show in the final empirical section.

The Pragmatics of Making Meaningful Distinctions

One way to bring these new terminologies to life is to create magic quadrants for them. However, during a presentation, a Gartner analyst notes some of the difficulties surrounding the pragmatics of doing this—particularly in separating out the Social Software category and making clear distinctions between the vendors operating within in it. The analyst notes how there were vendors producing software that could be counted as belonging to all three categories. Their fear was that there would be a great deal of overlap. However, he goes onto say, there turned out to be fewer overlaps than anticipated:

> It didn't actually turn out that the overlap was as far as we thought. It plotted exactly as reflected here... But the overlap we had in the final publication is really quite small. There is only a couple really that appear on several different ones. And parts of that is down to how we defined the criteria and what were the criteria and qualifications for being included in each magic quadrant. (Webinar, Gartner analyst)

The reason for this was how Gartner defined the evaluation criteria. Their setting and resetting of the criteria meant that the magic quadrants plotted "exactly" as they should do. This *pragmatics of making meaningful distinctions* can be seen more specifically in the creation of the Social CRM magic quadrant. Here an analyst describes the difficulty Gartner has had in producing this particular quadrant:

> We're in the process of creating a magic quadrant for this. There isn't one yet. Adam Sarner, Jean Alvarez, myself, and a couple of other folks... will create one in the second quarter of this year. It is a very onerous task because so many of these vendors are very new and hard to define. (Conference presentation, Gartner analyst)

To summarize, Gartner is faced with a large number of vendors working and claiming to work in the Social Software area. In order to compile its magic quadrants, Gartner embarks upon a number of strategies that help it actively create the beautiful pictures it desires, but performing this kind of pragmatics is complex. The analysts struggle to differentiate between vendors in these new groupings (they are imposing artificial boundaries onto the market and this can provide for difficulties). Many vendors could be included in more than one magic quadrant. Thus, deciding where a particular instance sits across a number of technology classifications requires a decision to be taken and this is often a complicated and ambiguous process.

Discussion and Conclusions

This chapter has brought to foreground the socio-material agency of rankings within markets (see also Scott and Orlikowski, Chapter 6). It has asked whether we can develop alternative agential explanations for the influence of rankings not yet identified by "social" accounts (Espeland and Sauder, 2007). Drawing on recent discussions of distributed material agency (Callon et al., 2007; MacKenzie, 2009), we have given attention to the material equipment that contain and surround rankings. Specifically, we have focused on how the "devices of ranking" do more than simply help communicate a rating conceived prior to its incorporation in a material form. Devices offer various affordances and constraints that the ranking organization need take into account. This can have a bearing on the nature of the ranking produced but also on how this ranking might constitute settings. The device *affords* a particular kind of ranking, but *realizing* this requires changes to the domain, a dynamic that we attempt to capture with the word "affordizing." The contribution of our chapter has been to show these twofold agential features of rankings and the intended and unintended consequences that stem from them.

We suggest, in this latter respect, that the magic quadrant has helped to create new visual and temporal dynamics within the IT domain. We say *visual dynamic* because the aesthetic and communicative aspect of the device turns out to be important. The ranking organization strove to produce a ranking that would allow everyone to see and compare how one vendor was performing in relation to the others, in the most straightforward manner, where there were neither *too many* nor *too few* dots. These experts had seemingly found the optimal number that could be represented (fifteen to twenty-five dots) and this was for them the *beautiful picture*.

What is the beautiful picture? We see the beautiful picture as a set of *socio-material* constraints. It was not simply the physical constraint of the graph by itself that limited how many vendors could be ranked. It would be possible, daresay, by shrinking vendor names and being creative with graphics, to include over a 100 dots, but this would adversely affect the aesthetic economy that was so important to the rankers. Thus, there were also *conventions* working in relation to the ranking that imposed constraints on the rankers. Moreover, the magic quadrant was also dependent upon and circumscribed by a large-scale *organizational apparatus*. It is for these reasons that we describe the affordances of the magic quadrant as both social *and* material (or simply *socio-material*).

We see the conventions surrounding the magic quadrant as an *ideal* (a type of "aesthetic economy"). In this respect, it resembles Garcia-Parpet's discussion (2007) of the "perfect market" (a market constructed specifically so that certain idealized conditions could be fulfilled). The conditions that most effectively shaped the market from the point of view of the ranking organization were those that kept them in "the game," so to speak. There were enough dots so that there appeared sufficient complexity in the market (i.e., people would call them) but never too many to lead to confusion within the figuration.

What we found was that these affordances could have adverse consequences for the shaping of domains. They were not *neutral* with regard to what constituted a market. For instance, magic quadrants appeared particularly ill suited to new, fast moving areas (i.e., where there were many players). While rankers were able to spot vendors entering a popular new category, they could not represent them *all* (the figuration lacked the affordances of a list in this respect). This issue resembles what Lynch (1985: 43), in his discussion of scientific graphs, has called the "problem of visibility." Scientists determine what is "natural" based on what their graphs are able to depict. Translated to our purposes, we suggest that rankers decided what a market "was" based on and what their devices were able to capture and communicate. This suggests that some technological fields—their boundaries, shape, number of players, etc.—were as much a product of the figuration's affordances as any other process of economization.

What is interesting about our study is the finding that, if the ranking organization could not capture beautiful pictures, then it would set about trying to *create* them. Because the graphs were seen to embody the feature of the markets under analysis, efforts were made to *intervene* in technological fields so that their characteristics became congruent with the affordances of the graph (Lynch, 1988: 226). From fieldwork, we saw rankers did this in one of two ways: through attempting to limit the number of vendors seen to be operating in a particular technological field (i.e., setting and resetting "inclusion criteria"); or, if the numbers could not be controlled, to create *new* pictures. This latter work they did through attempting to divide technological fields into wholly new designated areas of activity (with their own unique nomenclature, definition, inclusion criteria, magic quadrant, etc.).

Interestingly, one of the problems rankers now face in these markets-constructed-according-to-the-affordances-of-a-ranking is the pragmatics of making meaningful distinctions. Because these boundaries were imposed onto settings, individual rankers struggled to differentiate between vendors in the new groupings. This is evidenced by the fact that the products of certain vendors appeared in *all* three of the new magic quadrants (an outcome thought to be less than ideal because it suggested a lack of *distinction* within the rankings). These kinds of pragmatics were also visible when the rankers were forced to intervene when vendors clustered together on the graph. Clustering was thought to be a problem because it suggested that all the vendors had more or less the same qualities or status. It occurred because either the market was converging, or, over time, the vendors were conforming to the evaluative criteria, or, as in the case above, there was no meaningful distinction to be made between vendors. In all three cases, there would be little value to be found in the ranking device. Why would people contact the ranking organization, to paraphrase one respondent? Thus, another feature of this pragmatics was the process whereby rankers attempted to separate players out through exercising their organizational machinery (i.e., resetting the axes of the spreadsheet).

We say *temporal dynamic* because during interviews we were alerted to the fact that the affordances of the magic quadrants were not static but evolving over time. Espeland and Sauder (2007: 36) discuss how rankings end up becoming "moving targets": because people learn to "game" them and their producers are forced to update assessment criteria on a continuous basis. In our case, what we saw was that rankings were not only a moving target but also were surrounded by a "moving organizational apparatus." In particular, the magic quadrant had begun as a relatively informal and "subjective" device (Pollock and Williams, 2009*a*) but more recently has evolved into a formal ranking that could be audited and where those who produced it might be held to account for its judgments (Pollock and Williams, 2010). Thus, individual rankers could no longer rank vendors exactly as they wanted, or move dots once the graph had been generated by the spreadsheet. This not only constrained the discretion of these individuals but also the ranking organization's ability to respond to changes in the domain.

Because the production of rankings had become circumscribed by various kinds of organizational apparatus, this had significant effects on the ability of the ranking organization to produce "snap shots" of the market. It appeared that these experts could not react in time to capture specific innovations. Some beautiful pictures had disappeared even before these experts could mobilize their committees and spreadsheets. The picture (e.g., for Web Analytics) was there for a moment and then it was gone, to paraphrase a respondent. What this meant was that certain innovations could completely pass this ranker by. Pockets of the market could remain unranked in what was, in normal circumstances, a highly graded space. We think the instances where materiality creates these situations of "unrankability" deserve further attention. Were the markets for these technologies adversely or positively affected? Were the vendors who remain outside rankings "punished" or "rewarded" in some way?

Our fieldwork also showed that the affordances of the magic quadrant created cyclical pressures on the ranking organization to intervene at certain key moments. The beautiful pictures they sought were time limited. They were not there at the outset of a field (there were *too many* dots to be represented); nor were they there as it matured (either there were *too few* dots to allow anything meaningful to be said, or all the players had clustered and were now in the same box). This prompted rankers to make interventions not arbitrarily but at certain key points throughout the lifecycle of a technological field. This included the moment when a new area first appeared to be emerging and then again later as a field matured. It was through these cyclical interventions that the ranking devices became (or could remain) useful.

We conclude by considering whether the socio-material perspective set out here is helpful for considering the influence of rankings more generally. Does it help to draw attention to things not previously visible under social approaches? To return to where we began this chapter, does it tell us anything new about the extent to which rankings are "engines" within the economy (Espeland and Sauder, 2007; Karpik, 2010; Kornberger and Carter, 2010)? We speculate that this capacity to "intervene" is unequally distributed among rankings. This is because not all engines are the same. Where does this insight take us? Sauder and Espeland (2006) suggest that *competition* between industry observers is one of the major factors affecting the influence of a ranking. They evidence this by describing how US university law schools are rated by one sole body, and how owing to the monopoly this ranker enjoys it thus wields a high level of sway. Business schools, by contrast, find themselves assessed by multiple bodies and, because there is competition in this domain, the power any single commentator holds is diluted.

We suggest that influence is not simply mediated by competition. Domination can also stem from materiality: in particular, differences in the kinds of devices and equipment a ranking organization can gather for itself (Caliskan and Callon, 2010). It is not difficult to imagine how the machineries of ranking might be unevenly distributed among industry commentators. Nor is it impossible to imagine how rankings might have different affordances. Moreover, how many rankers could adjust a market to help their device better depict what is going on? Which organizations could name and launch a new technological field? The individual rankers studied in this chapter showed a sophisticated awareness of the affordances of their devices and their capacity and limitations in mapping the market. We might suggest therefore that the influence of a ranker, the extent to which their rankings can truly become engines, is related to these processes of "affordizing." In other words, it is divided unequally between those that can realize the affordances of their devices and those that cannot.

References

Akrich, M. and Latour, B. (1992). A summary of a convenient vocabulary for the semiotics of human and nonhuman assemblies. In W. Bijker and J. Law (eds.), *Shaping technology/building society*. Cambridge, MA: MIT Press.

Aldridge, A. (1994). The construction of rational consumption in *Which? Magazine. Sociology*, 28, 899–912.

Becker, H. S. (1982). *Art worlds*. Berkeley, CA: University of California Press.

Blank, G. (2007). *Critics, ratings, and society*. Lanham, MD: Rowman & Littlefield.

Burks, T. (2006). Use of information technology research organizations as innovation support and decision making tools. In Proceedings of the 2006 Southern Association for Information Systems Conference, Jacksonville, Florida, March 11–12. http://aisel.aisnet.org/sais2006/2

Caliskan, K. (2007). Price as a market device. In M. Callon, Y. Millo, and F. Muniesa (eds.), *Market devices*, Keele: Sociological Review Monographs.

——Callon, M. (2009). Economization, Part 1: Shifting attention from the economy towards processes of economization. *Economy and Society*, 38(3), 369–98.

————(2010). Economization, Part 2: A research program for the study of markets. *Economy and Society*, 30(1), 1–32.

Callon, M. (1998). An essay on framing and overflowing. In M. Callon (ed.), *The laws of the markets*. Oxford: Blackwell.

——(2007). What does it mean to say that economics is performative? In D. MacKenzie, F. Muniesa, and L. Siu (eds.), *Do economists make markets?* Princeton, NJ: Princeton University Press.

——Muniesa, F. (2005). Economic markets as calculative collective devices. *Organization Studies*, 26(8), 1229–50.

——Millo, Y., and Muniesa, F. (2007). *Market devices*. Keele: Sociological Review Monographs.

David, S. and Pinch, T. (2008). Six degrees of reputation. In T. Pinch and R. Swedberg (eds.), *Living in a material world*. Cambridge, MA: MIT Press.

Espeland, W. N. and Sauder, M. (2007). Rankings and reactivity. *American Journal of Sociology*, 113(1), 1–40.

——Stevens, M. L. (1998). Commensuration as a social process. *Annual Review of Sociology*, 24, 313–43.

————(2008). A sociology of quantification. *European Journal of Sociology*, XLIX(3), 401–36.

Free, C., Salterio, S., and Shearer, T. (2009). The construction of auditability. *Accounting, Organizations and Society*, 34, 119–40.

Garcia-Parpet, M. (2007). The social construction of a perfect market: The Strawberry Auction at Fontaines-en-Sologue. In D. Mackenzie, F. Muniesa, and L. Sin (eds.), *D. economists make markets?* Princeton: Princeton University Press.

Gibson, J. J. (1977). The theory of affordances. In R. E. Shaw and J. Bransford (eds.), *Perceiving, acting, and knowing*. Hillsdale, NJ: Erlbaum Associates.

Goody, J. (1977). *The domestication of the savage mind*. Cambridge: Cambridge University Press.

Henderson, K. (1999). *On line and on paper*. Cambridge, MA: MIT Press.

Hopkins, W. (2007). *Influencing the influencers*. Austin, TX: Knowledge Capital Group.

Hsu, G. (2006). Evaluative schemas and the attention of critics in the US film industry. *Industrial and Corporate Change*, 15(3), 467–96.

Ingold, T. (2007). *Lines*. Abingdon: Routledge.

Karpik, L. (2010). *Valuing the unique*. Princeton, NJ: Princeton University Press.

Kornberger, M. and Carter, C. (2010). Manufacturing competition. *Accounting, Auditing & Accountability Journal*, 23(3), 325–429.

Kwon, W. and Easton, G. (2010). Conceptualizing the role of evaluation systems in markets. *Marketing Theory*, 10(2), 123–43.

Latour, B. (1986). Visualization and cognition. In H. Kucklick (ed.), *Knowledge and society*. Greenwich, CT: JAI Press.

Law, J. (2002). Economics as interference. In P. du Gay and M. Pryke (eds.), *Cultural economy*. Gateshead, UK: Arthenacum Press.

Lynch, M. (1985). Discipline and the material form of images. *Social Studies of Science*, 15, 37–66.

——(1988). The externalized retina. *Human Studies*, 11, 201–34.

MacKenzie, D. (2006). *An engine, not a camera*. Cambridge, MA: MIT Press.

——(2009). *Material markets*. Oxford: Oxford University Press.

Muniesa, F, Millo, Y., and Callon, M. (2007). An introduction to market devices. In M. Callon, Y. Millo, and F. Muniesa (Eds.), *Market devices*. Abingdon: Blackwell.

Myers, G. (1988). Every picture tells a story. *Human Studies*, 11(2–3), 235–69.

Pollock, N. and Williams, R. (2009*a*). The sociology of a market analysis tool. *Information & Organization*, 19, 129–51.

————(2009*b*). *Software & organizations*. London: Routledge.

————(2010). The business of expectations. *Social Studies of Science*, 40(4), 525–48.

Quattrone, P. (2009). Books to be practiced. *Accounting, Organizations and Society*, 34, 85–118.

Sauder, M. and Espeland, W. (2006). Strength in numbers? *Indiana Law Journal*, 81, 205–17.

Schultz, M., Mouritsen, J., and Grabielsen, G. (2001). Sticky reputation. *Corporate Reputation Review*, 22, 24–41.

Shrum, W. M. (1996). *Fringe and fortune*. Princeton, NJ: Princeton University Press.

Strathern, M. (2000). The tyranny of transparency'. *British Educational Research Journal*, 26(3), 309–21.

6

Great Expectations: The Materiality of Commensurability in Social Media

Susan V. Scott and Wanda J. Orlikowski

Introduction

Social practices are necessarily bound up with the material means through which they are performed (Suchman, 2007; Leonardi and Barley, 2008; Orlikowski and Scott, 2008; Pinch, 2008). Whether human bodies, clothes, cell phones, computers, buildings, or infrastructures such as highways, electricity, and telecommunications, all such materiality shapes and defines the contours and possibilities of everyday life. This materiality is particularly evident in the fast-changing landscape of contemporary social media (e.g., Facebook, Wikipedia, and TripAdvisor), where recent technological innovations have facilitated the interconnection and interaction of large numbers of people across time and space (Surowiecki, 2004; Schroeder, 2009). The active creation of content by users is a central and distinguishing feature of social media websites. One aspect of this collective content creation is the online sharing of opinions, advice, reviews, and recommendations that discuss, assess, and rank the quality of a range of products and services, including books (Amazon), news (Digg), hotels (TripAdvisor), movies (Netflix), home services (AngiesList), healthcare (PatientsLikeMe), teachers (RateMyProfessor), and many more.

The mechanisms of online reviewing, rating, and ranking raise a number of interesting questions for organizational scholars. Questions of commensurability—how to measure by the same standard—become particularly salient. Commensuration is the process of comparing different entities in terms of a common metric (Espeland and Stevens, 1998: 313), and it makes up the core component of rating and ranking mechanisms (see Pollock, Chapter 5, this volume). For example, when buying a new refrigerator, we may find it useful to compare the different models on criteria such as price, performance, energy efficiency, capacity, and reliability. In so doing, we are engaging in a process of commensuration by imposing the same set of standard metrics on a set of items. This process both organizes and simplifies the information we have to analyze, while also reducing variability and eliminating nonstandard (often qualitative and contextual) information, for example, aesthetic look and feel, brand loyalty, etc. As Espeland and Sauder (2007: 16) note, the logic and process of commensuration is deeply consequential, it "shapes what we pay attention to, what things are connected to other things, and how we express sameness and difference."

We are interested in understanding what happens to the production of commensurability when it is performed online on social media websites. This is an interesting question because online user reviews, recommendations, ratings, and rankings constitute a significant aspect of social media activity, and are making a difference in what people are paying attention to and how. Compare for example, the rating and ranking of refrigerators on Consumer Reports (performed by experts in controlled test laboratories) with the user reviews of those same appliances on Amazon. Both provide qualitative descriptions of the appliances in free-text form, as well as ratings using quantitative scores. To produce their ratings, Consumer Reports employs six explicitly defined categories measured in their lab tests: temperature performance, energy efficiency, noise, ease of use, energy cost/year, and total usable capacity, while the Amazon website offers users a 1 to 5 rating option in response to the single question: "How do you rate this item?" These details make a difference to how things are comprehended, connected, and compared, and this changes the outcomes generated. These differences (and their consequences) in how commensurability is achieved online are the focus of our discussion in this chapter.

Espeland and Stevens (1998) argue that much more attention to commensuration is needed within the social science literature. They note (1998: 315): "Commensuration changes the terms of what can be talked about, how we value, and how we treat what we value. It is symbolic, inherently interpretive, deeply political, and too important to be left implicit in sociological work." We agree with Espeland and Stevens (1998) about the importance of examining the constitutive role of commensuration in everyday life, but want to offer a key amendment to their conceptualization: commensuration is not only symbolic, interpretive, and political, it is also thoroughly and inescapably material. This is the case for all forms of commensuration, but also and particularly so in the case of social media where commensuration is visibly and vividly performed through fast-changing and rapidly growing websites, databases, user contributions, algorithms, and networks. Focusing on the role of materiality in producing commensurability is thus central to our considerations here.

In this chapter, we explore the processes of materialization that are entailed in the production of commensurability, focusing in particular on the role and influence of rating and ranking mechanisms that have arisen in online social media. More specifically, our study focuses on a social media platform that has become especially dominant within the travel sector—TripAdvisor. A defining tag line of TripAdvisor's mission is "Get the truth, then go." We examine how the commensurability central to TripAdvisor's "truth" is performed through its website's distributed and dynamic materiality, and consider its influences for the expectations and encounters experienced by guests and hoteliers within the context of specific hotels.

Literature

In exploring the commensurability of social media, two streams of literature are particularly salient: the research on commensuration and that focused on materialization.

Commensuration

Commensuration is the process of transforming disparate forms of value into homogeneous units, which allows information reduction, uncertainty absorption, and simplification of decision-making. Through careful and detailed empirical work, Espeland and her colleagues (Espeland and Stevens, 1998; Espeland and Sauder, 2007; Sauder and Espeland, 2009) have shown how the commensurability performed by rating and ranking mechanisms changes both the responses of the institutions subject to them, as well as the institutions themselves over time. Drawing on a detailed study of law school rankings by the *U.S. News & World Report*, they find that such rankings normalize and construct law schools as particular standardized entities making them amenable over time to certain forms of manipulations and interventions (Sauder and Espeland, 2009). As they observe (p. 80):

> By imposing a shared metric on law schools, rankings unite and objectify organizations, reinforcing their coherence as similar objects. Commensuration strengthens the symbolic boundary that defines the field of legal education as comprised of the "same" organizations. It erodes the boundaries that define law schools' specialized niches, while at the same time establishing precise differences among schools based on an abstract, universal scale. Rankings have become naturalized and internalized as a standard of comparison and success. In changing how law schools think about themselves and pressuring schools toward self-discipline, rankings are now deeply embedded within schools, directing attention, resources, and interventions.

Espeland and Stevens (1998) argue that the process of commensuration is deeply embedded within and endemic to contemporary life. They further argue that its role and influence are typically overlooked by both everyday people and researchers (p. 315):

> Commensuration is often so taken for granted that we forget the work it requires and the assumptions that surround its use. It seems natural that things have prices, that temporality is standardized, and that social phenomena can be measured. Our theories presume that we commensurate when choosing and that values can be expressed quantitatively. Commensuration changes the terms of what can be talked about, how we value, and how we treat what we value.

Espeland and Stevens (1998) thus advocate that more explicit attention be focused on the process of commensuration in research studies. While we agree with Espeland and Stevens (1998) that commensuration is a central feature of contemporary life, we believe that their framing has overlooked the critical ways in which commensuration is also, always, materially constituted. For example, the production of law school rankings by the *U.S. News & World Report* involve people working with particular apparatuses, techniques, and protocols in certain ways at certain times and places to define and evaluate the different law schools that are included in the rankings survey. None of these specific assessments and measurements would be possible without the bodies, artifacts, and activities that accomplish them. They are material enactments.

A focus on commensurability would seem obligatory in studies of social media, where the raison d'etre for many user-generated content websites is the rating and comparing that is performed in the ranking of the products and services reviewed online. As we saw in the refrigerator review example, producing commensurability in social media websites typically takes a different form to that performed offline in labs, by experts, and through inspections and audits. It involves user opinions—personal, subjective, and experiential views—that are reduced to a single score or a few general criteria which are filtered and computed via algorithms to provide a hierarchical ordering or ranking.

To date, there has been limited study of commensurability within social media contexts. Research relating to online review sites has focused largely on the technology of online feedback mechanisms (Dellarocas, 2003), the form and quality of review comments (David and Pinch, 2006), the social networks of contributors (Pinch and Athanasiades, 2011), and the impacts of reviews on purchasing behavior (Gretzel and Yoo, 2008; Ye et al., 2011). As a result, we know relatively little about the nature and process of online commensuration and its influence for consumer and provider expectations. Additionally, we know little about how the materiality of online social media websites influences this process. We thus need to consider the role and influence of materiality in performing commensurability in social media.

Materialization

For Barad (2007), matter is "not a thing but a doing." We tend to think of matter and materiality in terms of objects and substances, yet Barad emphasizes that these are effects accomplished in practice. She draws on Judith Butler's work, which suggests that we should conceptualize matter as a process of materialization that stabilizes over time to produce the effect of boundary, fixity, and surface that we call matter. Thus, matter or materiality is not a fixed or static entity but "a dynamic articulation/reconfiguration of the world"; it is dynamically produced through a process that is both "stabilizing and destabilizing" (Barad, 2007: 151). She offers the metaphor of tree rings to emphasize that the making and marking of time is a lively material process of enfolding (Barad, 2007: 181). Such a conceptualization of matter may be regarded as sympathetic to the organizational process perspective that views actions as ordering an "intrinsic flux" and the world as in a constant state of "becoming" (Tsoukas and Chia, 2002).

We use the term "sociomateriality" (Mol, 2002; Suchman, 2002) to communicate the ways in which the social and the material are always in relation to and constitutive of each other. For Barad, any division between the social and the material is enacted, that is, brought into being, through a particular boundary-making practice. This cut, or "agential cut" as she refers to it, is integral to the process of materially defining objects and subjects in a specific time and place (echoing the theme of spacing/placing that other chapters have explored in this volume). One might imagine that once an object or a "thing" has been bounded and defined in this way, its meaning would detach and have a life of its own. But for Barad (2007: 151–2):

> ...matter and meaning are mutually articulated....Neither can be explained in terms of the other. Neither has privileged status in determining the other. Neither is articulated or articulable in the absence of the other.

Barad (2007) maintains that an important move toward understanding the relationship between meaning and matter in the production of knowledge is recognizing that theoretical concepts are not ideational in character but rather entail specific processes of materialization. She encourages us to consider "the conditions for the use of particular concepts" (p. 24) and to recognize that neither matter nor discourse can stand alone. These must be regarded as mutually constituted through practice. Our analysis thus focuses on the ongoing practices that enact specific material (re)configurings that produce particular boundaries, properties, and meanings in the world (p. 139).

We explore these theoretical ideas through an empirical examination of commensurability in the context of online social media. As discussed above, whereas Espeland and Stevens (1998) present a detailed critical discussion of "commensuration as a social process," we examine how commensurability is materially produced, and we do so by focusing on the phenomenon of hotel rankings. Of particular interest is the way in which commensuration collides with expectations through lived experience. Expectations are deeply relational and particular to experience (past, present, future), and they are conditioned by people's specific encounters with context. In this regard, hotels provide interesting and engaging "specific encounters with context" that call out the sociomateriality of expectations. Hotels are sociomaterial arrangements that encompass, among other things: the natural-cultural (e.g., a beach, lake, estuary, vale); through to the building (e.g., south-west facing, sunset capturing, site of a historic event); the practices of guests (e.g., golfers, night clubbers, shoppers, walkers); and staff (e.g., mixing signature cocktails, serving afternoon tea, producing flaming desserts at the table). Both hotels and expectations are highly performative; they are in a constant state of "becoming" (Tsoukas and Chia, 2002), producing particular "cuts" through the "intrinsic flux" of the world and differentially enacting different experiences and outcomes.

Research Site and Methods

Research Site

The growth of social media within the travel sector is having a substantial impact on the decision-making behavior of travelers planning and booking their trips (Starkov and Price, 2007; Xiang and Gretzel, 2010; Ye et al., 2011). Gretzel and Yoo (2008) report that almost 50 percent of travelers base their travel purchase decisions on user-generated travel content, using it during their travel planning to get ideas, narrow choices, and confirm their selections. Furthermore, most of these consumers believe that user-generated content is more likely to be relevant, reliable, and enjoyable, as compared to the travel information provided by traditional service providers such as travel agents and guidebooks (Gretzel et al., 2007).

We focus here on the TripAdvisor social media website, which aims to assist travelers in finding travel information (e.g., hotels, venues, etc.) posting reviews and opinions of travel-related content, and participating in interactive travel forums.[1] Most of the content on the website is provided by its users. Founded in 2000 by four software entrepreneurs with a mission to "Help travelers around the world plan and have the perfect trip," TripAdvisor's growth over the past decade has been rapid. It is currently purported to be

[1] For more information about TripAdvisor, see: http://www.tripadvisor.com/PressCenter-c4-Fact_Sheet.html

the largest online travel community, acting as repository for over 60 million user-generated reviews and opinions on approximately 1.5 million hotels, restaurants, and venues. The TripAdvisor website is visited by more than 50 million unique users per month. The company currently employs just over a thousand people located around the world, and operates websites for thirty countries with content available in twenty-one languages.

Trip Advisor has won a series of accolades recognizing its growing significance. For example, it was named one of the "Top 25 Travel Milestones" by *USA Today* in 2007—the only website included in the list—cited for being instrumental in changing the way in which consumers research travel (O'Connor, 2008). In 2009, TripAdvisor won the US Travel Association "Innovator of the Year" award, designed to honor companies whose innovations have had dramatic impacts on the larger travel landscape. And more recently, it won the "Favorite Site for Reviews" from *Budget Travel Magazine*.

Research Methods

Given the collective, distributed, and fluid nature of social media, they are a particularly challenging phenomenon to study. Where and how are we to direct our analytical gaze and what agential cuts does this entail for the phenomenon of study? We build on Newman's (1998) reconceptualization of the site of research as the dynamic and negotiated assembly and reassembly of actors and issues, and focus here on the fluid, negotiated (re)configurations of issues and interests involved in user-generated content on TripAdvisor. These are then examined in relation to the process of materialization entailed in performing commensurability.

We collected data on TripAdvisor by reviewing its English-language websites, examining hundreds of posted reviews on a number of select hotels, as well as the information provided on TripAdvisor procedures and policies. Interviews were also conducted with TripAdvisor employees in the US and UK offices (see Table 6.1). In addition, we reviewed articles, newsletters, and blog discussions about TripAdvisor published in various media outlets, and participated in industry events focused on social media in the travel sector.

Through a number of field visits, we collected data from several small to mid-sized hotels, conducting observations in situ, and interviewing owners and managers (see Table 6.1). For all the hotels we studied, we examined their available reviews, ratings, and rankings on both TripAdvisor and other more traditional hotel evaluation schemes (e.g., Frommers, the Automobile Association (AA)). Given the large amount of reviews and information available, we focused in particular on the reviews associated with two different hotels in a single region of the United Kingdom. The profiles and target markets of these two hotels are quite distinct. One—PubInn—is a small village hotel centered

Table 6.1 Number and type of interviews

	Number of interviews	Positions
Hoteliers	21	12 owners; 9 managers
AA employees	11	2 executives; 4 editors; 5 inspectors
TA employees	14	3 executives; 5 directors; 6 managers
Hospitality industry professionals	9	British Hospitality Association, English Tourist Board, Institute of Hospitality, Association of British Travel Agents, VisitBritain, VisitEngland
Total	55	

around a large pub, and the other—ManorHouse—is a country manor hotel situated some 30 miles outside of the nearest city.[2] The accommodation at both institutions has been assessed by established hospitality recognition schemes (AA, British Tourist Board) and both have been rated highly (four stars, silver award, etc.). On TripAdvisor, they are presented as the top two hotels in this region of the United Kingdom. We collected all the user reviews and their associated rankings for these two hotels (fifty-eight reviews for PubInn and fifty-two reviews for ManorHouse as of April 15, 2009). The ratings were tabulated in relation to user profile (age, gender, etc.) in order to provide us with a way of identifying patterns in the review data.

We also collected data on the hospitality industry, interviewing a number of professionals within the travel sector, including senior executives of tourist bodies, experts involved in hotel standards setting, and employees (e.g., inspectors, editors) of traditional offline hotel evaluation schemes (e.g., the AA, VisitBritain) (see Table 6.1). In addition, we reviewed hundreds of pages of online and print documentation relating to hotel evaluation schemes and/or the influence of social media, in such sources as governmental tourism reports, trade magazine articles, newsletters published by hoteliers' professional associations, and trade conferences on the hospitality industry.

Our study is exploratory, and thus our analysis is both inductive and iterative. We followed a grounded theory approach (Eisenhardt, 1989; Glaser and Strauss, 1967), beginning with a content analysis of user reviews/rankings of the two hotels (ManorHouse and PubInn), and proceeded to readings of the interview transcripts, observations, and other documentary materials, identifying threads associated with the nature, influence, and implications of TripAdvisor's user-generated content. We then drew on the conceptual lens of commensurability and specifically the notion of materialization to make sense of the emerging threads. This generated a central insight that focused on how TripAdvisor performs commensurability and how this differs from

[2] All hotel names are pseudonyms to protect confidentiality.

traditional hotel evaluation schemes such as those found in the AA or Visit-Britain schemes. More specifically, we found that the processes of materialization entailed in the production of TripAdvisor's user-generated ratings and rankings produce commensurability differently than the traditional schemes that rely on standardized criteria and annual inspections. We further found that these differences in what is paid attention to, by whom, when, and how have important implications for the kinds of knowledge produced and the expectations of both the hoteliers and travelers using TripAdvisor. We examine these findings below.

Performing Commensurability in Practice

Travel Guides to Social Media: The Manifestation of Travel Rankings

The practice of turning to those with prior knowledge before traveling to gauge one's expectations is as old as travel itself. Well-documented evidence for this comes from the letters, journals, and guides circulated among wealthy British travelers between the mid-sixteenth century and early nineteenth century detailing an "education and cultural circuit" called The Grand Tour (Towner, 1984: 215). Travel guides, as we know them, began in 1836 with the Murray Handbooks for Travelers and include specific details of transportation, accommodation, prices, suggested itineraries, as well as advice for travelers. At the turn of the twentieth century, formal schemes for rating hotels developed hand-in-hand with the rise of automobile touring clubs. For over 100 years, two accreditation schemes dominated the United Kingdom, the Automobile Association (AA) and the Royal Automobile Club (RAC). Seen in historical context, the rating and ranking of hotels performed by TripAdvisor is effectively the reconfiguration of an idea that is almost a century old.

Using TripAdvisor simply involves typing an Internet address into a connected computer to go to the website,[3] clicking on the hotel tab, and entering the name of a destination. TripAdvisor then lists the search results in order of "Traveler Recommendation." Like the AA hotel scheme, TripAdvisor structures the content in terms of symbols (stars, buttons, numerals) that score the hotel out of five however, unlike formal accreditation schemes, these are accompanied by user-generated reviews. These vary in length from a phrase or sentence to a short essay and are written in a range of styles from dispassionate accounts of specific issues such as room cleanliness to detailed and sometimes passionate accounts of the food, location, service, and the overall hotel experience.

[3] For example, tripadvisor.com tripadvisor.co.uk tripadvisor.fr tripadvisor.de tripadvisor.es tripadvisor.it tripadvisor.jp tripadvisor.in daodao.com

While the AA hotel rating focuses on operational issues and standardized assessments of certain types of facilities, services, and levels of cleanliness, etc., the reviews on TripAdvisor reflect individual users' personalized and situated experiences of the hotel (for further discussion of standards in practice, see Bowker and Star, 1999 and Timmermans and Berg, 2003). Whereas the AA and VisitBritain hotel evaluation schemes have well-defined and standardized criteria structured within a hundred plus categories of assessment (Table 6.2 offers an example of the criteria used to assess Room Service), TripAdvisor requests input on only six categories: "Value," "Rooms," "Service," "Location," "Cleanliness," and "Sleep Quality." The definition and meaning of these categories is unspecified and reviewers interpret them in their own

Table 6.2 Room service standards for the AA hotel grading scheme

One star	Two star	Three star	Four star	Five star
Optional Any room service provided may be limited in content, but nevertheless must be carefully presented.	Optional Any room service provided may be limited in content, but nevertheless must be carefully presented.	Provision made for room service meals to be eaten in comfort. The use of a dressing table or desk surface is acceptable.	Provision for room service meals to be eaten in comfort; it may be necessary to provide a dining table and chairs or a trolley service.	A formal mechanism for ordering room service, including menus provided in each bedroom and a dedicated room service staff.
		Room service provision with times of availability to be advertised and menus provided. Guests able to order without leaving the bedroom, normally by telephone or in-room order cards. Room service items well presented and served on a tray large enough to easily accommodate its contents. Service prompt and competent.	Trays, where used, well presented and easily able to accommodate all items. A procedure in place to arrange for the collection of trays or trolleys.	Presentation, whether by tray, trolley, or table of the highest standard. Service provided promptly, knowledgeably and professionally.

way as they score the hotel on a scale of 1 to 5 for each of the categories. TripAdvisor aggregates these ratings to determine each hotel's Traveler Recommendation score, and then compares the cumulative scores using a proprietary algorithm that ranks each hotel against others in its region to determine the hierarchical ordering of hotels within a region—what is known as the "Popularity Index."

Most of the reviews on TripAdvisor offer quite detailed descriptions and evocative accounts of reviewers' experiences at hotels, including their personal likes and dislikes. The postings may include images that demonstrate, often in graphic detail, critical points made in the review such as dirty linen and unpleasant or obstructed views. Below are two typical TripAdvisor reviews, one negative and one positive:[4]

> Booked a business class room for 4 days in May 2006. Asked for a quiet room and we got a room facing a construction site that starts work at 7am including Saturdays. In addition the room was very noisy at night as the clubs and bars nearby emptied out and drunken patrons staggered home, singing and yelling at all hours. The staff were not particularly sympathetic in helping to resolve the problem. In my opinion the stay was definitely not worth the expensive rate we paid.
>
> In the past six months, we have stayed in 17 hotels and have been disappointed with the quality and value of accommodation of all of them. So, we thought we would try something different on our recent trip to Amsterdam—an apartment. We heard about [name of establishment] and after reading some of the positive reviews on the TripAdvisor website we decided to give it a try. And we are so glad we did—it was by far the best accommodation experience we have ever had!!!

Many people use TripAdvisor without ever posting a review. As such, the social media site is a vehicle for accomplishing travel planning and booking. Engagement with formal guidebooks or social media websites configure practice; we are "using particular things in a certain way" as Reckwitz (2002: 252–3) notes. Most of the time when people check a hotel on TripAdvisor, they do so without giving TripAdvisor itself much thought because the purposefulness of their travel habits overcomes any interest in commensurability. However, their activities are neither neutral nor passive; such practices are part of the structuring process through which institutions and organizations are constituted over time (Giddens, 1984).

TripAdvisor is designed with the specific capacity to produce rankings of hospitality businesses, and this capacity entails specific materialization within some form (e.g., equipment, media, channels, bodies, buildings, spaces, etc.) that makes a difference to the outcomes produced. In the examples given here,

[4] In the interests of confidentiality, we do not quote from the reviews of any of the hotels we studied.

these practices enact a commensurability of hotels with specific consequences for the management of expectations and their materialization in practice. We consider this in the context of the cut instantiated by TripAdvisor's ranking of two hotels we studied within a region of the United Kingdom—ManorHouse and PubInn.

Apples and Oranges: The ManorHouse and the PubInn

As many guests have noted, arriving at the ManorHouse, located a few miles into the countryside, generates rather stately expectations; set at a slight angle, the heritage building peers down at guests from a gentle mount with a parterre garden around a circular drive. Although they have established a loyal following of leisure-seeking guests, the ManorHouse specializes in providing a venue for important occasions, particularly weddings (the hotel hosted over 170 weddings in 2008). Original medieval features such as inset stone window seats and garderobes have been preserved to create a distinctive historical setting. The hotel has thirty guest rooms and its current rates range from £140 to £250 per night.

The PubInn is located in a village near local tourist attractions; travelers can pull up directly from the road and park in front of the converted stone farmhouse. Guests are greeted by bar staff as they walk through the entrance of the PubInn and a regular flow of "locals" eat in the conservatory extension at the back of the drinking area. The pub is pet-friendly and there is generally a very relaxed atmosphere with trophies won by the village sports team proudly on show. Old photographs of bygone days hang on the wall side-by-side with stuffed animals. The inn has eighteen guest rooms and its current rates range from £90 to £125 per night.

Ostensibly, these two institutions would not be regarded as the same class of accommodation, and thus would not be seen to be direct competitors. Indeed, the AA lists them in two entirely separate guidebooks: ManorHouse is profiled in the *AA Hotel Guide*, while PubInn appears in the *AA Pub Guide*. On the British Tourist Board website, "VisitBritain," they are placed in different categories: "Hotel" and "Inn," respectively. By traditional forms of hotel evaluation and accreditation, these two establishments are incommensurable. Yet, within the reviewing and rating mechanisms of TripAdvisor, they are categorized as the same type of accommodation, assessed on the same rating criteria, and aggregated within the same algorithm. According to the TripAdvisor ranking algorithm—the Popularity Index—these two establishments are the top hotels in this region of the United Kingdom: out of a total listing of twelve hotels in the region, PubInn is ranked as the number one hotel and ManorHouse is ranked number two.

TripAdvisor does not reveal the details of its ranking algorithm but notes that it "incorporates Traveler Ratings to determine traveler satisfaction."[5] Both PubInn and ManorHouse have their relative merits and points of distinction, but in this instance it would seem that TripAdvisor is comparing "apples with oranges." The algorithm that computes the popularity ranking scores on TripAdvisor has a homogenizing affect, rendering these two hotels not only as commensurable but also effectively configuring them as rivals. Review practices on TripAdvisor intensify this material nullification of industry standards: both hotels receive excellent five-button reviews, and both are treated to similar hyperbole such as "fantastic hotel" and "best stay ever." Travelers encounter these hotels side-by-side, called up by place name and then ordered by popularity. The reference points for their knowledge practices are changed, making a difference to their expectations, and shifting the basis from which they make their hospitality choices.

The facilities of ManorHouse meet the sector standards through which an AA inspector experiences the accommodation, and its turrets, parterre, and tariff set high expectations among guests. Many of them are attending a wedding hosted at ManorHouse and might not normally choose to stay at this kind of accommodation. As the owner of ManorHouse put it:

> ...if you are a budget traveler, you've got, you know, £65 to spend and you've been to £65 hotels in the past, and this is the best £65 hotel you've stayed at. It may have nothing like what a £240 hotel has in it, but from your point view it is great.... I suspect the issue has a lot to do with the different people's perception of what they're comparing your hotel against. Maybe this was a budget hotel, but boy, that the best budget hotel I ever was in. It doesn't mean it's a better hotel than this other one is. It just means that there's no way to say what are you comparing it against. What is the frame of reference? And so I think that is a concern.

In contrast, PubInn performs according to a very different set of criteria, offering relaxed hospitality for a guest market predominantly populated by retirees and hikers hoping for a quiet night's sleep, value-for-money catering, and friendly personal service. When asked about how PubInn consistently achieves its number one ranking on TripAdvisor, the owner notes:

> It is the little touches... We have a system for remembering guests names, their preferences and past trips. If they have told us they like things a certain way, we make sure that we set up their rooms and the service how they like it. Simple really but it means they get what they want, leave us positive reviews on TripAdvisor, and come back to us for a repeat visit.

[5] For more information about TripAdvisor's ranking mechanism, see: http://www.tripadvisor.com/help/how_does_the_popularity_index_work

Table 6.3 Comparison of TripAdvisor reviews posted for ManorHouse and PubInn (as of April 15, 2009)

Postings/year	*2003*	*2004*	*2005*	*2006*	*2007*	*2008*	*2009*
ManorHouse (%)	1.9	11.5	13.5	9.6	9.6	40.4	13.5
PubInn (%)	1.7	0.0	0.0	20.7	41.4	25.9	10.3
Locations	*UK*	*USA/Canada*	*Europe*	*Middle East*	*Australia*		
ManorHouse (%)	71.7	26.1	2.2	0.0	0.0		
PubInn (%)	80.4	12.5	3.6	1.8	1.8		
Age ranges (years)	*18–24*	*25–34*	*35–49*	*50–64*	*65+*		
ManorHouse (%)	3.3	26.7	36.7	33.3	0.0		
PubInn (%)	0.0	2.6	28.2	61.5	7.7		
First postings	*Y*	*N*					
ManorHouse (%)	31.1	68.9					
PubInn (%)	61.1	33.9					

TripAdvisor's ranking engine gives these two hotels particular standing and a competitive position within their region which they would not have realized through most formal travel guides. Criteria that have been valued in the past, such as class of accommodation or particular number and type of facilities and services, are not included or considered salient. A closer examination of the TripAdvisor reviews relating to the ranking of these two hotels reveals some notable differences in the profile of travelers posting comments about their experiences (see Table 6.3). Most striking is the age range difference, with almost 70 percent of PubInn reviewers being over 50, while only a third of the reviewers for ManorHouse fall into this age group. Younger people are more likely to stay at ManorHouse (30 percent in the 18–34 range) than at PubInn (2.6 percent). This is not surprising given ManorHouse's emphasis on hosting wedding parties. Furthermore, a larger proportion of PubInn's reviewers are first-time posters (61.1 percent) as compared to ManorHouse reviewers (31.1 percent), possibly reflecting the age distribution of reviewers (with younger travelers more likely to be frequent contributors of online content). It is notable that while ManorHouse lays emphasis on in-house guest feedback cards, some establishments such as PubInn are more proactive and encourage customers to post TripAdvisor reviews.

A number of hoteliers and industry experts expressed confusion over the logic underlying the TripAdvisor ranking mechanism. One hotelier noted:

> I would say that there's no particular correlation between the reviews that we get and where we stand in the scheme of things and other places that any other person would say just aren't at the same level. So there is a sort of regional discrepancy, shall we say. And it's not just our site. I look at other places as well, really top properties nationally, that quite often come down and don't appear until they're maybe fifth or sixth or seventh in the local ratings. So you say, "Well,

> why is that?" Well, one is, I think that the higher spending customer that you have, for some reason, the least likely they are to actually write a review. Now, I don't say that with any scientific evidence. It's just a feeling I get. The second thing is that the places that are perhaps working at a lower level are the ones that probably work harder at trying to get the customers to write reviews, and in fact there are some locally who positively campaign to get people to write reviews.

A professional in the hospitality industry similarly observed that he could "pick any number of areas and just be in complete wonder at how certain places got to number one in the ranking in their area, for instance, and it just does make a mockery of the [TripAdvisor] ranking system."

Expectation Management: The Compulsion of TripAdvisor

The kind of reconfiguration of hotels' standing accomplished by TripAdvisor's rating and ranking mechanisms has caused consternation among some hotel proprietors. For example, one accommodation business in Australia has found itself excluded by how TripAdvisor's algorithms redefine the boundaries of regions. While TripAdvisor ranks Elizabeth's Place as the number one bed and breakfast establishment in a prestigious suburb of Sydney, a traveler would not find it while searching for accommodation within the main city listings for Sydney. Elizabeth's Place was listed by TripAdvisor and placed under the suburb name "Carrington" rather than Sydney. The owner has tried many times to get TripAdvisor to change her business' location information to "Sydney" or "Carrington, Sydney," as she noted in frustration:

> It is ridiculous! I'm fifteen minutes away from downtown Sydney and they [TripAdvisor] won't put me on the main City listing. I emailed them with a list showing how far each of the accommodations on the main list is from the centre of Sydney. One of them is on an island 50 miles away! But they won't budge! Good grief they won't even reply to my emails now.

TripAdvisor's response to this owner's request has alternated between automated replies, brusque refusals, and eventually silence. The apparent ambiguity about the geographic disposition of property locations within TripAdvisor contrasts with the precise specification for locations found in the AA Guide. In the section notifying readers "How to use this Guide," each edition states:

> Towns are listed in alphabetical order under each country: England, Channel Islands, Isle of Man, Scotland, Wales, Northern Ireland and the Republic of Ireland. The town name is followed by the administrative region. The map reference denotes the map page number, followed by the National Grid Reference. To find the location, read the first figure across and the second figure vertically within the gridded square.

By being the number one accommodation in a suburb offering only one establishment in which guests can stay, the owner of Elizabeth's Place believes that her business is not getting the marketing visibility or web traffic that benefits the hotels included on the main Sydney listing. She is thus concerned that she is being excluded from a potentially valuable revenue stream.

Hoteliers generally feel an increased vulnerability and loss of control associated with online reviews. Their expectations of being judged fairly and according to appropriate criteria have been challenged in the context of the reviewing and rating mechanisms afforded by social media. Two hoteliers noted:

> It's very difficult for a small operator because people who come automatically compare you to the big guys, so they often make what I would consider to be unfair comparisons [on TripAdvisor] and they often make very personal comments which I find quite insulting sometimes. I had one lady stay here who said that she thought we should get ourselves motivated and borrow some money and tart the place up. Now I find that quite offensive because she doesn't know what our circumstances are, and that is on TripAdvisor forever. Now since then, we've refurbished the room, we've done all sorts, but [the review] is still there. And I find that quite offensive.
>
> You just feel you've lost control. There's not much you can do about it. I think that's the issue. At the end of the day, you sort of feel like you could be doing the best job you can possibly be doing, put your bloody heart and soul into that, and then somebody who you've never heard of, or you probably can't remember, who didn't say anything to you at the time, goes and posts something really horrible that potentially thousands of people can see.

Some hoteliers have resorted to intervening actively with their guests to manage their expectations and thus preclude the possibility of negative online reviews. For example, the owner of a number one ranked establishment twenty minutes from a major airport said that he found out by talking to guests that they set a premium on being met in arrivals, helped with their luggage, and driven to the guest house. Set in a heritage property that was being renovated, the owner used the short drive from the airport to learn more about his guests' preferences, explain about the restoration work that was taking place, and manage their expectations about their visit. During the guests' stay, he and his partner would then configure services around this information, tailoring their guests' experiences so as to ward off any dissatisfaction that might achieve momentum and trigger a negative assessment on TripAdvisor, thus jeopardizing this hotel's number one ranking in the area.

Even when hoteliers do not want to engage with TripAdvisor, they find that they have become conditioned by it, and compelled to take it into account in their practices. One hotelier who resolutely claimed that he would not change his facilities or services based on an online review, went on to qualify this by saying:

> We were getting quite a number of [online] comments about TVs not working. Now, in every bedroom, we had a notice saying, "We're just about to go through digitalization. We're really sorry, but the TVs aren't working." But people would still write and say, "... the TVs didn't work." Even though this had absolutely been pointed out, and we had apologized, and so on and so forth. I mean, that was quite interesting. But it did make us be a bit more proactive about actually speaking to people, not just assuming that they would read something that was in their room.

The examples we have discussed highlight the highly personal, situated performance of both hosting and being a guest. When problems arise, their resolution is not generally achieved by instrumental or mechanical means at the click of an icon alone, rather it depends on context-specific expectation management, often requiring the reconfiguration of materiality (mending, fixing) and the re-making of boundaries (taking an item off the bill, awarding a future discount).

Implications

Analyzing social media phenomena through the lens of sociomateriality gives us a different perspective from conventional approaches that view technology as a stable substance, discrete entity, or passive mediator. The notion of materialization as processes of stabilizing and destabilizing helps us to understand how social media are entangled in the everyday practices of hoteliers and travelers.

Our study reveals not only how one sector is coming to terms with multiple forms and practices of commensurability but also how specific forms of commensuration become grounded materially. This helps us to understand Barad's assertion that meaning and matter are thoroughly entangled. Just as Fayard (this volume) illustrates entanglement through the practice of art—in which the materials are actively configured as form and content—so our study emphasizes that meaning is inseparable from its materiality. As our example of Elizabeth's Place shows, the material process of hotel rating and ranking makes a difference to who, what, and when things are included or excluded. Looking up a TripAdvisor entry for a hotel, we are reading *through* the processes of commensuration as they are materialized through TripAdvisor's rating and ranking mechanisms. Engaging with TripAdvisor enacts a different cut with regard to phenomena of hospitality than the cut enacted through encounters with schemes such as the AA.

Expectations are read—or experienced—through forms of commensuration. With the rise of social media, user experiences and opinions are related online instead of formal standards being compared as with AA star ratings. Accreditation schemes aim to manage expectations with a view to giving assurance and serving to reassure. If an AA guide awards a hotel four stars, then we have some confidence in expecting that we will get a good night's sleep with decent facilities. In contrast, user-generated content primes vigilance and opens possibilities for multiple accounts of worth (Stark, 2009). For example, a TripAdvisor reviewer notes that on Friday nights the rooms at the front of the hotel are noisy because of brawling revelers at a nearby bar, so when guests book they request a room at the back of the hotel. Another reviewer notes that the views of a historic landmark are better from the front of the hotel, so a guest booking mid-week asks for a room facing the landmark but books another hotel across town for the weekend. The hotelier checks the double-glazing on the front-facing rooms, and makes a note to visit the bar owner. The bar owner adds security and tightens the door policy. The streets are quieter, the pavement has less broken glass, and fewer bodies are bruised in fights.

We see from evidence in our study that there is increasing emphasis on proactive guest management and that the expectation of future reviews configures practices on the ground. Over time, the commensuration provided by TripAdvisor reviews and rankings will shift the materiality on the ground in hotels, so we may expect, for example, changes to room decor, policies, the types of dishes served on the breakfast buffet, the availability of room service and how it is served, and so on. While changes are also associated with traditional hotel evaluation schemes such as the AA, those are tied to a longer term strategy of hotel improvement. Responding to social media evaluation appears to involve a more immediate and spontaneous response by hoteliers.

There is recognition among TripAdvisor's management team that their algorithms reconfigure the phenomenon of hotels made visible on the website, and that this "has a material effect on their businesses" (Livingstone, 2007: 371). Nevertheless, the processes of materialization that are core to TripAdvisor's rating and ranking mechanisms are neither open nor subject to public scrutiny. In the past, hoteliers received private feedback from their guests on feedback forms, but through TripAdvisor's proprietary algorithms, hoteliers are now digitally exposed. This adds to the hoteliers' sense of "inequity" and "inappropriateness" with respect to commensurability, and is somewhat ironic when considered in the light of the charges of ratings manipulations that have been leveled at TripAdvisor (Kelly, 2009). Only if reviews can be proven to be fraudulent, can hoteliers request that they be removed.

The boundary between hoteliers' and travelers' opinions used to be indeterminate. The hoteliers provided a context in which travelers created their own moment in a special place, a unique experience that they would take away with them as a personal memory shared only with friends, family, and colleagues. Feedback requests were on the hotels' own questionnaires and handed directly to reception. TripAdvisor affords travelers an opportunity to publicly distribute what would previously have remained their own private sense of value. This contributes large amounts of information that influence traveler decision-making, but it does so without regulation or transparency about how ratings are produced or hotels are ranked. The phenomenon of user-generated content is thus involved in redrawing forms and practices of commensurability in online social media. The multiple and uncertain consequences of this reconfiguration in expectations and encounters for both hoteliers and travelers suggest that future research is required to further examine the ideas explored here.

Acknowledgments

We would like to thank Stuart Madnick and Leslie Willcocks for their support of this project, Vasiliki Baka for her research assistance, and the study participants for their contributions to the fieldwork. We appreciate useful discussions with participants in the Materiality and Organizing Workshop at Northwestern University, as well as Paul Leonardi's helpful editorial suggestions. We gratefully acknowledge the support of the Centennial Visiting Professors Program at the London School of Economics.

References

Barad, K. (2007). *Meeting the university halfway: Quantum physics and the entanglement of matter and meaning*. Durham, NC: Duke University Press.

Bowker, G. C. and Star, S. L. (1999). *Sorting things out: Classification and its consequences*. Cambridge, MA: MIT Press.

David, S. and Pinch, T. (2006). Six degrees of reputation: The use and abuse of online review and recommendation systems. *First Monday*. http://firstmonday.org/htbin/cgiwrap/bin/ojs/index.php/fm/article/view/1590/1505

Dellarocas, C. (2003). The digitization of word of mouth: Promise and challenges of online feedback mechanisms. *Management Science*, 49(10), 1407–24.

Eisenhardt, K. M. (1989). Building theories from case study research. *Academy of Management Review*, 14(4), 532–50.

Espeland, W. N. and Sauder, M. (2007). Ranking and reactivity: How public measures recreate social worlds. *American Journal Sociology*, 113(1), 1–40.

Espeland, W. N. and Sauder, M. Stevens, M. (1998). Commensuration as a social process. *Annual Review of Sociology*, 24, 312–43.

Giddens, A. (1984). *The constitution of society*. Cambridge: Polity Press.

Glaser, B. G. and Strauss, A. L. (1967). *The discovery of grounded theory: Strategies for qualitative research*. Chicago: Aldine Publishing.

Gretzel, U. and Yoo, K. H. (2008). Use and impact of online travel reviews. In P. O'Connor, W. Hopken, and U. Gretzel (eds.), *Information and communication technologies in tourism* (pp. 35–46). New York: Springer.

Hospitality News (2007). December 19. http://www.hospitalitynet.org/news/4034173.html

Keates, N. (2007). Deconstructing TripAdvisor. *Wall Street Journal*, June 1. http://online.wsj.com/article/SB118065569116920710

Kelly, M. (2009). TripAdvisor cleans out reviews for Fiji resort. *San Francisco Chronicle*, March 10.

Knorr Cetina, K. (1997). Sociality with objects: Social relations in postsocial knowledge societies. *Theory, Culture & Society*, 14(4), 1–30.

Latour, B. (1992). Where are the missing masses? The sociology of a few mundane artefacts. In W. E. Bijker and J. Law (eds.), *Shaping technology/building society: Studies in sociotechnical change* (pp. 225–58). Cambridge, MA: MIT Press.

——(2005). *Reassembling the social: An introduction to actor-network-theory*. Oxford: Oxford University Press.

Leonardi, P. M. and Barley, S. R. (2008). Materiality and change: Challenges to building better theory about technology and organizing. *Information and Organization*, 18, 159–76.

Livingstone, J. (2007). Stephen Kaufer: Cofounder, TripAdvisor. *Founders at work: Stories of startups' early days* (pp. 361–75). Berkeley, CA: Apress.

Media Metrix (2007). *Digital Calculator Report*, July comScore Inc. http://www.comscore.com/metrix/

Mol, A. (2002). *The body multiple: Ontology in medical practice*. Durham, NC: Duke University Press.

Newman, S. E. (1998). Here, there, and nowhere at all: Distribution, negotiation, and virtuality in postmodern ethnography and engineering. *Knowledge and Society*, 11, 235–67.

O'Connor, P. (2008). User-generated content and travel: A case study on tripadvisor.com. In P. O'Connor, W. Höpken, and U. Gretzel (eds.), *Information and communication technologies in tourism* (pp. 47–58). Vienna: Springer.

Orlikowski, W. J. and Scott, S. V. (2008). Sociomateriality: Challenging the separation of technology, work and organization. *Annals of the Academy of Management*, 2(1), 433–74.

Pickering, A. (1995). *The mangle of practice: Time, agency and science*. Chicago, IL: University of Chicago Press.

Pinch, T. (2008). Technology and institutions: Living in a material world. *Theory and Society*, 37, 461–83.

——Athanasiades, K. (2011). Online music sites as sonic sociotechnical communities: Identity, reputation, and technology at ACIDplanet.com. *The Oxford handbook of sound studies*. Oxford: Oxford University Press.

Reckwitz, A. (2002). Toward a theory of social practices: A development in culturalist theorizing. *European Journal of Social Theory*, 5(2), 243–63.

Sahlin-Andersson, K. and Engwall, L. (2002). *The expansion of knowledge: Carriers, flows, and sources*. Stanford, CA: Stanford University Press.

Sauder, M. and Espeland, W. N. (2009). The discipline of rankings: Tight coupling and organizational change. *American Sociological Review*, 74, 63–82.

Schroeder, S. (2009). The web in numbers: The rise of social media. *Rainer Digital*. http://www.rainierdigital.com/the-webin-numbers-the-rise-of-social-media-mashablecom/

Stark, David (2009). *The sense of dissonance: Accounts of worth in economic life*. Princeton, NJ: Princeton University Press.

Starkov, M. and Price, J. (2007). Web 2.0 vs. search engines: Are search engines becoming obsolete in the Web 2.0 frenzy. *Hospitality News*, September 17. http://www.hospitalitynet.org/news/4034173.html

Strauss, A. and Corbin, J. (1990). *Basics of qualitative research: Grounded theory, procedures, and techniques*. Newbury Park, CA: Sage.

Suchman, L. A. (2002). Located accountabilities in technology production. *Scandinavian Journal of Information Systems*, 14(2), 91–105.

——(2007). *Human-machine reconfigurations*. Cambridge: Cambridge University Press.

Surowiecki, J. (2004). *The wisdom of crowds: Why the many are smarter than the few*. London: Abacus.

Timmermans, S. and Berg, M. (2003). *The gold standard: The challenge of evidence-based medicine and standardization in health care*. Philadelphia, PA: Temple University Press.

Towner, J. (1984). The grand tour: Sources and a methodology for an historical study of tourism. *Tourism Management*, 5(3), 215–22.

Tsoukas, H. and Chia, R. (2002). On organizational becoming: Rethinking organizational change. *Organization Science*, 13(5), 567–82.

Urry, J. (2003). Social networks, travel, and talk. *British Journal of Sociology*, 54(2), 155–75.

Xiang, Z. and Gretzel, U. (2010). Role of social media in online travel information search. *Tourism Management*, 31, 179–88.

Ye, Q., Law, R., Gu, B., and Chen, W. (2011). The influence of user-generated content on traveler behavior: An empirical investigation on the effects of e-word-of-mouth to hotel online bookings. *Computers in Human Behavior*, 27(2), 634–9.

7

Digital Materiality and the Emergence of an Evolutionary Science of the Artificial

Youngjin Yoo

The Story of "Charlie Bit Me"

On May 22, 2007, Howard Peter Davies-Carr of England, a father of two toddlers—Harry and Charlie—uploaded a 56-second-long video clip on YouTube for their godfather who lived in the United States. The video shows little Charlie repeatedly biting his older brother Harry's finger. Although it was nothing extraordinary, the video went viral soon after it was uploaded. By October 2009, it became the most watched video in the history of the Internet with over 130 million hits. Simply known as the "Charlie Bit Me" video, it was viewed over 300 million times as of April 2011, with almost 600,000 "likes" and nearly a half million comments. What is particularly interesting is the emergence of numerous parodies of "Charlie Bit Me." Begun as simple imitations of Harry and Charlie, soon many creative adaptations of the original video started to emerge, featuring senior citizens, a group of kids, a gay couple, Lego blocks, and dogs, just to name a few. Some used the original audio track mixed with their own video, while others used their own sound and video and a slightly changed script. There were animations featuring unicorns and cartoon characters, such as Charlie Brown and Snoopy. There were even hip-hop dance music videos that remixed dance music with the edited version of original audio and video. By June 2010, there were over 5,800 different types of adaptations of "Charlie Bit Me" video available on the Internet.

Introduction

The story of "Charlie Bit Me" shows, among other things, the unbounded and generative aspect of digital artifacts. Digital artifacts rarely remain in their original form. They cannot be contained within an innovator's control: instead, they spill into more contested domains beyond their own boundaries (Harty, 2005). Be it products, contents, services, or processes, digital artifacts continue to evolve. We see such generative aspects of digital artifacts in the history of personal computers and the Internet (Zittrain, 2006), and more recently in digitally enabled consumer products such as the digital camera (Yoo et al., 2010*b*). Digital artifacts are generative because the underlying technology has an "overall capacity to produce unprompted change driven by large, varied, and uncoordinated audiences" (Zittrain, 2006: 1980). In this chapter, I explore how the materiality of digital technology enables such generative evolution of digital artifacts and its implication on our studying these artifacts.

Herbert A. Simon (1996) made an observation that the science of nature and the science of the artificial follow two different logics of inquiry. While the science of nature is built on the logic of *discovery* of what is *out there*, the science of the artificial, he argues, is built on the logic of *design*. The objects of the science of the artificial—management, engineering, law, medicine, and education—are not *out there* to be discovered. They are to be designed. In this chapter, however, I make an argument that we are now dealing with a new kind of artifacts, namely, digital artifacts, whose unique material characteristics make them generative. Although the use of technology, whether they are digital or non-digital, is always socially constructed, some of the digital artifacts are made deliberately incomplete and left to users to constantly remake, a property that Zittrain (2006) calls the "procrastination principle." I trace the generative nature of digital artifacts to the material constitution of digital artifacts. As a result, even though they are still being designed, digital artifacts evolve once they are designed, often beyond the intent and imagination of the original designers. I argue that we are at the cusp of a new type of artificial world that is inundated with digital artifacts that are continuously evolving and changing. The emergence of a new artificial world demands new approaches to digital artifacts and its materiality, with a focus on their evolution.

In what follows, I first discuss the nature of digital artifacts with four unique material characteristics. After that, I use an empirical study to illustrate how one can productively use tools from evolutionary genetics to study the evolution of digital artifacts. Based on the discussion of those empirical studies, I then conclude the chapter by discussing the implications of exploring a genetic basis of digital materiality.

Artifacts and the Sciences of the Artificial

From the stone hammers used by cavemen to hunt animals to the earliest Mesopotamian writing used for recordkeeping in crop inventory and taxation (Basu et al., 2009) and to ancient tribes organized for war and labor, human civilization cannot be separated from the artifacts it creates (Sennett, 2008). The agency that was exercised in creating these artifacts to serve human purposes and the artifacts that were created out of the process distinguish humans from other species on earth. The mastery of skills and social practices involved in the creation and the use of these artifacts, as well as the associated domestication of raw materials, largely define the developments of different cultures (Borgmann, this volume). Here, I follow a broad definition of artifacts as both material and immaterial objects that are man-made for subsequent use to accomplish certain goals by performing certain functions (Simon, 1996). Artifacts are man-made cultural objects that have enduring objectified qualities through their materiality. In defining artifacts, I follow the notion of materiality as practical instantiation and significance, as opposed to physical matters (Leonardi, 2010). Therefore, artifacts include a great many varieties of physical tools (stuffs), different forms of standardized routines, and different forms of representations (Buchanan, 1992).[1] Defined as such, artifacts indeed shape and are shaped by human agency over time, forming an intimate and complex web of sociomateriality (Orlikowski and Scott, 2008).

Seen this way, what makes a hammer a hammer and standardized accounting practices work is precisely the entanglement of physical matters and standardized practices that are often intangible.[2] In design, this combination is often referred to as *form* and *function*. To put it crudely, a designer transforms raw *materials* into a particular *form* in order to endow a certain *potential function*. I use the word *potential* deliberately to be sensitive to the shifting power from designers to users (Vargo and Lusch, 2004; von Hippel, 2005). Both Kallinikos and Borgmann, in this volume, provide powerful examples of how developments in science and technology make transformations of materials through design that seemed either impossible or inappropriate in the past now possible. Arthur (2009) notes that a technological artifact is created by capturing a set of natural phenomena in order to accomplish certain desired results. He also sees the progressive discovery of new phenomena that were

[1] I include both what Kallinikos (Chapter 4, this volume) refers to as technical objects and standardized routines as artifacts. Using the example he used, a *schedule* to be used as a schedule needs to be objectified either through norms, habituations, physical embodiments, or some type of combination of these.

[2] Here I suspend any moral evaluation of the word *entanglement*. I will come back to this point later in the chapter.

previously unavailable to appropriate as an important source of technological development.

What I am focusing on in this chapter is how the digitalization of these artifacts affords new forms of materiality to them, ultimately making them more generative than their analog counterparts. An important underlying force here is the historical shift from the physical and material to cognitive and immaterial (Kallinikos, 2011). Seen as some type of marriage between physical materials and immaterial ideas, the traditional analog artifacts were dominated by the physical materiality. Designers always attempt to negotiate with the material agency, appropriating the latest scientific and engineering knowledge (Pickering, 1995). Digitalization has brought a fundamental shift in the power balance between material and immaterial. As I will argue below, pervasive digitalization is decisively loosening the powerful grip of physical materiality on immaterial ideas. This shift sets up a condition for the highly generative and dynamic evolutionary patterns of digital artifacts. Drawing on a recent discourse on digital materiality, I will discuss why and how digital technology makes artifacts more generative.

Digital Materiality

Simply put, *digitalization* refers to the encoding of analog information into a digital format and the possible subsequent reconfigurations of the socio-technical context of production and consumption of the product and services. Digitization can happen at any of the three broad types of artifacts (physical objects, routines, representations) I mentioned above. For example, the invention of digital signal (as a representation) paved the way for digital information processing. When such digital signal processing capability was embedded into network switching devices and terminals in the early 1970s, a physical object—a phone—became digitalized. Although the digital switching technologies made phone calls a lot cheaper and a little bit better in quality and reliability, they did not fundamentally change the relationship between the producer and the customer. On the other hand, today's disruptions in telecommunication, mobile media, broadcasting, and publication industry caused by products like the iPhone, the iPad, and Android have led to the emergence of new forms of routines and relationships among consumers, content providers, manufacturers, and operators.

The rapid miniaturization of computer and communication hardware, combined with their increasing processing power, storage capacity, communication bandwidth, and more effective power management, has made it possible to pervasively digitalize previously non-digital artifacts (Kurzweil, 2006). As a result, a spectacular array of information is now digitally created, stored, and consumed (Kallinikos, 2006). All forms of contents such as books, music, photos, and maps, just to name a few, are now available in digital format. Furthermore, types of information that were simply impossible or impractical to capture are now routinely captured, stored, and analyzed (Yoo, 2010). Such digitalization of representations further enables and is enabled by small, yet increasingly potent, digital components such as sensors, screen, memories, and global positioning system (GPS) units that are becoming a standard part of previously non-digital artifacts such as books, cars, furniture, or buildings. These digitalized artifacts sense, generate, store, and communicate all types of information that were not available in the past (Kallinikos, 2006). Digitalization of artifacts not only promises innovative products, but also changes the way organizations operate. Digitalized artifacts are now routinely used to support various activities in organizations, often changing the ways individuals interact with each other (Boland et al., 2007).

In order to understand how digitalization is changing the nature of artifacts, we need to discern its essential characteristics. My point of departure here is the observation that digitalization of the physical world adds several unforeseen properties to previously non-digital, industrial-age artifacts. Drawing upon a growing body of theoretical work that discusses the unique characteristics of digital technology (Arthur, 2009; Kallinikos et al., 2010; Yoo et al., 2010*a*), I note four characteristics of digital technology that have played pivotal roles in facilitating recent spurts in an unbounded nature of innovations enabled by digitalization.

First, the most fundamental property of digital technology is the idea that all information can be structured as a series of binary digits of either 0 or 1, or simply "bits" (Shannon and Weaver, 1949). Unlike an analog signal, which maps changes in one continuously varying quantity onto changes in another continuously changing quantity, a digital signal represents analog signals as discrete bitstrings (Faulkner and Runde, 2010). Since analog signals are stored using the specific physical profile of a storage device (such as the pattern of a groove of an LP or magnetic variations on a tape) and transmitted through cables and air space, there is a tight coupling between analog data and analog devices, such as VHS tapes, vinyl records, books, and newspapers, for storing and transmitting them. Through digitalization, however, any type of data (audio, video, text, and image) can now be stored and transmitted using the same digital medium. Combined with the emergence of data and interface standards, the digital data allow different types of digital contents to be freely mixed and combined. Yoo et al. (2010*a*) call such mixing the *homogenization of data*.

Second, digitalization embeds microprocessors and software, which follow von Neumann Architecture, into the non-digital tools. Von Neumann Architecture is based on a stored-program concept, where both data and instructions (program) are temporarily stored in the memory instead of being permanently hardwired into the processing unit (Langlois, 2007). As a result, digital computers can shift between bits as data and bits as instructions. Unlike previous non-digital products where the hardware and the function were tightly coupled, digitalized products can now be flexibly re-programmed and re-purposed with very low cost (Faulkner and Runde, 2010; Kallinikos et al., 2010). Therefore, the integration of this computing architecture into a physical tool provides tremendous flexibility. Yoo et al. (2010*a*) call it the *re-programmability* of digital technology.

Third, digital technology is immaterial. Both digital data and software programs are all stored in immaterial bitstrings and temporarily stored in "bearers" (Faulkner and Runde, 2010). For example, a software program is a logical set of instructions written in texts. Likewise, digital data, as we have seen above, transform all forms of data into a string of bits, no matter what the actual contents were. Neither of them have tactile existence on their own (Leonardi, 2010). As Leonardi notes, "you can touch the screen... upon which data is displayed; but you can't touch data itself."[3] The immaterial characteristic of digital technology makes it extremely pliable, extensible, and recombinable and enables non-rivalry in use (Faulkner and Runde, 2010). Because analog data requires physical impression (whether it is a groove on plastic, invisible airwave, or inks on paper) of contents, it can have only as much fidelity as the materials will allow. To the contrary, digital data are a series of numerical snapshots of "on-and-off" status. Therefore, the fidelity of digital data is as flexible as the programmers allow. For example, if one wants to improve the quality of a digital recording, all one needs to do is to improve the sampling rate, that is, how frequently those snapshots of an analog sound stream are taken to create digital copy of the sound stream. Once it is captured, it can be infinitely reproduced, exactly. In the same way, one can repeat the same set of instructions coded in a software program. Recent developments in 3D printing technology that produces certain physical objects as dictated by an immaterial code of instructions (Anderson, 2010) takes the importance of the immaterial to a whole new level (Kallinikos, Chapter 4, this volume).

Finally, digital innovation relies fundamentally on various forms of digital technology as an enabling technology for its creation, diffusion, appropriation, and expansion. Yoo et al. (2010*a*) and Kallinikos (2006) call it the *self-referential nature* of digital technology. Combined with the ubiquitous availability of affordable digital devices—such as inexpensive PCs as a design and

[3] Ironically, the above text is from an article "published" on "web." Having a fixed page number is highly problematic as a web page is dynamically rendered, based on a number of different parameters of physical bearers such as monitor size, screen resolution, and the dimensions of the web browser. Thus, there is no page number for the quote. Anyone who tried to quote a page from Amazon's Kindle experienced the same problem.

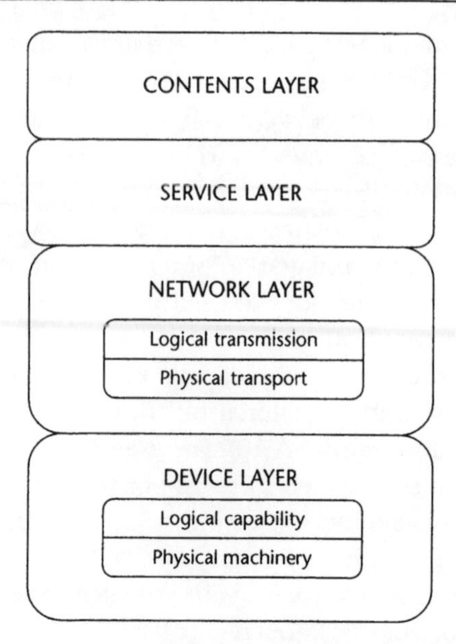

Figure 7.1 The layered architecture of digital technology (Yoo et al., 2010*a*)

production platform and the Internet as a distribution network—the self-referential nature of digital technology radically lowers the entry barrier for independent digital entrepreneurs. Unlike earlier physical resources, requiring extensive capital to acquire and operate, universal diffusion of PC and the Internet have largely democratized innovation, as individuals with little capital can participate in innovation activities (Zittrain, 2006).

These four characteristics of digital technology become the basis for making digital artifacts generative and highly evolving. Specifically, as I argued elsewhere (Yoo et al., 2010*a*), the homogenization of data and the re-programmability of digital technology have led to two critical separations: between physical device (i.e., form embodied in particularly materials) and service (i.e., function), and between contents and network. Furthermore, these two separations have led to the emergence of the *layered modular architecture*, which consists of four independent, loosely coupled layers of physical devices, networks, services, and contents (Benkler, 2006).

As shown in Figure 7.1, the device layer can be further divided into a physical machinery layer (e.g., computer hardware) and a logical capability layer (e.g., operating system). The logical capability layer provides the control and maintenance of the physical machine and connects the physical machine to other layers. The network layer is similarly divided into a physical transport layer (including cables, radio spectrum, transmitters, and so on) and a logical transmission layer (including network standards such as TCP/IP or P2P protocols). The service layer deals with the application functionality that directly serves users as they create, manipulate, store, and consume contents. Finally, the contents layer includes data such as texts, sounds, images, and videos stored and shared. The contents layer also provides meta-data and directory information about the content's origin, ownership, copyright, encoding methods, content tags, geo-time stamps, and so on. A digital artifact with a layered modular architecture is a result of the temporary binding of individual components in different layers. Such dynamic and flexible architecture is also enabled by modularity, granularity, and standardized interfaces of digital artifacts (Kallinikos et al., 2010).

Although the layered modular architecture shares many similar characteristics of the traditional modular architecture, they differ in a fundamental way. A *modular architecture* is characterized by its standardized interfaces between components (Baldwin and Clark, 2000). Modularity is a general characteristic of a complex system, referring to the degree to which a product can be decomposed into components that can be recombined (Schilling, 2000). A modular architecture offers an effective way to reduce complexity and to increase flexibility in design by decomposing a product into loosely coupled components interconnected through pre-specified interfaces.

Individual components and subsystems in a physical product with a modular architecture are product-specific (Ulrich, 1995). The product must be conceived before individual components and subsystems are designed. To the contrary, components in a layered modular architecture are product-agnostic. Most web APIs (application programming interfaces), for example, are written with no specific final products in mind. Furthermore, even if a digital component is developed for a specific product, because of the re-programmability and data homogenization, it can be easily repurposed for different products and services. Of course, users always appropriate artifacts in a *situated way* (Orlikowski, 1992). However, increasingly, digital artifacts with a layered modular architecture are deliberately designed to be open-ended without any specific purpose in mind. Many smart phones are incomplete, from a functional standpoint, when they are first purchased, as users must download and install their own apps. In fact, a smart phone remains incompletely designed as users continue to change its service profile as they install different apps throughout its lifetime. Finally, in a layered modular architecture, components in different layers do not necessarily belong within the same design hierarchy (Clark, 1985) or what Arthur (2009) refers to as *the same domain*. For example, design of Google map (contents layer) is independent from the design of smartphone hardware (physical layer) as well as the design of different applications that utilize the map (service layer). As components in different layers belong to different design hierarchies, a digital artifact with a layered modular architecture can evolve continuously through recombination of these components. Combined with the pervasive availability of affordable digital technology, such recombinatorial nature of digital artifacts makes their design process open and distributed (Yoo et al., 2008), which further amplifies the generative nature of digital artifacts as evidenced in the case of the Internet. The design of individual components at different layers is being carried out largely independent of each other. The *procrastinated binding* of these components at the time of consumption, not at the time of production, further makes the ontology of digital artifacts inherently generative and dynamic.

The emergence of digital artifacts with their generativity and the dynamic evolutions that produce innumerable varieties pose a serious challenge to research on materiality. Past research on materiality often focuses on idiosyncratic assemblage of sociomateriality (Orlikowski and Scott, 2008). Although such an approach provides a rich understanding on each instance of digital artifacts, it fails to explore the possible existence of generative and constitutive rules that "provide a wide space of possibilities" (Kallinikos, 2011: 135) whose realization is expressed in each local context. Below, I propose a genetics-based approach modeled after evolutionary biology. I provide a short description an empirical study as a way of sketching out what such a program of research might look like. It is being used as emerging sketches of what I hope to become an evolutionary science of the artificial.

A Sketch of an Evolutionary Science of the Artificial

An Overview of a Genetics-Based Approach

An important aspect of a genetics-based approach to studying generative digital artifacts is the cross-level analysis with focus on the complex interactions between an artifact and its "genetic" components. Such an approach will help us understand the complexity of the *phenotype* (i.e., distinguishing characteristics of living organisms) through its *genotype* which is the basic genetic building block of those artifacts. In the case of digital artifacts, the phenotype is the diverse appearance and usage of digital artifacts such as applications and smart phones, while the genotype is their internal logical structure.

Recently, the field of biology is gradually shifting to system-level understanding, mostly referred to as system biology. The recent advance in computing capabilities in modern biology such as genetics or genomics allows researchers, for the first time, to study from the very bottom, gene level, to the top, population level. The goal of system biology is: "to understand biology at the system level, we must examine the structure and dynamics of cellular and organismal function, rather than the characteristics of isolated parts of a cell or organism" (Kitano, 2002).

Similarly, I argue that the study of digital artifacts and their "ecology" should not be separated from how lower level components are constituted and recombined. Changes in higher level digital artifacts are often triggered by changes at a lower level; at the same time, changes at the higher level will also influence the lower level. By decomposing an artifact into the basic elements in a similar way that genetics does, and by looking at the system as whole built by combining such elements, we are able to better understand the evolution and changes of those artifacts.

What do we gain in our exploration of evolution of digital artifacts and their materiality by using the idea of genes? In order to explore that question, let me first briefly explain the roles of genes in nature. Genes perform two key functions. First, genes carry genetic information coded in a set of four nucleotides from one generation to the next. It is through genes that basic genetic information is preserved in DNA and traits are inherited across generations within a species via a process called DNA replication. In a genetics-based study of digital artifacts, "genes" of digital artifacts perform this role. As organizations refine the design of digital artifacts, the basic genetic information of those artifacts is preserved when these genes are passed on unchanged. When they are altered, however, changes or "mutations" in artifacts take place.

However, genes perform another important role. They act as basic instruction codes to produce proteins—the basic building blocks of living organisms. This process is performed through what biologists call DNA transcription, which consists of a series of transformations. During this process, a triplet of nucleotides (known as a codon) is mapped into a particular kind of amino acid. Given that there are four nucleotides,[4] there are 4^3 or 64 different possible triplets. These sixty-four triplets of nucleotides are mapped onto twenty different possible amino acids that are found in nature. This relationship between sixty-four triplets of nucleotides and twenty amino acids is known as the *universal genetic code*. Therefore, each gene consists of a string of triplet codons, which is then translated into a set of amino acids.

It is important to note that DNA sequences in genes do not carry any physical properties that we see in organisms. Instead, they merely carry abstract *immaterial* information. Genetic code translates the immaterial genetic information encoded in DNA into physical traits—eye colors, heights, and the shape of the cheekbone. But there is no one-to-one relationship found between the genetic information (i.e., *genotype*) and the distinguishing features of living organisms (i.e., *phenotype*). Combinations of the amino acids

[4] Nucleotides are molecules that form the fundamental structural unit of DNA.

coded in these triplet codons produce proteins that determine physical traits. In addition to the genetic component that influences traits or phenotypes, however, the expression of this phenotype is also determined by environmental factors. In a genetics-based approach, therefore, we are translating building blocks of digital artifacts, such as software subroutines, physical components, and subsystems, or *performative routines* (Pentland and Feldman, 2005) into *genetic* building blocks of those digital artifacts characterized with a fixed set of alphabet.

In the past, scholars of materiality of artifacts have been drawn to study seemingly unlimited possibilities of the phenotype. Each organization seems to produce and consume unique and idiosyncratic artifacts. The potential power of genetics-based approach is, however, to allow scholars to go beyond the immediately observable phenotypes of materiality and begin to look at the underlying "genetic" principles that give birth to these phenotypes. Below, I outline an empirical study that I am carrying out with my colleagues as an illustration of such genetics-based study of digital artifacts.

Evolution of Web APIs and Mashups[5]

In this study, we explore the evolution of Web Services as a digital artifact. Many popular applications that are available on the web as well as on mobile platforms are combinations of existing web-based applications with specific technology resources provided by the hardware. These new applications, or simply "apps," are often created by mixing and matching existing web APIs and other digital technologies that belong to different layers. Following the definition by World Wide Web Consortium, we define Web Service as "a software system designed to support interoperable machine-to-machine interaction over a network."

We are particularly interested in the evolution of a class of Web Services called web APIs that enable web-based applications also known as mashups (or often referred to as Web 2.0). A *mashup* is an application that combines data and functionalities provided through web APIs to create new services. It is normally a web page or smartphone application and often includes more than one web API. Firms like Google, Amazon, Facebook, and Twitter are aggressively using web APIs to expand their presence on the web. For example, more than 60 percent of eBay listings come from its APIs. Also, Twitter gets ten times more inbound traffic through its APIs than its website. More importantly, however, these APIs are deliberately used to attract heterogeneous third-party

[5] This study is being carried out together with Rob Kulathinal and Zhewei Zhang (both at Temple University). The project is supported in part by a research grant from the Fox School of Business.

developers to reimagine the products and services and to transform them into new ones (Bush et al., 2010).

According to Programmableweb.com, there are more than 5,600 mashups in the format of a web page. These mashups range from digital maps to social networking to music downloading. Web APIs are published using Web Services Descriptive Language (WSDL) and there are several publicly available directories of various types of web APIs. We explore how we might be able to characterize each API with a fixed set of generative elements (again akin to four DNA nucleotides), and how the combination and recombination of those APIs form the basis of continuous and seemingly unbounded innovations among mashups. Web APIs provide an ideal context to study the evolution of digital artifacts, as the combination and recombination of commonly available APIs produce a highly dynamic and evolving assortment of mashups.

We see infinite possible combinations of these APIs as formative elements of mashups. Therefore, in order to capture and analyze innumerable APIs and apps and their evolutions in a meaningful way, we must be able to characterize them and their elements with a fixed set of possible values. Our approach was inspired by modern genetics, which uses a fixed set of four nucleotides (represented as G, A, T, and C) to code for all the varieties of living organisms we see in the nature. As shown in Figure 7.2, a mashup can be modeled as a syndication of web APIs with their own data and functionalities, combined with its own client-side logic and data representation to offer new services. A web API, in turn, can be further characterized by a set common base elements that defines its function and performance.

We find a great similarity between the structures of a mashup and protein, as shown in Figure 7.3. Protein is structured as two-level structure by subunits, called protein domains, and basic elements, amino acids. Protein domain is also constituted by amino acids. Therefore, it can be characterized as nested multilayered structure. Protein domains evolve, function, and exist independently without being a part of a protein. Our conceptualization of mashups and APIs allow an almost identical structure. A mashup is constituted by web APIs connected by its own elements. Like a protein domain, web API can evolve, function, and exist independently without being part of a mashup and it is also constituted by basic elements.

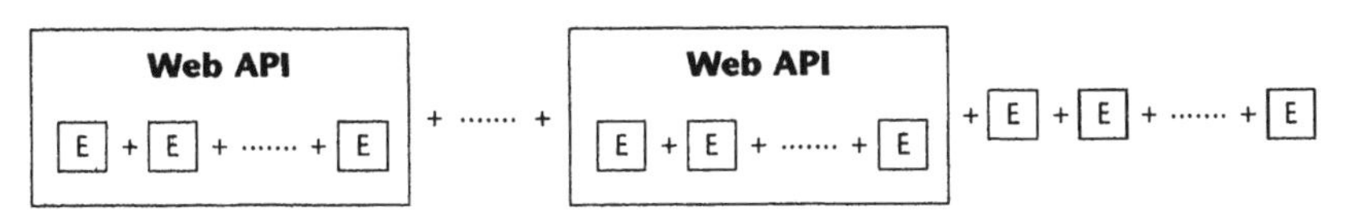

Figure 7.2 Multilevel structure of web API and mashup

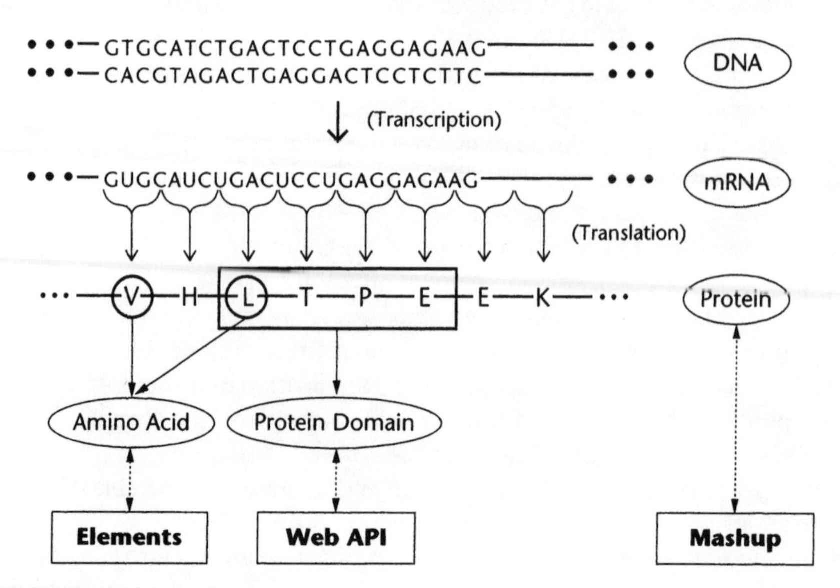

Figure 7.3 Analogy to biology

As a digital artifact, a web API needs to be represented as an ensemble of both social and technical components. In this study, we characterize a particular web API using five base elements: (*a*) an *owner* who owns the API can be a individual firm or an alliance; (*b*) *openness* of an API denotes whether the source code is open or not; (*c*) *licensing* indicates if it needs to be paid for use; (*d*) *service context* represents the domain this API is designed for; and (*e*) *service function* represents the software function that API provides. Each of these categories has a limited set of possible values as shown in Table 7.1. According to Progammableweb.com, there are fifty-four different categories of service context. Based on similarities of purpose, we grouped them into nine service contexts, which cover almost all the web APIs commonly used while keeping the computational complexity manageable. The four types of service function are drawn from the most basic database syntax: select, add, update, and delete, as we believe the relationship between an API and a user is often analogous to that between a user and a database. Taken together, we have a fixed set of nineteen possible categorical values for five basic components for web APIs. Using this "alphabet," each API's internal structure can be characterized as a simple "string" by concatenating a set of categorical values, each describing a unique aspect of the API. A mashup, in turn, is characterized as a sequence of APIs.

Using this taxonomy, we can characterize each API as a sequence of values for these base elements in the order of owner–openness–service

Table 7.1 A taxonomy of web API DNA

	Possible values	Amazon A2S	OpenLayers	Twitter
Owner	I (individual), A (alliance)	I	I	I
Openness	O (open), C (closed)	C	O	C
Service context	N (news), H (shopping), V (media), N (social network), M (map), S (advertising), E (search), U (utility), T (communication)	H	M	N
Service function	R (retrieve), L (upload), D (update), V (remove)	R	R	RL
Licensing	F (free), P (paid)	P	F	F

context–service function–licensing. For example, from Table 7.1, the sequence of Amazon A2S is [ICHRP], OpenLayers is [IOMRF], and Twitter is [ICN(RL)F]. Once each web API is characterized, we can further characterize a particular mashup as a concatenated sequence of APIs that are utilized by the mashup. For example, a mashup that consists of Amazon A2S, OpenLayers, and Twitter can be characterized as [ICHRP IOMRF ICN(RL)F]. This sequence represents the basic code, or DNA, of the mashup. A change of a mashup can occur through additions or deletion of APIs or through the mutation of an API, as it might add new functions or change its licensing structure.

Once APIs or mashups are sequenced, we can then analyze them with a sequence analysis tool. Genetics researchers seek to analyze the configuration of nature's elements in the DNA. Similarly, we seek to discover patterns in mashups and APIs and their "mutations." We use a method similar to the one used by Gaskin et al. (2010). Using genetics software, we perform a pairwise alignment of the sequences in order to construct a similarity matrix which is then converted into distance scores. The distance score matrix can be then used to identify the variety among mashups and APIs. Using these results, we can identify distinct groups of mashups that are similar in their composition. Similarly, we can group APIs based on their similarities. Each instance of mashups and APIs within the same group can be understood as a mutation of the species, while different groups of API and mashups represent different species.

Using sequence analysis, evolutionary geneticists explore relationships among different species such as: What are the evolutionary relationships among different species? When did primate ancestors split? And what were the specific events prior to and subsequent to their separation? We can ask similar questions in the context of evolution of digitalized artifacts. For example, Figure 7.4 shows a phylogenetic tree that we constructed from our preliminary analysis of top 100 most popular APIs since 2008. Phylogenetic trees allow us to estimate the rates of evolution, identify the evolutionary pattern, and infer the ancestral sequences. From these results, we can explore

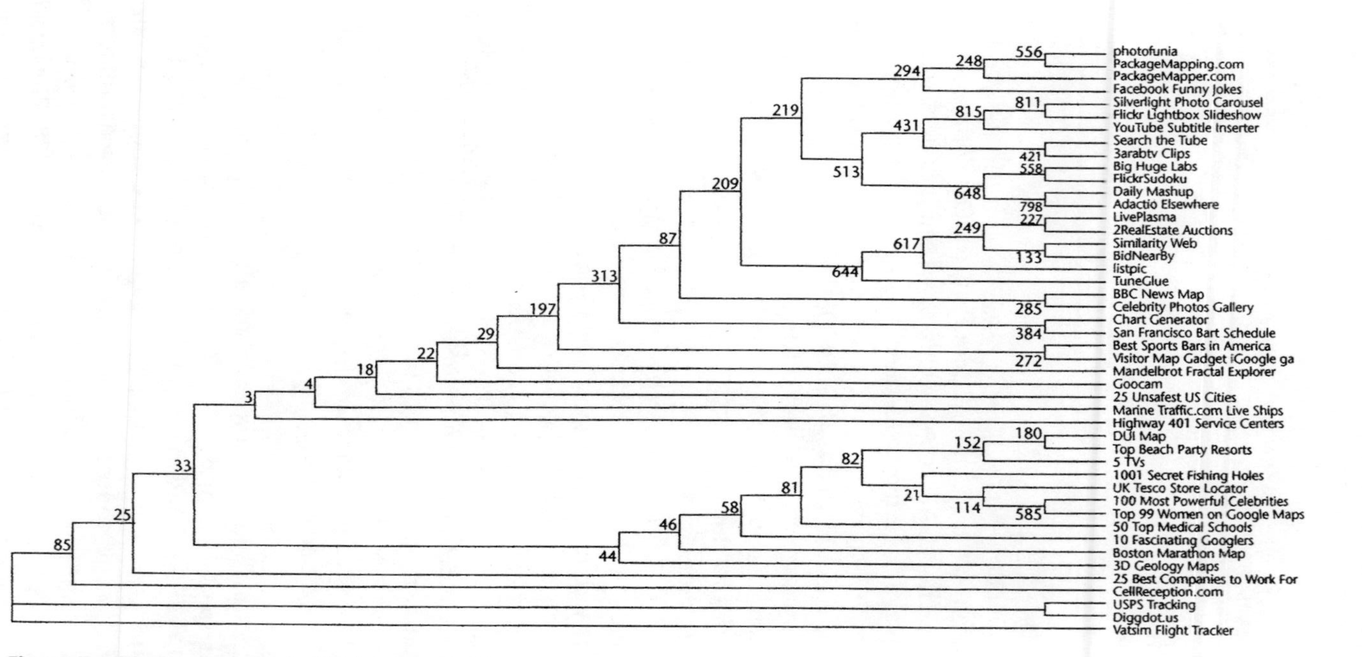

Figure 7.4 Phylogenetic tree after bootstrapping

whether evolutionary patterns of mashups that use Google APIs are different from those that use Yahoo APIs, for example. We can also explore whether APIs that use open source produce more "mutations" than proprietary APIs.

Furthermore, one can complement the sequence analyses with other research methods. For example, from a circular view of the phylogenetic tree shown in Figure 7.5, we can clearly see that most tree branches highlighted by the red line are derived from a single mutation that happened at the node highlighted by the yellow circle. In biology, this means that a mutation that happened at that node has better performance in terms of reproduction. In the case of digital artifacts, the artifact from that particular "mutation" was more generative than others. With this empirical insight, we can conduct an in-depth case of that particular artifact to gain better understanding on how to design generative digital artifacts.

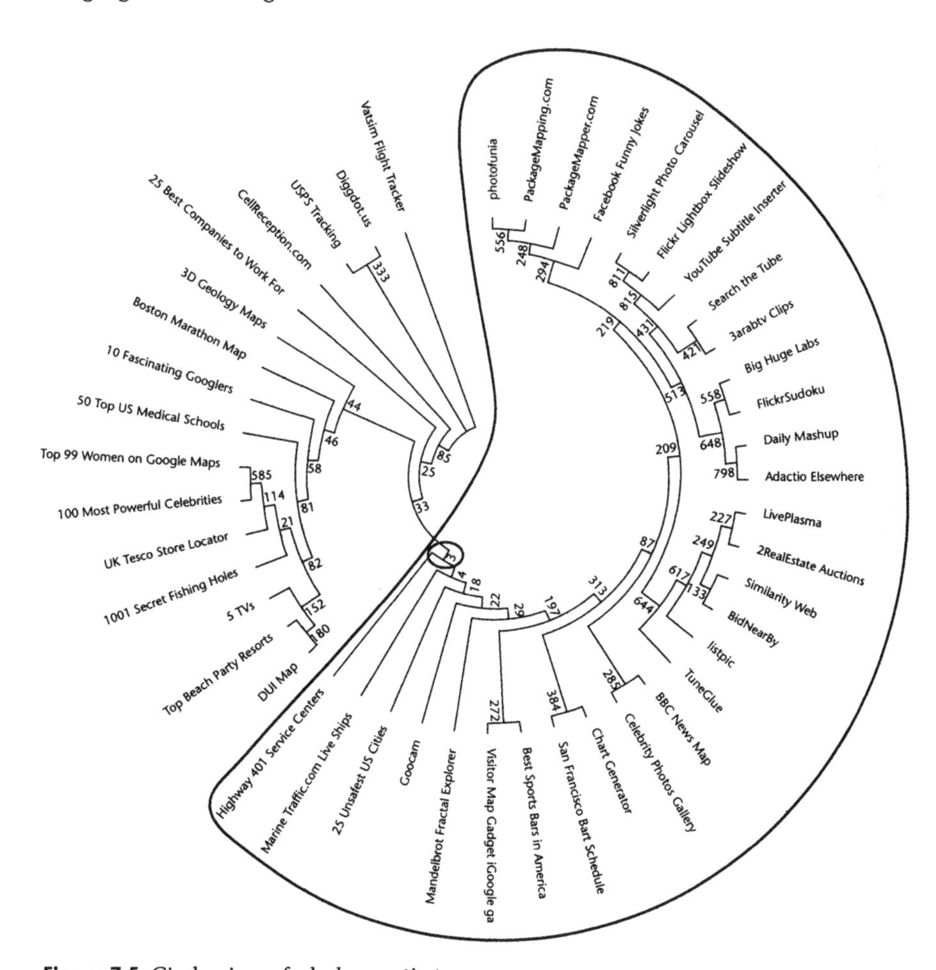

Figure 7.5 Circle view of phylogenetic tree

Since we are still at a very formative stage of our research, there are many theoretical and methodological obstacles that we must overcome. However, the preliminary results that we have gained so far have shown promising avenues going forward in understanding the evolutionary patterns of digital artifacts. One can conduct similar analyses on routines and representations. We are carrying out separate studies in those domains (Gaskin et al., 2010).

Toward a Genetic Basis In the Study of Digital Materiality

In this chapter, I proposed a genetics-based approach to the evolution of digital artifacts and their materiality. To illustrate how the evolution of digital artifacts can be analyzed through a genetics-based approach, let me explicitly discuss how a study of digital artifact can be mapped to genetics-based biology. First, at the most basic level, digital artifacts need to be characterized by a fixed set of base elements that act like the four nucleotides that constitute the DNA. In the case of web APIs, we have nineteen of those. Second, these base elements make up the basic units (web APIs) that carry "genetic" information of digital artifacts (mashups). In a way, these basic units act as *genes* of digital artifacts. Third, varieties of digital artifacts come from either the changes in the sequence of these "genes" or through the mutation in those genes. Therefore, one can study the evolutionary patterns and dynamics of both of these digital artifacts by studying the sequence of these "genes" and their changes. I argue that these three similarities provide a genetic basis for studying the evolution of digital artifacts.

Arthur (2009) cautions the idea of using genetics and genes as the basis of understanding combinatorial innovation with technology. One of his key concerns is that genes in nature are stable, where new human artifacts are constantly being created. However, an important point to note here is that his approach to combinatorial innovation is at the phenotype level. That is, when he talks about combinatorial innovations through mix and match, he is talking about subsystems and components as they are already expressed. In that sense, his approach is similar to the traditional biological approach before the discovery of DNA and its role as the carrier of genetic information. Since Darwin, biologists have tried to explain the varieties in the nature through physical properties, which has turned out to be an insurmountable problem. Both the natural and artificial worlds seen at the phenotype (or performative) level are always dynamic with infinite varieties. Every person on earth looks different and every routine in organizations is unique. Yet, modern genetics shows that beneath the seemingly infinite varieties on the surface, there is remarkable consistency across all living organisms at the genetic level.

My proposal here is to complement Arthur's evolutionary approach (2009) to technology innovation with a genetics-based approach. The combination and recombination are happening at the levels of both the phenotype and the genotype. Just as in nature, it might be the case that much of the recombination at the phenotype is directed by the changes at the genotype. If that is the case, we might be able to understand the seemingly intractable problem of evolution of digital artifacts by focusing on its genetic foundations.

The proposed genetics-based approach to the evolution of digital artifacts has an analog in social sciences known as structuralism in cultural anthropology, linguistics, semiotics, and art theory (Lévi-Strauss, 1963; Kallinikos, 2011).[6] The complementary, yet possibly contentious, relationship between the messy, chaotic material reality (phenotype) and the simple, abstract genetic signs (genotype) that I note here is akin to the structuralism as developed by Lévi-Strauss (1963) who saw the all ancient myths in different cultures as combinations of a few universal oppositions.

So, what does this mean for research on materiality? While the currently dominant approach to materiality highlights the unique and situated nature of the sociomaterial ensemble (Orlikowski and Scott, 2008), the approach that I outlined here emphasizes the universal common structures that give birth to those varieties that we observe. In the words of Lévi-Strauss, a genetics approach demands a shift from *conscious* material phenomena to *unconscious* infrastructure. Or, in the words of Kallinikos, it is a transition from object-mediated discourse with context-embedded knowledge to the standardized world of constitutive rules with decontextualized representation. In order to truly appreciate the material richness of the artificial world with all its expressed varieties, we need to step back and peel off the superficial surface of phenomena and look for the genetic instructions that produce varieties.

Given that these digital artifacts must be designed by humans who exercise agency, the evolutionary perspective that I propose here raises a question about the role of human agency. I suggest that there are two different places where we must carefully consider the role of human agency. First, the basic genetic information of digital artifacts must be designed. In biology, DNA is represented by the four bases A, C, G, and T, and the triplet combinations describe the basic twenty amino acids found in nature. They were not man-made. However, in both studies, the basic genetic information we *discovered* were man-made. They were deliberately designed by designers. Therefore, unlike biology, these basic genetic elements can be deliberately altered in order to generate fundamental shifts in the evolutionary dynamics.

[6] I am indebted to Jannis Kallinikos for this point.

Second, much of the variations that we see in the digital artifacts are again results of human actions (either deliberate or accidental). In digital artifacts, as in the case of mashups, designers create new "mutations" by mixing and matching components, taking advantage of the loosely coupled nature of the layered modular architecture of digital artifacts. In biology, such changes take place through random mutations, gene flow (exchange of genes from one species or population to another), or recombination via sexual reproduction. One might be able to make similar categorizations of variations in digital artifacts. Therefore, one can see that the outcomes from the evolutionary science of the artificial might in fact help us better design these man-made digital artifacts. Nevertheless, the evolutionary study of artifacts demands us to decenter human actors from our scientific investigation and instead move the artifacts right to its center. Some may find it uncomfortable or unacceptable.

One must not rush to blindly borrow models and tools from the field of evolutionary biology to understand the evolution of digital artifacts. We must carefully explore the theoretical and methodological assumptions that are embedded in these models and tools to understand the possible implications of making certain adaptations to those models as we move them from the natural world to the artificial world. We may end up finding that the application of evolutionary biology to the study of the evolution of man-made digital artifacts simply does not work. The idea of genes and genetic code may not translate well into the domain of man-made artifacts for the reasons that I cannot envision at this point. We may also find that the degree to which we need to change the tools and methods is too great to meaningfully make any associations between the two.

Despite this caveat, I believe contemporary evolutionary biology serves as the best known source of theoretical and methodological inspiration to study the evolution of digital artifacts and their materiality. So far, the similarities between the two worlds, as I have found in these two studies, are quite striking and the outcomes promising. Just as modern evolutionary biology has opened up a new way of understanding the natural world, I hope the evolutionary sciences of artifacts will open up new ways to understand man-made digital artifacts, which are increasingly becoming the most important part of our civilization. From a philosophy of science standpoint, the epistemological argument should not be confused with the ontological position that there *are* actually digital artifact genes *out there*. Whether they *are* actually out there or not is not the focus here. The focus is whether the idea of genes and genetic codes of digital artifacts will help us to understand the phenomenon better than other scientific devices. Just as the current ideas of genetic variation and evolution in biology will remain true until another idea is proven to be more effective in explaining the varieties that we see in nature, the idea of genes and the genetic code of digital artifacts needs to be deployed and tested to explain the varieties we see in the artificial world: that is, until we find a better scientific device to study the incredibly dynamic varieties in man-made digital artifacts.

Acknowledgments

This chapter is based on work supported, in part, by the National Science Foundation under grants *VOSS-0943010* and *VOSS-1120966*. Any opinions, findings, and conclusions or recommendations expressed in this material are those of the author(s) and do not necessarily reflect the views of the National Science Foundation.

References

Anderson, C. (2010). In the next industrial revolution, atoms and the new Bits. *Wired*, 18, Vol. 18. http://www.wired.com/magazine/2010/01/ff_newrevolution/

Arthur, W. B. (2009). *The nature of technology*. New York: Free Press.

Baldwin, C. Y. and Clark, K. B. (2000). *Design rules, Vol. 1*. Cambridge, MA: MIT Press.

Basu, S., Kirk, M., and Waymire, G. (2009). Memory, transaction records, and *The Wealth of Nations*. *Accounting, Organizations and Society*, 34(8), 895–917.

Benkler, Y. (2006). *The wealth of networks*. New Haven, CT: Yale University Press.

Boland, R. J., Lyytinen, K., and Yoo, Y. (2007). Wakes of innovation in project networks. *Organization Science*, 18(4), 631–47.

Buchanan, R. (1992). Wicked problems in design thinking. *Design Issues*, 8 (2), 5–21.

Bush, A. A., Tiwana, A., and Rai, A. (2010). Complementarities between product design modularity and IT infrastructure flexibility in IT-enabled supply chains. *IEEE Transactions on Engineering Management*, 57(2), 15.

Clark, K. B. (1985). The interaction of design hierarchies and market concepts in technological evolution. *Research Policy*, 14, 235–51.

Faulkner, P. and Runde, J. (2010). The social, the material, and the ontology of non-material technological objects. Working Paper.

Gaskin, J., Lyytinen, K., Thummadi, V. et al. (2010). Sequencing DNA design. Paper presented at the 2010 International Conference on Information Systems, St Louis, MO., December 12–15. http://aisel.aisnet.org/icis2010_submissions/202

Harty, C. (2005). Innovation in construction. *Building Research & Information*, 33(6), 512–22.

Kallinikos, J. (2006). *The consequences of information.* Cheltenham, Glos, UK: Edward Elgar Publishing.

—— (2011). *Governing through technology.* New York: Palgrave Macmillan.

—— Aaltonen, A., and Marton, A. (2010). A theory of digital objects. *First Monday,* 15(6–7). http://firstmonday.org/htbin/cgiwrap/bin/ojs/index.php/fm/article/view/3033/2567

Kitano, H. (2002). Systems biology. *Science,* 295, 5560.

Kurzweil, R. (2006). *The singularity is near.* New York: Viking Penguin.

Langlois, R. N. (2007). Computers and semiconductors. In B. Steil, D. G. Victor, and R. R. Nelson (eds.), *Technological innovation and economic performance* (pp. 265–84). Princeton, NJ: Princeton University Press.

Leonardi, P. M. (2010). Digital materiality? *First Monday,* 15(6–7).

Lévi-Strauss, C. (1963). *Structural anthropology.* New York: Basic Books.

Orlikowski, W. J. (1992). The duality of technology. *Organization Science,* 3(3), 398–427.

—— Scott, S. (2008). Sociomateriality. *The Academy of Management Annals,* 2, 433–74.

Pentland, B. T. and Feldman, M. S. (2005). Organizational routines as a unit of analysis. *Industrial and Corporate Change,* 14 (5), 793–815.

Pickering, A. (1995). *The mangle of practice.* Chicago: University of Chicago Press.

Schilling, M. A. (2000). Toward a general modular system theory and its application to interfirm product modularity. *Academy of Management Review,* 25(2), 312–34.

Sennett, R. (2008). *The craftsman.* New Haven, CT: Yale University Press.

Shannon, C. E. and Weaver, W. (1949). *The mathematical theory of communication.* Chicago, IL: University of Illinois Press.

Simon, H. A. (1996). *The sciences of the artificial.* Cambridge, MA: MIT Press.

Ulrich, K. (1995). The role of product architecture in the manufacturing firm. *Research Policy,* 24, 419–40.

Vargo, S. L. and Lusch, R. F. (2004). Evolving to a new dominant logic for marketing. *Journal of Marketing,* 68(1), 1–17.

von Hippel, E. (2005). *Democratizing innovation.* Cambridge, MA: MIT Press.

Yoo, Y. (2010). Computing in everyday life. *MIS Quarterly,* 34(2), 213–31.

—— Boland, R. J., and Lyytinen, K. (2008). Distributed innovation in classes of network. Paper presented at the 41st Hawaiian International Conference on Systems Science, Big Island, Hawaii.

—— Henfridsson, O., and Lyytinen, K. (2010*a*). The new organizing logic of digital innovation. *Information Systems Research,* 21(5), 724–35.

—— Lyytinen, K., Thummadi, B. V., and Weiss, A. (2010*b*). Unbounded innovation with digitalization. Paper presented at the 2010 Annual Meeting of the Academy of Management, Montreal, Canada, August 6–10.

Zittrain, J. (2006). The generative Internet. *Harvard Law Review,* 119, 1974–2040.

IV
Materiality as Assemblage

8

Inverse Instrumentality: How Technologies Objectify Patients and Players

Hamid Ekbia and Bonnie A. Nardi

Popular narratives of technologies portray them as tools in the service of human welfare and well-being. These portrayals derive from an instrumentalist view which considers technology as something that can be mastered, a means to an end (Heidegger, 1949). They are also compatible with the modernist framework of a "service economy," which conceives of all natural and man-made systems as calculable resources in the service of human well-being (Suchman, 2007). On close scrutiny, however, these portrayals are untenable. Indeed, in examining certain large, complex technologies, we observe the inverse: a move to strategically insert human beings within technological systems in order to allow the systems to operate and function in intended ways. Such is the case, for instance, with massively multiplayer online video games and certain health information technologies (HITs) such as electronic health records. We argue that some large, complex technological systems cannot meet their goals without instrumentalizing *end users* as indispensable mediators in the system. Such systems, which we call "technologies of objectification," stand in contrast to "technologies of automation" such as banking and financial systems. Technologies of automation are designed to disallow human intervention at nearly all points within the system, excluding carefully controlled access to simple input and output (such as a bank teller accessing a customer account) or professional information management conducted by specialists who configure, troubleshoot, and repair systems. These systems represent different ways of regulating human behavior, some of which are more novel than others. However, they also represent something more—namely, how certain technologies objectify individuals and turn them into particular kinds of subjects. While all technologies may entail some degree of objectification through enframing—enabling and allowing certain types of behaviors and constraining or disallowing others (Heidegger, 1982)—objectifying systems are distinctive in that they push critical tasks down to the very lowest level, that is, to end users themselves. In so doing, they objectify users as critical mediating elements in networks of interaction.

Objectifying technologies affect human beings in various ways, particularly in how they turn an individual into a particular kind of *subject*. By "subject" we mean the person as a social agent in interaction with other persons and objects. Rather than a fixed entity given a priori, the subject as such is constituted in interaction with these other entities. In other words, subjects and objects emerge in networks of interaction through a process of mediation. This conception of subjects, and their relationship to objects, is in line with post-structuralist thinking of Foucault (1982), Serres (1982), and others (cf. Ekbia, 2009; Day, 2010), but also with certain interpretations of Vygotsky (1978)'s cultural-historical psychology and activity theory (Kaptelinen and Nardi, 2009). In discussing power relationships in modern societies, for instance, Foucault suggests "three modes of objectification which transform human beings into subjects" (1982: 208). Foucault delineates practices of objectification that result in particular kinds of human subjects:

1. The modes of inquiry that try to give themselves the status of science, for example, objectification of the speaking subject in linguistics and philology, of the productive subject (labor) in economics, of the sheer fact of being alive in biology and life sciences.
2. Dividing practices that divide the subject inside themselves or from others, for example, the mad and the sane, the sick and the healthy, criminals and the "good boys."
3. The way a human being turns him/herself into a subject, for example, how people have learned to recognize themselves as subjects of "sexuality."

These modes of objectification are sometimes clearly discernible in modern technological systems. With respect to the first mode of objectification, for instance, it has been argued that dominant theories of information, particularly the Shannon–Weaver Model of Communication, provide a narrow view of language and communication as having solely to do with the transfer of information (Day, 2010). One could argue that technologies that are built on the basis of these theories enable and embody this view of language. Our interest here, however, is in the other two modes and in the technologies that support and enable them. For reasons that will become clear throughout this chapter, we identify these as, respectively, "fragmenting" and "totalizing" technologies. As an example of the first, we discuss electronic health records, especially what has come to be known as "Personal Health Records," and for the second, we focus on massively multiplayer online video games, in particular "World of Warcraft."

The two realms of computer gaming and health technology are interesting to us for a number of reasons. First, as we elaborate below, they provide salient examples of the two modes of objectification mentioned above. At the same time, they lead to the curiously common consequence of using objects to construct subjects—they both subjectify through objectification. Second, they instantiate distinctive ways of mediating and organizing large assemblages of technologies and people (players in one case; patients, health providers, administrators, technologists, and others in the other case). They show how human actions and human bodies are inserted into these assemblages in order to bridge functional gaps in technological systems. The dynamics of functionality and materiality manifested by these systems is of deep interest to us (Kallinikos, this volume). Third, these two systems embody a defining feature of modern societies—namely, the transition from direct ownership of bodies, as practiced in earlier historical periods such as slavery and serfdom, to one of uninterrupted, constant, subtle coercion. This transition marks, in Foucault's poignant terms, the attention to the body "that is manipulated, shaped, trained, which obeys, responds, becomes skillful and increases its forces" (1978: 136). We would like to explore the meaning of materiality inherent in these manipulations of the human body. Lastly, these two realms clearly illustrate the relationship between power and materiality in areas that are generally underexplored. While this relationship is studied and relatively well understood in organizational and work contexts (e.g., Zuboff, 1988), it is scarcely articulated in broader, non-work-related settings such as healthcare and game play. With the introduction of digital technologies into our daily lives, we believe that this transition is taking on new shapes and meanings, which we seek to understand as the interplay between processes of materialization and dematerialization. Power is deeply, but subtly, implicated in these and other similar activities of modern societies that are not evidently disciplined or hierarchical and we would like to understand the meaning of these implications for patients, players, and, through them, other social actors of contemporary societies. The active complicity of social actors, increasingly sought out in current designs of technological systems, supports and enables the functioning of such systems. How can we meaningfully conceptualize this role and its consequences for such modern values as human freedom and autonomy without rushing into normative judgments about technology?

Digital Docile Bodies: The Fragmentation of the Patient

> A body is docile that may be subjected, used, transformed and improved. (Foucault, 1978: 136)

The healthcare situation in the United States is rather gloomy, as is the state of HIT. It is widely believed that these two are closely related, and that the former can be largely explained in terms of the latter, that is, that poor technology accounts for poor healthcare. According to some commentators, "The U.S. healthcare industry is arguably the world's largest, most inefficient information enterprise" (Hillestad et al., 2005: 1103). According to others, this state of affairs has largely to do with "a medical care environment that provides insufficient help for clinicians to avoid mistakes or to inform their decision-making and practice" (Stead and Lin, 2009: 14).

It is also widely believed that broad adoption of electronic medical records (EMRs) and personal health records (PHR) can "[lead to] major healthcare savings, reduce medical errors, and improve health" (Hillestad et al., 2005). These beliefs are supported by casual observations of daily encounters between patients and health providers, for instance, and perhaps most conspicuously, in the anachronistic paper forms attached to clipboards that patients fill out again and again in doctor's offices, clinics, hospitals, and elsewhere. Currently, 15–20 percent of US physicians' offices and 20–25 percent of hospitals have adopted EMR systems (RAND, 2005). Most medical records, however, are still stored on paper, hampering the coordination of care, the assessment of quality, and the reduction of medical errors. Similarly, many people are not familiar with PHRs, and only a small percentage of those who know about these technologies actively use them (California Healthcare Foundation, 2010). To understand the reasons for this underutilization and, more importantly, the potentials and implications of their use, a closer look at the technology is in order.

The origin of EMRs goes back to attempts in 1970s at creating a standardized medical file that would allow physicians to act *scientifically*. Through the problem-oriented record, the doctor "is able to organize the problems of each patient in a way that enables him to deal with them systematically" (Weed, 1968; Hurst, 1971). The dream was to create an automated record that would turn all narrative data into a structured and retrievable form of information. EMRs serve an important function. They replace an inefficient, byzantine system of disconnected paper records. But, as Diamond and Shirky (2008: 383) point out, "If you computerize an inefficient system, you will simply make it inefficient, faster." They advocate "transforming the U.S. health care system as a whole, rather than simply computerizing the current setup."

The interest in PHRs, on the other hand, largely derives from a growing trend toward a preventative and "patient-centric" model of healthcare. Although the concept of "patient-centered medicine" is not new (Slack, 1972), the resurgence of interest in it can be arguably attributed to the affordances of new electronic technologies, which allow the recording, transfer, and sharing of information at increasingly lower transaction costs. As such, these technologies can potentially enable the engagement of patients in the management of their own health at lower effort and economic cost (Bodenheimer et al., 2002). The patient-centric approach seeks to put patients at the center of healthcare, not only as recipients of health service but also as "managers" of their health and health information. This approach favors a collaborative health model that would "promote effective self-management of health habits that keep people healthy through their lifespan" (Bandura, 2004). To maintain healthy lives, minimize risks, and effectively achieve optimal health outcomes, individuals need to understand risks and to make significant changes in lifestyle, such as increasing exercise, dietary modifications, weight loss, and cessation of unhealthy habits. These activities, referred to as self-management, require both patient–provider collaboration and the education of self-management (Bodenheimer et al., 2002). Self-management education is meant to help support people in attaining the best possible quality of life. The principle of self-efficacy, or the confidence to carry out a behavior necessary to reach a desired goal, is key to self-management.

On a broader scale, the interest in PHRs and preventative health derives from the financial burden on national healthcare systems of the increasing healthcare costs of an aging, chronically ill population. According to current policy trends in Europe and the United States, sick people are expected to monitor their risk factors more carefully in order to contribute to cost reduction. These policies are partly enabled by the explosion of epidemiological knowledge that allows the screening of "at-risk" populations in advance, but they are also driven by a moral order informed by a neoliberal philosophy. This perspective considers healthcare largely a matter of individual responsibility and, by implication, regards illness as a matter of individual choice (Mathar and Jansen, 2010). Altogether, these perspectives, policies, and practices reconfigure not only the relationship between patients and providers, but also between the state and citizens. In this process of bio-medicalization (Clarke et al., 2003), the Foucaultian medical gaze (Foucault and Seitter, 1973) is not only at work in the clinic and hospital but also in daily life, disciplining people to take more responsibility for their health (Mathar, 2010: 172).

According to the proponents of the patient-centered approach, to allow people to manage their health, we must provide them with tools and technologies that enable and encourage informed decisions concerning their short- and long-term health. Self-management support is defined by the Institute for Healthcare Improvement as the care and encouragement provided to people to help them understand their central role in managing their health, make informed decisions about care, and engage in healthy behaviors (IHI, 2009). Information transfer by itself has been found insufficient in impacting outcomes, as greater patient knowledge does not amount to greater patient engagement (Bodenheimer et al., 2002). Furthermore, web-based information resources are limited in that they cannot be personalized to an individual's specific health conditions and health risks. Therefore, this approach advocates PHRs, which can help patients manage, maintain, and exchange their health information with support from medical practitioners to become "co-pilots" of their own care (Tang et al., 2006).

The patient-centric approach to EMRs, however, is not aimed at the kind of thorough, top-down transformation Diamond and Shirky have in mind. With no top-down change, the PHR approach mandates patients—the end users of the system—to manage, maintain, and exchange their health information with support from medical practitioners. The implications of assigning critical tasks of mediation to patients have not yet been explored, but would seem to be significant and should be taken into account when designing PHR systems.

What is more important perhaps is the way these technologies play into the unfolding of the illness trajectory. On one hand, earlier technologies (e.g., machinery, drugs, and procedures) have been shown to change or lengthen trajectories, often in unpredictable ways; on the other, technological innovations generate, and are generated by, medical specialization in a tightly coupled manner (Strauss et al., 1985: 4). Most importantly, however, these interactions have resulted in the fragmentation of chronic care, "with increasing possibility that continuity of care will go awry, accompanied by accusatory cries of dehumanization" (ibid.). To compensate for these effects, the health profession has responded to this by creating new workers or roles such as "primary nurse" or "patient advocate," particularly in hospitals. The emergence of such roles indicates that the patient is not always optimally suited to handle information tasks as a node in a technological system, and may require further human mediation. The advocates, along with the patient, become technical workers administering the system. This situation suggests that it is important to pay attention to the instrumental, regulative role of technology in the organization and conduct of health work.

A key concern would be for patients' bodies to acquire an ephemeral nonmaterial existence in these networks of information flow, fragmenting the patient as a subject from their body as the object of treatment. In the case of PHRs, the embodied human person (the patient) transforms, through the objectifying technology, to a dematerialized mediator acting within the technical system. The eminent risk is for patients to turn into a set of data points, abstracted away from their experiences as particular individuals in particular situations, and fed into these networks.

We can see further glimpses of such fragmentation in what is called "patient-centered cognitive support"—that is, a kind of technology that would support the physician by providing a "conceptual model of the patient" (Stead and Lin, 2009). In this vision of EMR technology, "the clinician interacts with models and abstractions of the patient that place the raw data into context and synthesize them with medical knowledge in ways that make clinical sense for that patient. These virtual patient models are the computational counterparts of the clinician's conceptual model of a patient" (ibid.). The implications of this cognitivist approach to healthcare and HIT should be of concern to those who design and use such technologies.

Active Bonds and Bodies: The Totalization of Players

Massively multiplayer online video games also rely on mediation by end users to enable system function. We analyze "World of Warcraft" (WoW), a popular game with 11 million players worldwide. "WoW," as it is known, is a medievally themed fantasy world organized around slaying monsters, practicing medieval crafts, and trading at an Auction House (Nardi, 2010). It is a rich, diversified gaming environment rather than a single game. WoW is structured into a set of win–lose contests among which players choose, playing as many or as few contests as they like. The game is produced by Blizzard Entertainment, Irvine, California.

WoW is a highly social game. Players gather in "guilds," which are relatively stable groups that play together consistently. Players may also have a "friends list" of others with whom they play. Game utilities support sociality: multiple text chat channels, mechanisms for tracking in-game and real-life friends, and indicators of who among guildmates and friends are currently online. These capacities foster social bonds, encouraging players to come to know one another quite well, and to develop ties of friendship. Many game activities require a group, and the guild and friends list are handy sources of readily available teammates (Figure 8.1).

WoW is, in our parlance, a "totalizing technology"; players make a strong connection to the game in terms of hours played, affective engagement, and participation in game-related activities outside the game.[1] Unlike patients who are caught up in an unfortunate situation they would absent themselves from if they could, WoW players autonomously seek out gaming activity, often prioritizing it so that they coordinate other life activities around periods of play (see Nardi, 2010).

A key part of engaging with this totalizing technology is gaining levels of skill sufficient to play well. Players are caught up in the function of the technological system as they train others to play. Kallinikos (this volume: 14)

[1] Our usage of the term "totalize" here is somewhat unusual relative to contemporary literature (e.g., in critical or political theory), and more akin to the dictionary definition of the term, meaning "to combine into a total" and "express as a whole" (Merriam Webster). We think of (good) gaming as a *totalizing experience* because it engages the person as a whole, allowing them to participate with the medium in a tight and interactive manner.

Figure 8.1 A playable World of Warcraft character

emphasizes that computation "addresses primarily perception and cognition . . . [such that] user understanding of the functionality of software . . . is *sine qua non*, the gate, as it were to the software." A technological system remains out of reach, inert, until its users pass through a gate of understanding. WoW, as a complex software artifact, presents a steep learning curve as a prerequisite for participation. Arcane constructs such as "aggro," "hit boxes," and "enrage timers" infuse game mechanics. In order to play, it is necessary to extract knowledge of game mechanics from other players (or to have already learned them from other games, in which case a different set of players has served the instructional function). Because the game constantly changes through software updates and expansions, learning is "lifelong" in WoW.

Despite the significant learning needed for the game, it is delivered with minimal documentation. No manual exists apart from a small introductory booklet, and there is little in the way of online tutorials or guides outside simple decontextualized comments presented at the login screen. Blizzard "game masters" interact in-game with players, but game masters are forbidden to offer what Blizzard refers to as "game hints." (The second author has, through participant-observation research, pushed the game masters to offer such help, which was politely but staunchly refused.) Instead, game masters assist with technical difficulties, harassment, and other matters outside of learning to play.

Instruction, rather than being supplied by the purveyor, falls to players. Because they love the game, players willingly serve as instructors. Teaching and learning occur within the game itself, as well as outside in forums, blogs, wikis, and chatrooms (Nardi et al., 2007; Kow and Nardi, 2009; Nardi, 2010). Players write and publish how-to and strategy guides, some very stylish and sophisticated. These guides demonstrate deep knowledge of the gaming software, and often deploy effective presentation techniques including pictures and diagrams as well as clear, crisp text. Players also answer questions on forums, offering a multitude of tips and hints to fellow gamers.

The most skilled players, many adept at other similar video games, lead the way in authoring guides and answering questions. As new content becomes available in updates and expansions, expert players populate forums with lively, nuanced discussions of game mechanics. An online search "definition aggro Blizzard" yields multiple sites (e.g., wowwiki.com and many others) containing elaborate discussions of aggro, in contrast to the one line definition at Blizzard's official site. Yet, aggro (a quantity specifying the probability that a monster will attack a particular player) is one of the most fundamental concepts of play that every WoW player must master.

Player-teachers answer questions in real time in-game (often within 30 seconds of being asked; Nardi et al., 2007), and they group together for game activities to show each other the ropes. The most serious players experiment with the game by playing it, and, through patient trial and error, learn how to win the various contests. They share this knowledge with others. In this fashion, experienced players act as both enablers and gatekeepers of the overall system. These players themselves are, even as they experiment, recursively searching the forums for useful information to inform their own play.

The work of teaching, then, is delegated to the leaf nodes, as it were: the end users. The technological system would fail to function without human mediation between less experienced players and the gaming software. (As a small comparison, a product such as Microsoft Word is also supported by user forums. However, it is possible to successfully use Word without consulting forums, or asking for help. But essential to WoW play are sources beyond the technological system of the game itself.)[2]

If we envision the capacity to play the game as a network of resources, the network is configured with gaps to be filled. Through human mediation, gaps are closed, completing the network, yielding the capacity to play. We can say that the system as it is designed and merchandised is not wholly "off the shelf" or "shrinkwrapped," in being self-contained or complete. Rather, it is organized *expectantly*, anticipating its functions to be actuated via external

[2] Other similar games such as EverQuest organize instruction in the same way. See Steinkuehler (2005) and Taylor (2006) for descriptions of learning to play.

mediation from end users. Such actuation may be voluntary, for example, in the case of totalized subjects, or it may be prompted by urgent needs such as health, as for patients.

This configuration of resources in some ways turns artificial intelligence (AI) on its head. Instead of programming humanlike intelligence into a software artifact, the software expectantly leaves gaps to be closed by human intelligence. The very complexity of WoW is possible because of its power to generate totalizing experience. If Blizzard were required to document every feature in its complex game, to provide elaborate tutorials and instruction, it would probably be economically infeasible to charge players $14 for a monthly subscription—a price point at which millions of players can maintain subscriptions, sometimes for years (Debeauvais et al., 2011).

Let us turn to a second instance of objectification within WoW, one that is arguably less positive than teaching-learning, and more like patient as mediator. In this case, we want to argue that design changes have resulted in the objectifications of human players as "NPCs," or non-player characters (Figure 8.2).

Human-like computer-generated characters are central elements in many video games. NPCs provide services and perform as actors in the game narrative. In WoW, NPCs act as the monsters killed in contests, as quest givers assigning quests (narrativized tasks), as vendors selling items in the game economy, auctioneers staffing the Auction House, guards in the capital cities providing directions to destinations within the city, and seasonal characters purveying holiday events (such as the Midsummer Fire Festival, Brewfest, Pilgrim's Bounty).

Players do not, however, play with NPCs as teammates in the group contests. Composing a team can be difficult because of the requirement to assemble a precise mix of specific skills. These skills are distributed according to a "trinity" of tank–heal–damage (Gamasutra, 2009). Each character type can perform one function of the three. Without worrying too much about further details of the trinity, it specifies that successful groups contain players from each of the categories. Although usually a player looks to his guild or friends list for teammates, sometimes a proper group cannot be assembled unless other players are recruited (guildmates and friends may be busy, etc.). Thus, players organize ad hoc pick-up groups or "pugs," similar to pick-up basketball games.

From the game's inception in late 2004 until late 2009, pug members came from a player's own server. Server-only groups often required long wait times until the correct mix of players could be found. Blizzard redesigned its servers to make cross-server pugs possible, drawing group members into "dungeons" (sites of group contests) from the much larger pool of players available across multiple servers.

Figure 8.2 World of Warcraft non-player character

Under the new technical regime, pick-up group players treated each more like NPCs than human players. Players served primarily to fill out the roles in the trinity so that contests could be completed and items of equipment won; the social dimension transmogrified to minimal instrumental interaction, and sometimes no interaction at all. The friendly sociality of pugging on a player's own server—making new friends and behaving well in the expectation of future contact—disappeared from the game.

In the past, when playing in a pick-up group on a player's own server, players knew they might, in the future, be grouped with the same people. They were generally friendly and well-behaved. They attempted to make conversation, used game "emotes" such as dance to lighten the atmosphere, and exhibited (for the most part) good manners. Players often got to know one another, adding each other to their friends lists, perhaps even eventually joining a new guild based on friendships forged in pugs.

Cross-server pugs changed all that. Players knew that they would not be playing with the same people again because the pug population had expanded to thousands of players across hundreds of servers. There was little if any conversation in pugs, unlike the own-server pugs in which people usually made an effort to be pleasant or funny, to tell little stories, or provide brief amusing commentary. When players in the new pugs did interact, often it was to comment derisively on the play skills of another player—skills deemed detrimental to a player's narrow goal of obtaining an item of equipment. With no social controls in place, players were unconstrained in expressing hostile feelings. (There is a study to be done on why players of games such as WoW are often so undeniably crabby.) Players would sometimes leave a pug without a word after receiving the "loot" they desired from a dungeon. Instead of finishing the encounter, killing all the key monsters—some of which might have loot other players needed—miscreant players simply vanished. A further small but noticeable technical change eroded sociality: players could no longer trade items with one another, such as the shared temporary enhancements to play that commonly smoothed social interaction and cooperation in normal play. These changes to the gaming software removed key means of socializing and spreading good will—leading to "players as NPCs."[3]

Cross-server pugs generated a system that objectified human players as technical components to enable (other) players to meet their goals in the logic of the game. Pug members provided the necessary trinity-derived skills for the contests, but in a new context in which their subjectivity as social beings was nearly extinguished. A game designed for lively socializing altered to meet a demand that could not, for technical reasons, be filled without instrumentalizing human subjects—transforming them to NPCs.

This turn of events suggests that large, evolving computational systems may transform over time such that gaps appear in the network of resources required for system function. The system may respond to new demands by objectifying human participants. Gaps, then, are not always designed into a system; they may emerge in time as the software expands and changes. On the other hand, gaps may be planned, as in the case of patients deliberately assigned tasks they will undertake as "co-pilots." Gaps are expected from the outset, and work is done to ensure that human subjects are capable of filling the gaps, for example, the provision of primary nurses in patient-centered care.

The case of players as NPCs lies somewhere between fragmented patients managing their own illnesses and the totalized player happily laboring to instruct others in the ways of the game. Pug members are still meeting their

[3] We are indebted to T. L. Taylor for the phrase "players as NPCs" (personal communication 2010). Taylor is developing a different set of examples of the phenomenon.

own goals in an activity of play. But game forums are replete with discussions of how to deal with the miseries of the altered pugs. An unpleasant sense of being used is particularly strong when members abandon a group after they have attained their goals, with no regard for players who have not completed their own.

It should be noted that players often remarked that they liked the new system because it enabled them to advance in the game more quickly. Many players enjoyed an active social life within their guild, and because pug play was seen as convenient for certain goals, it was acceptable, even with the deterioration of social experience. But acceptance of, and acclimation to, an ethos and practice of objectifying people, unwillingly, to function as components in a technical system may have far-reaching effects.

The Reconfiguration of Patients and Players

As pointed out earlier, our interest in the two domains of video gaming and HIT derives from the conceptual and practical contrasts that they provide, particularly in regard to issues of technology, organizing, and materiality. Kallinikos (Chapter 4, this volume) argues that modern technology evinces a "growing emancipation from the materials with which they are entangled." While this observation is no doubt correct in the overall trajectory it sketches, we feel it is nonetheless important to examine a diverse ecology of technological systems because, just as in biological evolution, a particular mutation, even if not pronounced, can, over time, shake up the gene pool. Our examination here brings to forth certain aspects of this ecology, which we highlight here on four dimensions of organization, mediation, materiality, and power.

Expectant Organizing

Our focus on the objectification of subjects under two technological regimes (PHR and gaming) hints that large technological systems may evolve as "expectant" systems that await human interventions from end users. End users are thus at once beneficiaries of the systems—"users" in the classic sense—but also subjects bound into the system as necessary functional components. The capacity of software systems to insert human subjects, deriving labor from them as evidenced in our empirical examples, speaks to the systems' growing regulative power (see Kallinikos, 2011). However, it also points to the enormous potentials created by such systems in engaging, enabling, and perhaps even empowering end users. As such, we are interested in both aspects of these technologies.

Of course it is not as simple as asserting that software systems straightforwardly regulate and objectify humans in particular ways. Behind every PHR system is a medical establishment in which vulnerable patients go along with what they are told by doctors, nurses, and insurance companies in the hopes of getting well or at least not getting sicker. Some games are so inviting that they create and sustain totalizing experience, but most fail to do so. Game designers constantly wrack their brains to bring forth the elusive "fun" to lure players and keep paying customers coming back for more. These human elements of power and imagination indicate that "assemblages" are in play.

Our discussion may bring to mind processes of translation as enumerated in actor-network theory (ANT) (Callon, 1991). However, the slippage inherent in ANT's metaphorical deployment of concepts such as "enrollment," "mobilization," "spokespersons," etc., applied to nonhuman entities, hides certain questions that interest us, questions we do not wish to treat metaphorically. Rather than asserting that nonhuman actors "enroll" or "persuade" human subjects (a common analytic in ANT), we would like to know exactly how such relations can be established between a software system and a patient or player. It is important to our understanding of systems of inverse instrumentality that we keep the question before us as a problematic to be addressed rather than backgrounding it as a theoretical axiom. The question bears on continuing struggles to apprehend such matters as the structure and evolution of complex technological systems (or assemblages), the regulative power of software (Kallinikos, 2011), and the meaning of terms such as "affordances" in the context of digital technologies (see Robey et al., this volume).

These discussions illustrate exemplary processes through which individuals are drawn into, or withdraw from an activity, as the case might be, through their relationship with a system. Video games such as WoW seem to attract individuals by enabling a total experience with a strong aesthetic and sensual thread (Dewey, 1938; cf. Nardi, 2010). A PHR, on the other hand, largely lacking in such an aesthetic component, might be unlikely to have such an attraction (Ekbia, 2010), especially for individuals such as the younger adults with no clear sense of immediate health issues (Kutz and Ekbia, 2011). Given the appeal of games to the same population, an interesting practical question would be whether and how the techniques of gaming can be applied in engaging individuals with their health. This is the approach pursued under the general rubric of serious games for health, which has thus far had limited, but promising, success (Kato, 2010). Of specific interest to us here is the possibility of designing integrated game–health systems with the affordances of a totalized experience.

Double Mediation

Voluminous research spanning literatures in human–computer interaction, computer-supported collaborative work, media studies, and organizational studies informs us of the many ways in which human subjects act as end users of technologies. But the rhetoric of "users" has perhaps obscured other relations which systems define for us (see Day, 2010, for related discussion). PHR and gaming indicate that end users may be objectified into the operation of software as functional parts without exactly being cognizant that they are performing system labor beyond their own goals, or labor that might be automated into, or executed by, a system (such as the provision of instruction in a game). There is a kind of double mediation at play in systems of inverse instrumentality; subjects act as intentional beings in pursuit of object-oriented activity using technologies to mediate relations with reality (Leontiev, 1978), while at the same time, subjects are (partially) objectified through interactions with the technology to become particular kinds of transformed subjects (Day, 2010).

We suggested two kinds of technological systems—fragmenting and totalizing—that bear on subjects' experiences of objectification. These two systems represent examples of technologies of objectification, but the manner by which they objectify is rather different. The fact that gaming is a pleasurable activity while patient treatment is not may account for part of the difference, but it is not the key differentiator. The key difference is in how the person as a subject–object totality is distinctly reconfigured in these two systems through processes of fragmentation or totalization. All technologies fragment and integrate individuals as subject–objects through their mediations, but the degree by which they do this is what is at issue here.

(De)Materialization as a Process

Where does this leave us with the question of materiality, especially as it relates to embodied human beings? In thinking about patients and players, we find the material–nonmaterial dichotomy of limited conceptual utility. The way this dichotomy is sometimes set up reminds us of the classic mind–body dualism in Descartes's philosophy of mind, which has engaged scientists and philosophers in an ongoing intellectual pursuit with no clear end (Ekbia, 2008). An alternative approach, and one that we would like to pursue, is to adopt a processual perspective that would examine how things (human bodies, in this case) become materialized or dematerialized in human–artifact assemblages (Ekbia, 2009). In fragmenting technologies, the process of objectification appears to dematerialize the human subject, reducing him or her to a set of functions relative to the technical system. The fully embodied human is attenuated. In PHR, the patient's needs to recover a healthy body become separate from his function as information manager. In WoW pugs, the player's social needs are driven underground as she is compelled to act as a computer-generated character. But the flip side is that a technological system can objectify a totalized subject whose material embodied intelligence and creativity are summoned in the generative production of useful outputs such as instructional materials and interactions. A key point, then, is the immense power of technological systems to objectify. As Foucault noted, subjects are objectified in variable ways. In our examples, we perceive two ends of a spectrum of materiality: systems that fragment, objectifying subjects such that they dematerialize as they divide from their material bodies; and, on the other hand, totalizing systems advancing a fully realized human person materialized to willingly use her capacities to serve others in collaborative relations of mutual interest, even joy. In both cases, the technological system extracts a technical capability for system function.

Partly because WoW is a leisure activity that players can freely choose, and the consequences of playing or not playing are low, such labor as goes into teaching, which players enjoy, ends up on the positive side of a moral ledger. Whether such cost savings in medical care are morally defensible is another question. The fact that PHRs seem to have worked successfully in Europe suggests that socio-technical differences in terms of how they function, the kinds of support patients have, and the kinds of relationships between patients and providers do matter.

Our account of inverse instrumentality leaves us with key questions about the notion of materiality as it plays into the systems of mediation such as those that we discuss here. On one hand, traditional notions of the material as coextensive with the "physical" (e.g., Stoljar, 2009) seem too narrow and inadequate here. On the other, the more cultural understandings of the term as having to do with "possibilities and resistances of specific social, cultural, and physical forms" (Day, 2010) sound too broad to be able to capture and explain the *mechanisms* by which the material reveals its potencies. Drawing from both Stoljar and Day together, we have hopefully indicated at least something about these mechanisms in formulating notions of objectification and inverse instrumentality. The empirical examples suggest that the embodied person as a material object is subject to manipulations by expectant technological systems awaiting completion through human cognition. Kallinikos (Chapter 4, this volume, p. 81) speaks of "the immateriality of cognition and communication, supported underneath by increasing selectivity in the use of matter." Here we see our patients as information managers, and the hapless pick-up group players. Bodies are set aside in the technological systems of medical records in favor of the cognitive labor of managing information; with gaming, fully embodied persons transform to virtual nonplayer computer characters.

Stoljar points to the physical as central to materiality, and that is apropos for our argument as we foreground the human body in drawing attention to transformations of embodied human subjects. However, the complexities of virtual identities suggest that Day's notion of materiality as harboring social, cultural, and physical aspects is also pertinent to understanding processes of objectification.

For now, we can say that the human person as materially realized in a physical body is important to a concept of inverse instrumentality, and that conceptions of materiality that incorporate the physical must continue to shoulder the burden of preserving a notion of the material as matter that has substance and occupies space (as against, say, the material conceived as only what "matters" to someone at some particular time (see Cooren et al., Chapter 15, this volume) or is of consequence in some human activity (Burrell, this volume)). Materiality/physicality continues to do important work in characterizing the embodied person—an essential category we cannot do without. Patients struggling with illness, being slotted into technological PHR systems, denote and crystallize the inescapability and frailty of the human body, a frailty all too purely physical. We can further say that, in this digital age, embodiment may at times exist at one remove; the avatar can represent the embodied human in the welcome grip of a totalizing technological system that stimulates his intelligence, creativity, and deep engagement. But the very same pixels may also visualize the diminished "player as NPC," regulated by the goals of the technological system. Day enjoins us to consider a cultural interpretation of materiality, and it seems this is necessary if we are to understand how modern technology produces its complex effects of objectification.

Our contribution has been to explore a curious property of large digital systems—their capacity to harness human cognitive activity to cover their own shortcomings—and to suggest that a reason to find these systems of inverse instrumentality interesting is that they powerfully objectify subjects in variable ways. These systems reconfigure subject–object relations, producing "docile bodies" in some cases and "active bonds" in others. These varied versions of the human subject appear to skew our natural human materiality away from full embodiment in some instances, and to enhance and deepen it in others.

Digital Discipline: Active Complicity of the Modern Subject

Technologies of inverse instrumentality raise interesting questions in regard to the flow of power in modern societies, especially with respect to how digital technologies enable and produce such flow. The "voluntary" character of game play, for instance, suggests that individuals are driven toward this activity by their own desire and motivation. While this is the case in a serious sense of the term, it is important to understand how the technology of the game, broadly understood, disciplines the player.

Foucault (1978) describes how modern societies employ a whole set of techniques, methods, and plans to discipline individuals and their bodies. In particular, he identifies spatial distributions (enclosures, partitions, functional sites, etc.), anatomo-chronological control of activity through time tables, gestures, and technically defined articulations, and a techno-political machinery that inserts the body in a whole ensemble in order to accomplish efficiency, arranging a positive economy in the process. With the advent and infusion of digital technologies, some of these methods and techniques (e.g., temporal control through schedules and plans) have been refined and enhanced, while others (e.g., spatial distributions) have been, by and large, abolished, giving their place to what can be characterized as "digital distributions." The kind of discipline enabled by these technologies is less spatial and more temporal, less anatomic and more cognitive, but they are by no means less techno-political. What makes current arrangements different is the disappearance, or rather the invisibility, of spatial enclosures, and at the same time the spatio-temporal extension of Foucaultian gaze, whether we consider the medical gaze and its broadening to include "healthy" individuals or we examine video gaming and the "guilded" gaze of leader-teachers and other co-players. The outcome in both cases is a subject who is disciplined, digitally if you wish, through the "obligatory syntax" (Foucault, 1978: 153) of software. The difference is that in gaming a positive economy is, in fact, attained; whereas in the case of PHRs and other HITs such an outcome is not only not ensured, it seems unlikely to be accomplished, especially given the increasingly dominant conceptions of these technologies. The issue at stake is whether and how we can develop and design HIT that can produce a positive economy for patients, providers, and the state. This is a question that should be answered in practice.

References

Bodenheimer, T. and Grumbach, K. (2003). Electronic technology: A spark to revitalize primary care? *JAMA*, 290(2), 259–64.

California HealthCare Foundation (2010). *Consumers and health information technology: A national survey*. Retrieved from www.chcf.org

Callon, M. (1991). Techno-economic networks and irreversibility. In J. Law (ed.), *A sociology of monsters: Essays on power, technology and domination. Sociological Review Monograph*, 38.

Clarke, A. E., Shim, J. K., Mamo, L., Fosket, J. R., and Fishman, J. R. (2003). Biomedicalization: Technoscientific transformations of health, illness, and U.S. biomedicine. *American Sociological Review*, 68(2), 161–94.

Day, R. E. (2010). Death of the user: Reconceptualizing subjects, objects, and their relations. *Journal of the American Society for Information Science and Technology*, 62(1), 78–88.

Debeauvais, T., Nardi, B., Schiano, J., Ducheneaut, N., and Yee, N. (2011). If you build it they might stay: Retention mechanisms in world of warcraft. Submitted to *Foundations of Digital Games*.

Dewey, J. (1934). *Art as experience*. London: Penguin Books.

Diamond, C. and Shirky, C. (2008). Health information technology: A few years of magical thinking? *Health Affairs*, 27(5), 383–90.

Ekbia, H. (2009). Digital artifacts as quasi-objects: Qualification, mediation, and materiality. *Journal of American Society for Information Science and Technology*, 60(12), 2554–66.

——(2010). Personal health records for self-management support. Paper presented at the Wellness Informatics Workshop, CHI2010, Atlanta, GA, April 10.

Fonkych, K. and Taylor, R. (2005). The state and pattern of health information technology adoption. RAND Corporation. www.rand.org

Gamasutra (2009). http://www.gamasutra.com/view/feature/4219/rethinking_the_trinity_of_mmo_.php retrieved April 11, 2011.

Heidegger, M. (1982). The question concerning technology. In M. Heidegger (ed.), *Basic writings* (pp. 282–316). New York: Harper Collins.

Hillestad, R., Bigelow, J., Bower, A., Girosi, F., Melli, R., Scoville, R., and Taylor, R. (2005). Can electronic medical record systems transform health care? Potential health benefits, savings, and costs. *Health Affairs*, 25(5), 1103–17.

Institute for Healthcare Improvement (2009). *Self-management support*. http://www.ihi.org/IHI/Topics/PatientCenteredCare/SelfManagementSupport/

Kallinikos, J. (2011). *Governing through technology: Information artefacts and social practice*. London: Palgrave MacMillan.

Kato, P. M. (2010). Video games in health care: Closing the gap. *Review of General Psychology*, 14(2), 113–21.

Kow, Y. M. and Nardi, B. (2009). Culture and creativity: World of warcraft modding in China and the U.S. In B. Bainbridge (ed.), *Online worlds: Convergence of the real and the virtual*. Heidelberg: Springer.

Kutz, D. O. and Ekbia, H. R. (2011). Designing for the invincible: Health engagement and information management. Paper presented at the 44th Hawaii Conference on System Sciences, Kauai, Hawaii, January 4–7.

Leontiev, A. N. (1978). *Activity, consciousness, and personality*. Englewood Cliffs, NJ: Prentice-Hall.

Nardi, B. (2010). *My life as a night elf priest: An anthropological account of world of warcraft*. Ann Arbor, MI: University of Michigan Press.

——Ly, S., and Harris, J. (2007). Learning conversations in world of warcraft. In Proceedings of Hawaii International Conference on Systems Science, Big Island, Hawaii, January.

Serres, M. (1982). *The parasite.* Translation Lawrence R. Schehr. Baltimore, MD: Johns Hopkins University Press.

Slack, W. (1972). Patient power: A patient-oriented value system. In J. Jacques (ed.), *Computer diagnosis and diagnostic methods: Proceedings of the second conference on diagnostic process* (pp. 3–7). Held at the University of Michigan. Springfield, IL: Charles C. Thomas.

Stead, W., Lin, H., Informatics NRCCoEtCSRCiHC, et al. (2009). *Computational technology for effective health care: Immediate steps and strategic directions.* National Academies Press.

Steinkuehler, C. (2005). *Cognition and learning in massively multiplayer online games: A critical approach.* Madison, WI: University of Wisconsin-Madison.

Stoljar, D. (2009). *Physicalism. Stanford encyclopedia of philosophy.* http://plato.stanford.edu/entries/physicalism/ retrieved March 24, 2010.

Strauss, A., Fagerhaugh, S., Suczek, B., and Wiener, C. (1985). *Social organization of medical work.* Chicago: Chicago University Press.

Suchman, L. (2007). *Human-machine reconfigurations: Plans and situated actions* (2nd ed.). Cambridge: Cambridge University Press.

Tang, P. C., Ash, J. S., Bates, D. W., Overhage, J. M., and Sands, D. Z. (2006). Personal health records: Definitions, benefits, and strategies for overcoming barriers to adoption. *Journal of Medical Informatics Association*, 13(2), 121–6.

Taylor, T. L. (2006). *Play between worlds: Exploring online game culture.* Cambridge, MA: MIT Press.

Weed, L. (1968). Medical records that guide and teach. *New England Journal of Medicine*, 278(11), 593–600.

9

Space Matters, But How?

Physical Space, Virtual Space, and Place

Anne-Laure Fayard

> Space is "a living system, a collection of interacting, and adjacent patterns of events in space." (Alexander, 1979: 74)

Introduction

Virtual organizing in various forms—distributed teams, online communities, etc.—has become commonplace, and virtual space[1] is now part of our imagination and organizational life (Schultze and Orlikowski, 2001). Organizations are becoming increasingly distributed and technology ubiquitous (e.g., Fulk and DeSanctis, 1995). The development of tools such as Skype, Second Life, social network applications such as Facebook, and the use of mobile devices that allow people to be "always on" (Nardi and Whittaker, 2002; Baron, 2008) are merging and redefining the boundaries between physical and virtual spaces in organizational contexts. Online forums and virtual teams in organizations engage participants, and team members interact and collaborate with people they have never met or will never meet, while still feeling a sense of belonging. Global souls (Iyer, 2001) and neo-nomads (Abbas, 2010), these new citizens of our digital and global world, moving through the airports of the world's cities and surfing the web while hanging out with their friends and

[1] I chose the phrase "virtual space," although it can be confusing sometimes, because it is frequently used and referred to in the literature and in everyday conversations. Virtual is sometimes associated with "unreal." For example, in the Oxford dictionary, one meaning of "virtual" is "not physically existing as such but made by software to appear to do so." In this chapter, "virtual" does not mean "immaterial." On the contrary, "virtual" encompasses "online," "distributed," and "digital."

families, claim their freedom from locations and nationalities but still long for a sense of home. Hence, people must increasingly negotiate multiple layered spheres of interaction that occur in different social worlds via various modes of interaction (e.g., face-to-face and technologically mediated).

Voluntary or forced exile, nomadism, and distributed collaboration are not new notions, yet technology and increased mobility have dramatically changed our perception of space and led to contradictory imaginations of space. Increasingly, more spatial connections and longer distances are involved in the understanding of any organization or community because of the possibility of remote collaboration or of being part of the same organization—or even the same team—without being colocated. Meanwhile, there is an implicit belief that space—because of fast travel times, as well as improved communication technologies—no longer matters. In other words, there is a belief that distance can be abolished (Massey, 2007) and that physical space, seen as a limitation and constraint, is subsumed by virtual space, described as endless and free (Mitchell, 1995; Negroponte, 1995; Davenport and Pearlson, 1998). The references to virtual space are grounded by a definition of space as an extension, a matter of *xy* coordinates. However, a body of literature questions this geometric or traditional definition of space and proposes a new definition of space as constantly constructed, a product of relationships and sociomaterial practices (Alexander, 1977; Malpas, 1999; Lefebvre, 2000; Massey, 2007). Although it is assumed that virtual space is a different kind of space (Kitchin, 1998), neither distance nor materiality can be abolished. For example, the instantaneity of a Skype exchange (textual or audio) does not annihilate the differences in locations, and the participants are still two different entities.

In this chapter, virtual space is understood as deeply material, that is, relying on a lot of "stuff"—technical infrastructure, as well as organizational infrastructure, tools, interfaces, etc. (Orlikowski, 2007; Leonardi and Barley, 2008). My aim is to explore perspectives on space developed by philosophers, geographers, and sociologists and to see how they frame our understanding of virtual space. I argue that space is not an empty extension to be filled in, but instead one that is constantly constructed and which emerges from the relationships and practices of people living, working, and interacting in space. Hence, I understand space as an "entanglement" (Orlikowski, 2007) of materials—rooms, walls, buildings, routes, etc.—and social practices and narratives. More deeply, I argue that space matters, much like Cooren et al. (Chapter 15) and Burrell (Chapter 16) argue that discourse and language matter. More specifically, I build upon Malpas's concept of place, a condition for our embodiment and for our ability to think, to suggest a definition of virtual space. Finally, by describing a multimedia installation I designed with an artist, I illustrate how practices construct virtual space.

Perspectives on Space

Space is an essential element of organizational life whether we think of physical environments, such as office buildings, offices, science parks, or of geographical locations where distributed teams, suppliers, and customers might locate. A review of the organizational and management literature shows a relative paucity of studies of space, despite calls such as those made by Kornberger and Clegg (2004) to bring space back into organization theory. When space is mentioned in organizational studies, it is usually as a material constraint and/or as a distance that affects interaction and communication, as per Allen's seminal study (1977).

The French philosopher and sociologist Lefebvre in *The Production of Space* (2000) notes that the word "space" is often used without an understanding of what is meant. Developing a better understanding of the relationship between space and materiality, as well as of the notion of virtual space and its implications for organizational studies, is therefore crucial.

The concept of space is particularly interesting for understanding materiality, because space and materiality are often associated. Space is seen as the location where objects, buildings, and people are situated. It is interpreted as the material boundaries of such locations and the structure in which they allow things to happen. In organizational studies, space is often assumed to be the physical distance separating people at work and as a material constraint that can prevent interaction and communication (Allen, 1977; Keller and Holland, 1983; Davis, 1984). In such an interpretation, space and materiality are interpreted as constraints: "You don't have enough space" or "You would like more space," that is, to get rid of the material constraints of space. Thus, space is often conceived as a structure opposed to everyday practices (DeCerteau, 2002)—it is seen as the frozen or "dead" that opposes the quotidian associated with the temporal (Bergson, 1907; DeCerteau, 2002). Space, which is typified by measurable, geometrical space that can be located, and materiality are also often conceived as "objective" as opposed to social and subjective.

Such definitions of space are grounded on a deeper opposition: against time, which is associated with life and activity. Also linked with creativity, flow, and movement, time thus comes to oppose space, which is associated with fixation and representation (e.g., Bergson, 1907). The tension between the two sits at the heart of many concepts of space (Malpas, 1999; Massey, 2007).

Yet, as soon as one begins to consider how human beings organize their activities (e.g., communication, interactions, work) in the world, the limitations in viewing space as it is "objectively" conceived become apparent. One way to address these limitations is to distinguish between the "objective" space of geometry or physics and the "subjective" space as experienced and constructed through our practices. For example, Merleau-Ponty (1976) distinguishes between objective space and subjective or experiential space but presupposes an "original" definition of space as geometrical extension and structure. Subjective space, in which time is integrated so that practices can be enacted, comes in second place, however. Thus, similar attempts to acknowledge the experiential and social nature of space lead to the introduction of the concept of place. Take, for instance, the geographer Tuan (1977), who introduced the concept of "place" or experienced space. For Tuan, place is an experiential construct encompassing the affective responses of human beings to their environment. It is seen as the space that gives rise to experience, and space is conceived as "more abstract" than place (Tuan, 1977: 6), that is, abstracted from various experiences of places.

Massey challenges these attempts because they continue to presuppose life on the side of time. Indeed, place is seen as "meaningful, lived, everyday" (2005: 6), as opposed to space, which becomes the "outside, the abstract, the meaningless" (p. 6). Massey rejects such a vision of space. Criticizing the politics implied by such imaginations of space, she proposes a conception of space as a process and as "the product of the interrelations; as constituted through interactions" as well as "the sphere of the possibility of the existence of multiplicity in the sense of contemporaneous plurality" (p. 9).

While I agree with Massey's agenda to rehabilitate space as not necessarily opposed to time, but as intertwined and influencing each other, I am not sure her argument goes far enough in rejecting the distinction between space and place as meaningless. The route suggested by Malpas (1999) in his development of the concept of place seems more productive. Rather than rejecting place and reclaiming space as temporal, he proposes the concept of place as one that integrates time and space:

> The idea of place encompasses both the idea of the social activities and institutions that are expressed in and through the structure of a particular place (and which can be seen as determinative of that place) and the idea of the physical objects and events in the world (along with the associated causal processes) that constrain, and are sometimes constrained by, those social activities and institutions. (Malpas, 1999: 35)

While he acknowledges the importance of space and temporality, Malpas is critical of attempts such as Heidegger's in *Being and Time*, where what Heidegger calls "existential spatiality" is nearly reduced to a form of temporality. Massey's attempt to reintroduce temporality into space is similar to Heidegger's: it leads to the hegemony of one dimension—space. Malpas offers another way to integrate space and time with the concept of "place," which offers a structure in which both subjective and objective spaces are interconnected and interdependent. Malpas's interest in place is not so much as it is experienced but as "a structure within which experience (and action, thought and judgment) is possible" (1999: 71).

Malpas' concept of place offers us a productive perspective to think about practices and organizations because it acknowledges the role of space as the "place" through which, in the enactment of multiple relations between individual agents and the circumstances within which they act, the social is constructed. Such an approach and a concept of place belong to a philosophical perspective whereby human thought and experience is essentially grounded in spatiality, locality, and embodiment (Heidegger, 1962; Merleau-Ponty, 1962). Hence, not only are all our encounters with persons and things "taking place" in place, but the very possibility of being human and engaging with the world (and in particular its objects and the events within it) and thinking about the world is tied to place (Davidson, 1980; Cavell, 1993; Malpas, 1999). We must understand ourselves as already "in" the world to be able to think and understand. Space and materiality are therefore intrinsically connected not only because space is material (made of "stuff") but also because space or, more specifically, place as defined by Malpas (1999), is the condition for and the structure within which human beings as embodied can engage with objects and artifacts. Such a definition of place is closely connected to the conceptualization of matter as a process (Barad, 2007) framing Scott and Orlikowski's study (Chapter 6, this volume).

Underlying Imaginations of Virtual Space

The literature on virtual space, whether referring explicitly to "virtual space" or discussing virtual teams, organizations, distributed work, and collaboration, is also based on an objective and geometrical conceptualization of space, with virtual space referred to as a vast expansion allowing for the free flow of information and as a surface with fewer constraints than physical space. Hence, Handy (1995: 42) argues that virtual organizations "exist as activities not as buildings." Some, such as Thomas (1991) and Castells (1996), suggest that there are two spaces, with virtual space overlapping real space, "allowing organizations to be more flexible in relation to real-space geographies" (Kitchin, 1998: 387).

Thus, work can be abstracted and disembodied from its particular circumstances, and virtual organizations can operate "everywhere and nowhere in space" (Schultze and Orlikowski, 2001: 57). Space is imagined as a "uniform and infinite expanse" (p. 58) across which people and ideas can move freely (Tuan, 1977; Casey, 1997). Such a conception presupposes an objective space, one understood as a structure, a container separated from time and place, place being where people and things are copresent and interact face-to-face

(Giddens, 1991; Casey, 1993). Yet, as discussed above, such a view of space is limited, as it does not recognized everyday practices and presupposes a distinction between time and space. It implies that virtual space, which annihilates distance, also allows us to annihilate time since time is associated with place, interactions, and practices. In that construct, the world becomes "flat" (Friedman, 2005), a "global village" (McLuhan, 1962; Iyer, 2001), or even erased. Acknowledging differences in the nature of virtual space (Kitchin, 1998) does not mean that we can or should abolish distance or materiality. Virtual organizations still operate with people in physical offices. Moreover, the instantaneity of a text message or an email does not annihilate the differences in locations, and the participants remain two different entities.

Exploring Materiality, Space, and Place through an Interactive Installation

In previous works, I looked at people's interactions in physical spaces in organizations (Fayard and Weeks, 2007); in virtual spaces, such as public online forums (Fayard and DeSanctis, 2005, 2010); and in hybrid spaces, such as video-mediated contexts that include two connected physical spaces and a third that I described as one constructed by the participants, which I called "the virtual stage" (Fayard, 2006). All these studies showed how neither an objective nor a subjective definition of space was sufficient for understanding space and interactions within space.

I wanted to explore the construction and materiality of virtual space through the use of communication technologies. Communication technologies embody an ambiguous and unwieldy form of "materiality," however, as their materiality represents only a minuscule fraction of their "function." Moreover, the materiality of software is difficult to grasp (Leonardi, 2010). Also, virtual and physical spaces, as well as online and colocated interactions, are intertwined and thus difficult to investigate. Therefore, it is hard to gain insights into how individuals experience the materiality of virtual space, as well as how they develop and enact practices through which virtual space is constructed.

Following Weick's (2005) advice, I chose to study the practices through which virtual space is constructed in a context less "opaque" than an organizational one can be. The growing interest in organizational studies for art and design (e.g., Weick, 2007; Barry, 2008), as well as the argument that the arts and arts-based practices provide different ways of describing and relating to the complexity of our world that are distinct from those offered by traditional logic and rationality, prompted me to look at other forms of exploration. While many studies of art, design, and management often use the arts to look at organizations metaphorically—for example, via Calder mobiles (Barry and Rerup, 2006) or the improvisation of a jazz quartet (Lewin, 1998)—I decided to take a more action-research approach (Gayá Wicks and Rippin, 2010) by collaborating with Aileen Wilson, an artist and art education scholar at Pratt Institute in New York.

Together we embarked on an exploratory project and designed a multimedia interactive installation, *building_space_with_words* (Fayard and Wilson, 2010). This site-specific installation responded to the possibilities and limitations of a large space in a historic building on the NYU-Poly campus in Brooklyn, NY, open to the public for three-and-half weeks in March 2009. The installation, which involved nine LCD projectors and nine computers, including two touch-screen computers, offered a number of modalities meant to engage a diverse public in exploring the matter of building space with words. It had an ambient soundtrack and a maze made of fabric panels on which posts from a blog (created for the installation and on which the visitors could post using two computers located in the center of the maze) were dynamically posted.

By dramatizing "virtual space" and making it stark, we were able to see clearly what is normally subtle or hard to discern. For instance, by inviting visitors to post on the blog and then projecting their posts onto the maze, we gave them the opportunity to create space. They could then experience and consider what online interactions and "virtual space" mean, especially as it relates to physical space. Thus, it is possible to describe the installation as a critical and evocative object qua critical design (Dunne, 2005).[2]

Building the Installation

The Maze: The Physical Space

Chiffon panels created a maze (22 ft × 37.8 ft) onto which digital text was projected. We created a wire grid from which we "floated" the maze structure. The structure and its paths reflected the geography of online forums and possible online interactions, with participation characterized by a core of regular activity and intermittent participation with significant variations and lurkers (Gray and Tatar, 2004). The maze had several entrances, all with passages large enough for one person to go but too narrow for two people to walk through them together or stand in them side by side. The paths represented people checking information or posting a question. Lurkers were those

[2] Critical design as popularized by Anthony Dunne and Fiona Raby uses designed artifacts (and their subsequent use) and the process of designing to generate a reflection on existing values, mores, and practices in a culture.

Figure 9.1 View of the maze and text, with computers in the center. © Anne-Laure Fayard and Aileen Wilson. Photo © Pratt Institute.

who chose to stay out of the maze. The central area, where at least four people could stand and where two computers allowed them to access the blog, invited several people to idle, chat, and interact. This area represented online communities where people develop relationships and share a sense of identity (Figure 9.1).

The physical space of the maze was not just physical; it also consisted of the projected blog text. Its projection made the text readable, but that was not our primary goal. We wanted visitors to consider the text's texture. It was dynamic: it updated every minute and featured the last twenty entries on the blog. The discourse enveloped people as they entered the space, suggesting the materiality of words and the idea of living in language. The participants were both defining the discourse and being defined by it. By posting on the blog, visitors constructed the space, which was constantly "under development" as new posts were created (Figure 9.2).

Sound as a Sociocultural Dimension

We added the dimension of sound[3] to evoke another community, another space where people asked questions and shared their thoughts. We combined

[3] An excerpt of the soundtrack can be found as supplement content 2 at http://www.mitpressjournals.org/doi/suppl/10.1162/leon.2010.43.3.257

Figure 9.2 Projected text and materiality of discourse. © Anne-Laure Fayard and Aileen Wilson. Photo © Pratt Institute.

sound heard in public and virtual spaces (e.g., clicks and typing) and recordings of people reading online forum posts in different languages. The sound was diffused throughout the space through four speakers. Its intensity varied from barely audible to very loud and included many different voices. We wanted to evoke a sense of place, such as a busy network or a train station, and thus create a bridge between virtual and physical spaces. We hoped to underscore the distinction between public spaces, which are full of sounds, and virtual spaces, which are often perceived as silent, and remind visitors that our online interactions often occur in a public space, such as a coffee shop or airport lounge. We also intended to call attention to the increasing interweaving of the virtual and the physical. Lastly, through languages and accents, the soundtrack highlighted the diversity of narratives and geographies involved in online interactions.

The Blog: The Virtual Space

The blog launched in October 2008, and we invited artists, sociologists, designers, architects, etc. who had an interest in space—physical or virtual—to contribute. While some blog members were part of our personal and professional networks, many were experts we invited to the discussion. Most out-of-network experts accepted, with the number of members totaling thirty, including us. Our invitation provided them with a general description of the project's aim: "to deepen our understanding of how virtual public space can trigger interactions and (potentially) relationships through an interactive installation *building_space_with_words*, which aims to prompt reflections on communication and community (or 'we-ness') in both physical and virtual spaces."

Figure 9.3 Visitors interacting in the maze and "building" the space. © Anne-Laure Fayard and Aileen Wilson. Photo © Marian Goldman Photography, NYC.

The blog was originally conceived as a companion to the installation and as the source of all the projected text, but it took a life of its own and is still active. Visitors could post and see their posts projected on the maze walls after a minute or two. This provided visitors with the opportunity to interact with the maze and "build" the space. Adding the blog to the installation allowed us to explore the blurred boundaries between physical and virtual space: the projected text made the maze partly "virtual," while the virtual space of the blog became physical through its projection on the maze panels (Figure 9.3).

Reflecting on the Installation: Physical Space, Virtual Space, and Place

The Installation as a Critical Object

The installation was a success[4] in that it led people to reflect on their interactions in physical and virtual spaces and on the ideas and the research underlying the work. In that sense, it became an evocative and critical object (Dunne, 2005). Visitors often indicated (mostly through their posts during

[4] About 500 people visited it in three-and-half weeks.

Figure 9.4 Visitors posting on the blog: online interactions and colocation. © Anne-Laure Fayard and Aileen Wilson. Photo © Marian Goldman Photography, NYC.

their visit, but also through verbal comments made during their visit[5]) that they interpreted the installation as an embodiment of the virtual: "Your exhibition makes tangible this intangible concept" (March 5, 2009).

Many also commented on how experiencing the installation space led them to reflect on the nature of virtual space and their own experiences of online interactions (Figure 9.4). For example, some visitors addressed the differences between physical and virtual spaces, as in this visitor's definition of privacy:

> We are in a cultural shift in terms of our definition of private space. The notion of private may or may not have ever existed in "reality." Despite this, the idea of a private space is becoming more and more ambiguous in an electronic context... (March 4, 2009)

The quote highlights the difficulty of enacting privacy in an online context, an observation the installation illustrated when it displayed the text of any "private" post on the panels that composed the installation itself. The excerpt

[5] I was in the space every day for at least 90 minutes during the three-and-half weeks of the installation. A research assistant was present on site during open hours (five hours for five weekdays). She counted the number of visitors, answered their questions, and took some notes of the behaviors of the visitors, for example, their comments, questions, and ways to interact with the space. I also read visitors' comments posted on the blog to understand how they experienced the space and what reflections this experience generated.

further speaks to how the online experience leads us to question our definition of private space. Thus, the experience of writing a comment in the exhibition space differed from the act of posting on a blog or a forum in a more "private" setting, such as at home or work. So, although the experience could be perceived as private, the product—for example, the text found in blogs, forums, and, in the case of the exhibit, the installation—is displayed and shared online where it can be accessed by a larger, and more public, audience. The installation "materializes" by making explicit that which is tacit, thus highlighting the thin line between private and public.

In conclusion, visitors experience the space as constructed by their posts, and for most of them it became an evocative object that triggered reflections about the nature of virtual space and the interactions within it.

Social Practices Frame Our Interactions in the Virtual Space

In one comment, a visitor linked space and social interactions, implicitly referring to her subjective experience or what some define as subjective space or place (understood as a psychological and experiential concept):

> The somewhat relaxed sense that developed in me in the exhibit seemed related to a sense of being in a community as well as a space. People at the show could post to the blog, the words moving through the visual area in a rhythmic fashion, the sound of voices in the background. It was like being able to work or just be, with a sense of people around you whom you could consult when you wanted to, but whose voices and words were accessible and connecting... (March 24, 2009)

By linking space with a sense of a community, the visitor gets at how space might be constructed through social interactions and encounters, yet the title of her post "Is this space?" and her way of associating community with space, "as well as space" itself, suggest that, for her, space is more of an extension, a container wherein social interactions might exist. Moreover, her experience of place—following Malpas's definition of place—involves the physical boundaries of the space, as well as the moving text, voices, and style of interactions.

Some discussions on the blog pointed to important issues in how participants' interactions within space—both physical and virtual—are framed by social practices. For instance, blog members Jim and Mark[6] underlined the importance of the social designation—that is, the expected behaviors and interactions—of a space and the differences between a physical and an online space. They both felt that "this space may not have walls that divide those who are inside from those who are outside, but it has memory and it is instructive that memory can serve so effectively as a boundary" (Mark, April

[6] Names have been changed to protect identity.

14, 2009). They perceived the fact that previous interactions were available as an expectation, if not an obligation, to read. They interpreted the expected practices based on their behavior in public (physical) spaces: "certain expectations and behavioral scripts from other spaces that we find similar" (posted by Mark on April 14, 2009). In fact, Mark's perception of the blog space and the expectations he associated with it changed when he thought of it as a party rather than a discussion panel: he could just be "there" discussing topics he wanted.

Our observations indicate that people did not perceive virtual space as an infinite expansion, but as constantly constructed through specific practices. As Malpas notes, "we understand a particular space though being able to grasp the sorts of "narrative of actions that are possible within that space" (1999: 186). While studies of organizational practice (Pentland, 1992; Orr, 1996) have shown that practice is always situated in the sociomaterial environment and remind us that we need to take into account the influence of the material environment and artifacts used in a practice to understand it, these quotes highlight the importance of the practice and its social expectations in a virtual context. Hence, space is always interpreted as a specific place with a certain set of practices tied to it.

The Intermingling of Physical and Virtual Spaces

The installation was an environment where the physical and the digital were interwoven and where discourse was what linked the two. This intermeshing of the physical and the virtual was also illustrated by the soundtrack of merged recordings of public spaces and of quotes from posts on online forums, highlighting how one could be located in a physical space while simultaneously engaged with people in many different spaces. The texts projected on the panels alerted visitors to the presence of others with whom they could communicate and who were as "present" as, and sometimes more present than, the visitors standing or colocated with them in the installation space. For example, a consultant for a big technology company who worked with several teams in India and the United States visited the installation and said that, although he was based in New York during the workday, he often felt closer to colleagues and clients in India. His experience echoes Turkle's finding that individuals can feel more "present" when online than when interacting face to face (Turkle, 1997) and suggests that developing a sense of shared place with teammates might matter more than being colocated. Several visitors evoked a similar experience—the blending of physical and digital environments and the social practices they involve (within and across these environments), as well as the creation of a "hybrid cultural ecology" (Lindtner et al., 2008).

The subtle and complex ways in which physical and virtual space interacted were seen in the reactions of blog members when we created a visitor account to the blog that could be accessed only from the installation space. By allowing access to the blog in a physical location—an access that was open to all visitors and not just invited contributors—we unintentionally changed some of the material characteristics of the blog. It became more "open." It was as if the installation had become an extension of the blog, a door to its conversation space.

The change had an impact on the blog's discursive practices. Indeed, with an increase in the diversity of authors and discursive practices, the blog space and therefore the space of the installation constructed by the projected texts itself changed. Messages posted became shorter. A "chat" style emerged on the day of the opening night; the maze had by then become a giant bulletin board. Worried that the visitor account might kill the discussion between the original contributors, one of the original blog members suggested displaying the visitors' posts on another page in order to avoid mixing them with the rest of the blog's conversations. The participant viewed the blog as a semiprivate space despite its public nature, with the visitor account seen as a threat to that privacy.

Interestingly, after the visitors' account was introduced, conversations by blog members continued but mostly as comments to previous messages rather than posts. These comments can be interpreted as "corridor" conversations avoiding the main public space, which had lost some of its privacy. When the installation came down, the visitors' account disappeared, and a few members posted again.[7]

Learning from the Process

Working with an artist was particularly illuminating for my reflection on materiality, especially because materials are central to artistic practice. Yet, it is really more production—as Scott and Orlikowski (Chapter 6, this volume) argue—that is central. Art practice involves materials and content, but in a complex and entangled fashion: the materials might be the form *and* the content; in many instances, they are inseparable.

True to the definition of an installation, *building_space_with_words* responded to the exhibition site, and although its essential form had been

[7] Although the blog is still "live" and accessible, invited members stopped posting and commenting a few weeks after the closing of the installation.

imagined before we had access to the site, the final form emerged "from the act of doing." The content (the notion of virtual space and its relation to physical space), the form (a maze, floating or free-standing), and the material (paneled fabric, wires, etc.) became central to our work and reflection. The whole process was a conversation with materials—we responded to the materials and their possibilities, as well as to the space. While there was an underlying aesthetic idea, we also had technical or practical issues that created constraints to which we responded. Materiality might seem constraining, but it affords options and the response itself is creative.

My exploration of the art practice led me to revisit the distinction Kallinikos (this volume) makes between matter, form, and function.[8] It also made me deeply aware of how embodied we are and how matter (stuff, artifacts, etc.) and space are valuable. As I moved from conceiving the installation to making it—materializing it in a specific space with tools and engaging in physical activities such as measuring, sketching, or pulling wires—I was confronted with the material as stuff, as process, and as something that matters, through the constraints and limitations it imposed, as well as the options it offered. All of this also happens, but in a more subtle and less obvious fashion, as one writes a paper—by this I mean, not only typing "the final version" on your computer but also reading, annotating, taking notes, presenting to various audiences, thinking about it in various contexts and spaces, and of course drafting, revising, and editing. The whole experience made me profoundly aware of embodiment, being in the world, as suggested by Malpas's notion of place is indeed constitutive to our ability to think.

Virtual Space and Place

The installation was designed to reflect the intermingling of physical, virtual, and social components present in online interactions. It allowed us to "embrace paradox, contradiction and duality" (Schultze and Orlikowski, 2001: 67) and thus investigate the tensions at the core of virtual space. Indeed, we were able to go beyond the fixed opposition between space, time, and place on which virtual space is typically understood and to attend to the dynamism at the heart of the notion of "virtual space." *building_space_with_words*, including the interactions of blog members and visitors with and within the installation and their reflections, highlighted how practices in space and time constitute places. It illustrated how neither virtual nor physical space could be equated to a dead structure, but are instead continuously enacted. It also

[8] Contemporary artists, especially since Marcel Duchamp, have been questioning the relationship between objects and functions.

suggests how in today's world, digital and physical environments are increasingly mixed, offering us a hybrid space in which to interact.

Place was articulated though the narratively structured activity of people within an objective, physical environment, that is, a paneled maze featuring projected online discourse. Through the various posts and interactions, the installation highlighted the multiple narratives (suggested to a certain extent by the soundtrack) that constitute our experience of space and place. *building_space_with_words* was endlessly built, dismantled, and rebuilt via the posted text projected onto the maze, as well as through the movements of visitors in the space. Finally, the installation showed how people rely on sociomaterial practices enacted in physical spaces to help them interpret virtual space and to guide their actions within it, thus highlighting how materiality, agency, and space are intrinsically connected.

The installation was inspired by a need to revisit the distinction between space as the room we move in—its location and geometrical characteristics—and place as the experiences and social practices associated with that room or space. Building the installation space, experiencing it, and observing others interacting with it, confirmed this need. Rather than erasing the distinction between space and place and making it minimal, I adopted Malpas's concept of place as a simultaneously objective and subjective structure. This concept allows us to recognize the material and social dimensions of virtual space, thus acknowledging the multiplicity of practices and narratives and rejecting unidimensional visions of globalization, virtual organizing, and distributed work. As Lefebvre's work highlights, the introduction of virtual forms of organizing affects mental and social ordering. For example, if we assume, as many studies on virtual organizing do, that work takes place "in space" (understood as an infinite extension), how can we reconcile the situated performance of work at specific locations with the theories of work in virtual space, understood as an abstract and general space (Schultze and Orlikowski, 2001)?

Lastly, the concept of place suggests understanding the various elements of practice and experience (artifacts, locations, people, events) not in terms of an underlying structure to which they can be reduced, but in terms of their own interrelations. It thus allows us to recognize spatiality and materiality while avoiding both determinism (space is a continuous process, always "in construction" and including multiple narratives) and voluntarism (we are embodied "in the world").

Acknowledgments

I would like to thank Aileen Wilson, who joined me in this adventure and without whom *building_space_with_words* would not have happened. She offered me the opportunity to explore space and materiality in novel and exciting ways. Guilhem Tamisier, Liz DiNapoli, Ardis Kadiu, and Don Diesche were all essential in the design and development of this project. I would like to thank the members of the *building_space_with_words* blog for their participation and thoughtful posts. John Weeks provided me with useful feedback on previous versions of this work. I gratefully acknowledge the useful conversations with the participants of the workshop on *Materiality and Organizing* organized at Northwestern University as well as Bonnie Nardi's helpful editorial suggestions.

References

Abbas, Y. (2010). Neo-nomads and the practice of re-location: Designing for mobilities. In D. Tsigaridis and J. Jungclaus (eds.), *Tracing mobilities: Designing ubiquities, Cambridge. Design and technology report series*. Cambridge, MA: Harvard Graduate School of Design.

Alexander, C. (1977). *A pattern language: Towns, buildings, construction*. Oxford: Oxford University Press.

—— (1979). *The timeless way of building*. New York: Oxford University Press.

Allen, T. (1977). Managing the flow of technology: Technology transfer and the dissemination of technological information within the R&D organization. Cambridge, MA: MIT Press.

Barad, K. (2007). *Meeting the university halfway: Quantum physics and the entanglement of matter and meaning*. Durham, NC: Duke University Press.

Baron, N. (2008). *Always on: Language in an online and mobile world*. Oxford: Oxford University Press.

Barry, D. (2008). The art of leadership and its fine art shadow. In D. Barry and H. Hansen (eds.), *The SAGE Handbook of new approaches in management and organization*. London: Sage Publications Ltd.

—— Rerup, C. (2006). Going mobile: Aesthetic design considerations from Calder and the constructivists. *Organization Science*, 17(2), 262–76.

Bergson, H. (1948) (1907, 1st edition). *L'évolution creatrice*. Paris: Presses Universitaires de France.

Cavell, M. (1993). *The psychoanalytic mind: From Freud to philosophy*. Cambridge, MA: Harvard University Press.

Casey, E. (1993). *Getting back into place: Toward a renewed understanding of the place world*. Bloomington, IN: Indiana University Press.

—— (1997). *The fate of place: A philosophical history*. Berkeley, CA: University of California Press.

Castells, M. (1996). *Rise of the network society*. Oxford: Blackwell Publishers, Inc.

Davenport, T. and Pearlson, K. (1998). Two cheers for the virtual office. *Sloan Management Review*, 39, 51–66.

Davidson, D. (1980). *Essays on actions and events*. Oxford: Clarendon Press.

Davis, T. R. V. (1984). The influence of the physical environment in offices. *Academy of Management Review*, 9(2), 271–83.

DeCerteau, M. (2002). *The practice of everyday life.* (2nd edition). Minneapolis, MN: University of California Press.

Dunne, A. (2005). *Hertzian tales: Electronic products, aesthetic experience, and critical design*. Cambridge, MA: MIT Press.

Fayard, A.-L. (2006). Interacting on a virtual stage: The collaborative construction of an interactional video setting. *Information Technology and People*, 19(2), 152–69.

—— DeSanctis, G. (2005). Evolution of an online forum for knowledge management professionals: A language game analysis. *Journal of Computer-Mediated Communication*, 10(4), article 2. http://jcmc.indiana.edu/vol10/issue4/fayard.html

—— —— (2010). Enacting language games: the development of a sense of "we-ness" in online forums. *Information Systems Journal*, 20(4), 383–416.

—— Weeks, J. (2007). Photocopiers and water-coolers: The affordances of informal interaction. *Organization Studies*, 28(5), 605–34.

—— Wilson, A. (2010). Building_space_with_words: An interactive, multi media installation exploring the relationship between physical and virtual space. *Leonardo: The Journal of the International Society for the Arts, Sciences and Technology*, 43(3).

Friedman, T. L. (2005). *The world is flat: A brief history of the twenty-first century.* New York: Farrar, Straus and Giroux.

Fulk, J. and DeSanctis, G. (1995). Electronic communication and changing organizational forms. *Organization Science*, 6(4), 1–13.

Gayá Wicks, P. and Rippin, A. (2010). Art as experience: An inquiry into art and leadership using dolls and doll-making. *Leadership*, 6(3), 259–78.

Giddens, A. (1991). *Modernity and self-identity: Self and society in the late modern age.* Stanford, CA: Stanford University Press.

Gray, J. and Tatar, D. (2004). Sociocultural analysis of online professional development: A case study of personal, interpersonal, community, and technical aspects. In S. A. Barab, R. Kling, and J. H. Gray (eds.), *Designing for virtual communities in the service of learning* (pp. 404–35). New York: Cambridge University Press.

Handy, C. (1995). Trust and the virtual corporation. *Harvard Business Review*, 73, 40–50.

Heidegger, M. (1962). *Being and time*. New York: Harper and Row.

Iyer, P. (2001). *The global soul: Jetlags, shopping malls and the search for home*. New York: Vintage Departures.

Keller, R. T. and Holland, W. E. (1983). Communicators and innovators in research and development organizations. *Academy of Management Journal*, 26(4), 742–9.

Kitchin, R. (1998). Towards geographies of cyberspace. *Progress in Human Geography*, 22 (3), 385–406.

Kornberger, M. and Clegg, S. (2004). Bringing space back in: Organizing the generative building. *Organization Studies*, 25(7), 1095–114.

Lefebvre, H. (2000). *La Production de l'espace* (4th edition). Paris: Anthropos.

Leonardi, P. M. (2010). Digital materiality? How artifacts without matter, matter. *First Monday*, 15(6). http://firstmonday.org/htbin/cgiwrap/bin/ojs/index.php/fm/article/view/3036/2567

—— Barley, S. (2008). Materiality and change: Challenges to building better theory about technology and organizing. *Information and Organization*, 18(3), 159–76.

Lewin, A. (1998). Jazz improvisation as a metaphor for organization theory. *Organization Science*, 9(5), 605–22.

Lindtner, N., Nardi, B., Wang, Y., Mainwaring, S., Jing, H., and Liang, W. (2008). A hybrid cultural ecology: World of Warcraft in China. In *Proceeding of CSCW '08*. San Diego, CA: ACM.

Malpas, J. E. (1999). *Place and experience: A philosophical topography*. Cambridge: Cambridge University Press.

Massey, D. (2007). *For space* (3rd edition). London: Sage Publications.

McLuhan, M. (1994) (1st edition, 1964). *Understanding media*. Cambridge, MA: MIT Press.

Merleau-Ponty, M. (1976). *La Phénoménologie de la Perception*. Paris: Gallimard.

Mitchell, W. (1995). *City of bits: Space, place, and the infobahn*. Cambridge, MA: MIT Press.

Nardi, B. and Whittaker, S. (2002). The place of face-to-face communication in distributed work. In P. Hinds and S. Kiesler (eds.), *Distributed work: New research on working across distance using technology*. Cambridge, MA: MIT Press.

Negroponte, N. (1995). *Being digital*. London: Hodder and Stoughton.

Orlikowski, W. (2007). Sociomaterial practices: Exploring technology at work. *Organization Studies*, 28(9), 1435–48.

—— Scott, S. (2008). Sociomateriality: Challenging the separation of technology, work and organization. *The Academy of Management Annals*, 2(1), 433–74.

Orr, J. (1996). *Talking about machines: An ethnography of a modern job*. Ithaca, NY: Cornell University Press.

Pentland, B. T. (1992). Organizing moves in software support hot lines. *Administrative Science Quarterly*, 37(4), 527–48.

Schultze, U. and Orlikowski, W. (2001). Metaphors of virtuality: Shaping an emergent reality. *Information and Organization*, 11(1), 45–77.

Thomas, D. (1991). Rituals for new spaces: Rites de passage and William Gibson's cultural model of cyberspace. In M. Benedikt (ed.), *Cyberspace: First steps* (pp. 31–48). Cambridge, MA: MIT Press.

Tuan, Y. (1977). *Space and place: The perspective of experience*. Minneapolis, MN: University of Minnesota Press.

Turkle, S. (1997). *Life on the Screen: Identity in the age of the internet*. New York, NY: Simon & Schuster.

Weick, K. E. (2005). The experience of theorizing: Sensemaking as topic and resource. In I. K. Smith and M. Hitt (eds.), *Great minds in management* (pp. 394–413). Oxford: Oxford University Press.

——(2007). Drop your tools: On reconfiguring management education. *Journal of Management Education*, 31(1), 5.

10

Socio-material Practices of Design Coordination: Objects as Plastic and Partisan

Jennifer Whyte and Chris Harty

Introduction

The importance of the material to social organization is mundanely obvious, but often absent from accounts of interaction. The development of sociology as a discipline has tended to position the social realm as analytically distinct from other domains: as an area of inquiry apart from the scientific or technological. Materiality has had parts to play, but arguably passive ones. They are the backdrop for social interaction (Goffman, 1971), a reflection of social distinctions (Bourdieu, 1984), or material levers employed by knowing actors (Giddens, 1984). Objects have often remained secondary to the understanding of the social, whether positioned as the product of human agency, effects of extraindividual social structures, or some combination of both.

But alternative conceptions have been emerging in recent decades, where the material is attributed a more active role in understanding interaction, practice, and social organization. The area of Science and Technology Studies (STS) especially has provided a number of approaches that more substantively incorporate the material into accounts of the social. Concepts such as the actor-network (e.g., Callon, 1986), the boundary object (Star and Griesemer, 1989), and socio-technical system building (Hughes, 1983) as well as discussions of the politicized nature of objects (Winner, 1985) challenge this passive view of the material.

Management and organization scholars have also recognized and contributed to this debate, with materiality becoming a more central explanatory factor in understanding organizations and the processes of organizing

(Orlikowski, 2007, 2010; Leonardi and Bailey, 2008; Nicolini, 2009; Leonardi and Barley, 2010). But there are alternative perspectives on whether to analytically separate the social and the material and to look for the relations and connections between them (Leonardi and Barley, 2010), or to see them as "entangled" (Pickering, 1995; Orlikowski, 2010), where practices are positioned as socio-materially hybrid (Harty and Whyte, 2010) and embodied or situated within and across social actors and material artifacts (Gherardi, 2006; Lanzara, 2009; Yanow and Tsoukas, 2009).

In this chapter, we take an explicitly socio-material view through analyses of design work and its coordination across a large construction project. In this empirical setting, an integrated software system or "single model environment" is mobilized in the coordination of design across the delivery project as professional teams work to design and construct a new airport terminal. In the process of theorizing our empirical data on this coordination work, we have returned to and reread Star and Griesemer's (1989) seminal study of coordination in the research museum, finding their work more tentative, more ambiguous, and also less of a generic explanation of coordination than is portrayed in later studies. They examine the coordination of volunteer inputs from collectors, trappers, and other nonscientists to the activities of the research museum. We draw on our empirical work and our rereading of this earlier study to highlight the plasticity of objects and their partisan nature in coordination, drawing out the relevance for contemporary debates. While the process of developing this argument has led us to work iteratively between data and theory, in the chapter we first discuss Star and Griesemer's study, and then briefly outline our empirical work, discuss our findings, and develop their interpretation. Our analytic focus, like that of Star and Griesemer, is on the interactions between people and objects in achieving organizational goals. We conclude by discussing the implications of the plasticity of objects and their partisan nature for recent debates on socio-material practices and suggesting some directions for researchers of management and organization.

Revisiting Coordination: A Seminal Study

In their seminal study of scientific work in the professional museum, Star and Griesemer set out to explain, within an ecology of institutions, how: "*consensus is not necessary for cooperation nor for the successful conduct of work*" (Star and Griesemer, 1989: 388). The explanation Star and Griesemer develop identifies methods of standardization and the development of boundary objects as two of the major factors in developing a coherence between radically different meanings within intersecting social worlds. The work of translation, which is required to meet the scientific goals of the museum, involves *methods* to discipline information and *objects* to maximize both the autonomy and communication between social groups.

While later scholars have focused most attention on objects, for Star and Griesemer the methods and objects are used together. In their analysis, methods are described as emphasizing "*how*," that is, a set of standardized procedures for action; not "*why*," the reasoning and motivation for that set of actions. The methods created common ground in their case by setting out "clear, precise, manual tasks" (Star and Griesemer, 1989: 407). The social groups answered to different audiences and pursued different sets of tasks, hence for example:

> the trick of translation required, first, developing, teaching and enforcing a clear set of methods to "discipline" the information from collectors, trappers and other non-scientists and generating boundary objects that would maximise the autonomy and communication between worlds. (Star and Griesemer, 1989: 404)

Star and Griesemer are not consistent in their terminology and also refer to "boundary objects" as "coherence objects" and "marginal objects." The disparate list of objects that are of interest to actors within the multiple social worlds that they describe include species and subspecies of mammals and birds; the terrain of California; physical factors in California's environment (such as temperature, rainfall, and humidity); and the habitats of collected animal species (Star and Griesemer, 1989: 392). Elsewhere in the chapter, the museum itself is included in their list of boundary objects. The list of objects also included standardized forms that might ensure that the collectors, trappers, and other nonscientists all follow similar methods for data collection.

Star and Griesemer's study describes multiple examples and intersecting categorizations of boundary objects. They explain that boundary objects may be abstract or concrete and are:

> ...plastic enough to adapt to local needs and constraints of several parties employing them, yet robust enough to maintain a common identity across sites. They are weakly structured in common use and become strongly structured in individual-site use. (Star and Griesemer, 1989: 393)

It is the extent to which the object allows both the autonomy of and communication between social groups that marks it out as a useful boundary object. Thus, in the production of boundary objects, the intersection of information focused on those parts of the work that were essential to maintaining coherent information while ignoring others.

Star and Griesemer are, however, cautious with their claims and do not rule out alternative explanations. They note that their analyses are intended as: "*suggestive rather than conclusive outlining an approach to case studies as well as a partial analysis of the case in hand*" (Star and Griesemer, 1989: 388). The question of power informs their analyses, and they position boundary objects as one of a number of means of addressing the fundamental tension between heterogeneity and cooperation in science, writing that:

> The production of boundary objects is one means of satisfying these potentially conflicting sets of concerns. Other means include imperialist imposition of representations, coercion, silencing and fragmentation. (Star and Griesemer, 1989: 413)

They also note that such materials: "*contain at every stage the traces of multiple viewpoints, translations and incomplete battles*" (Star and Griesemer, 1989: 413). The production of boundary objects is part of a set of activities of coordination, not the only mechanism through which coordination happens, and this production activity is socio-material, rather than social or material in nature.

While it is a cooperative mode of coordination, which is of interest to Star and Griesemer and which they expand on in their paper, they acknowledge other forms of coordination. Indeed, they articulate different strategies and divisions of labor for situations where there is conflict:

> via a 'lowest common denominator'... via the use of versatile, plastic, reconfigurable (programmable) objects... via storing a complex of objects from which things... can be physically extracted and configured... each participating world can abstract or simplify the object to suit its demands... work in the worlds can proceed in parallel except for limited exchanges... or work can be staged so that some stages are relatively autonomous. (Star and Griesemer, 1989: 404)

These indicate forms of socio-material practice: use of common denominators; programmable objects; storage, extraction, and reconfiguration; and abstraction and simplification. They also indicate rearrangements of broader divisions of labor: the use of parallel and limited exchanges and autonomous stages. The tenor of their writing is exploratory and tentative, and these strategies and divisions of labor are reworked slightly in the later reprinting of this chapter (Star and Griesemer, 1999 [1989]: 515). They might become variously used (in concert or separately) to achieve coordination in the case of conflict.

Since this foundational work, the concept of the boundary object has been mobilized across a broad range of work discussing the processes of knowledge production and exchange across epistemologically distinct groups. However, we would argue that some of the original intention has been lost through this expansion (see also Zeiss and Groenewegen, 2009; Star, 2010). Boundary objects may, in some cases, function as a device to gradually bring about uncontested interactions between diverse groups, but this naturalization of the object into different communities of practice is only one potential outcome—deletion of controversy (through standardization) is just as, if not more, important (see also Contu and Wilmott, 2003). The scale and scope of the boundary object often remains unexplored, especially in the way specific objects function as part of larger systems of more or less stable, more or less structured objects and activities, and how boundary objects become part of boundary infrastructures (Bowker and Star, 1999). Finally, the issue of other forms of bringing about cooperation, such as coercion and the power and influence differentials between groups and actors, is often overlooked (Oswick and Robertson, 2009). In this process, the recent wider debates that use Star and Griesemer as a starting point have lost the tentative nature of this analysis and discussion, and the mobilization of an explanation of coordination through boundary objects as one of a number of possible explanations of how coordination is achieved.

In the following empirical descriptions and discussions of coordination work in a large construction project, we use this reexamination of the boundary object concept to analyze the interactions between people and objects in achieving organizational goals. We observe how objects are produced and circulated between groups, with a particular focus on their plasticity and on issues of their partisanship or neutrality.

Setting and Methods

The setting for our study is a large construction project in Europe, the design and construction of a new airport terminal: Heathrow Terminal 5, in London. The project is a "mega project" in Flyberg et al.'s terms (2003)—a multi-billion dollar project. At the time of construction, it was one of the largest projects worldwide. There are significant organizational challenges associated with such a project-based setting (Boland et al., 2007). For example, the nature of the organization changes fundamentally through different stages of the project process—with different individuals and firms joining and leaving the overall project team in the planning, design, and construction phases, both in the operational teams and in the strategic management of the project. As such, it provides a good context for studying the coordination of design work.

Aside from its size, one of the key aspects of the project was a commitment from the client to introduce a set of information technologies for producing, storing, sharing, and representing information. This innovative way of working would allow all project information—plans, spreadsheets, models, written documents, and so on—to be coordinated and distributed digitally. Information exchange would be mediated through these technologies and ensure all information was consistent and up to date. The term used for this digital coordination was the "single model environment," but the integrated software system that linked a data repository and computer-aided design

(CAD) software was largely consistent with the type of system now termed the "building information model" (BIM). This new set of technologies to design and coordinate work was introduced with other innovations in contractual arrangements and organization of the work.

Both the present authors have conducted significant empirical work on the technologies and practices within this large construction project. Data were primarily collected through two separate empirical studies focused on this setting; but we also maintain ongoing relationships with the key participants having first developed a research interest in the project in 2000. The first study involved ethnographic research by the second author, involving blocks of three or four days observing staff at work, engaging in informal discussions, and attending team and project meetings with data collection over several weeks between October 2001 and April 2002. To clarify observations from this immersion within the site, toward the end of the period the second author conducted twenty-three semistructured interviews (Harty, 2005). The second study was largely interview-based and conducted between October 2005 and November 2006: the first phase involved the first author and her colleagues interviewing at a senior level within the project, and the second phase, which was conducted by the first author alone, collected additional data on the roof subproject as well as on the use of the single model environment (see Whyte, 2008; Harty and Whyte, 2010).

It is the commonalities across these studies that led us to work together. The separate studies share more than just the empirical case; they are closely matched in terms of theoretical interests and approach. As we jointly discussed and analyzed our data, we conducted further interviews and informal conversations to check facts and test interpretations in conversation with key participants, and we shared our readings of the Star and Griesemer study as we worked to theorize the interactions of people and objects in achieving organizational goals in our setting.

Findings: Plasticity of Objects and their Partisan Nature

We observed material and digital objects being used in shifting ecologies of practice (Harty and Whyte, 2010). Objects that are productive in terms of coordination of ongoing work have a dual epistemic and boundary-spanning role, allowing participants in the project to maintain conventions and legibility of objects, across different locations and times, while allowing flexibility and partiality to enable the development of new ideas and innovation. They are not neutral but are partisan: bridging boundaries between some groups while creating and sustaining others. This makes the roles and functions of objects more complex; not only are there multiple interrelated objects, and

Table 10.1 Examples of information technologies that can be described as objects at Heathrow Terminal 5

Object type	Nature	Actors that become central	Actors that become marginal
Computer hardware	PC-based systems	Computer literate	Computer illiterate
Standards	Standards documentation	IT managers	Engineers
Repositories	Ordering devices	IT managers	Engineers
Computer-aided design software (CAD)	AutoCAD software tool	AutoCAD users	Nonusers and users of other software systems
CAD representations and models	Drawings manipulated and viewed in CAD	Professional with expertise to read the drawings	Nonexperts
Product models	Physical and digital models and visual representations	Designers	Central IT managers

not only are these of different digital and material forms, but each potentially performs different roles. In Table 10.1 we outline some key examples—both hardware and software—relating to "information technology" used in the project, showing how they make some actors central while marginalizing others.

The plasticity of objects and their partisan nature are further explored below, using examples from across the studies to describe in detail how particular people and technologies come together to develop and deliver organizational aims.

The Plasticity of Objects in Socio-material Practice: Multiple Models on the Roof Subproject

The idea that objects are plastic comes out of our observations in the project of objects that are worked on collectively and evolved over time. This finding also resonates strongly with previous work on epistemic and technical objects (Knorr Cetina, 1997, 1999, 2001; Rheinberger, 1997; Ewenstein and Whyte, 2009), where objects are described as *epistemic* where they are only partially instantiated in the materials that scientists or designers engage with, question and explore, and *technical* where they are seen as already complete, are unquestioned, and are used instrumentally to frame ongoing inquiry. In the large construction project, material objects only partially capture aspects of the final building that is to be built, and hence the designers work across multiple objects to achieve this goal.

The roof subproject was seen by project managers as a good example of the integrated team using the single model environment, and its associated software and processes; yet the research shows that on the roof subproject there was also prolific use of physical models in a variety of different media. As the team developed the design and the construction plans, it worked across multiple models at a range of scales to make decisions. By moving from across different scales, the designers could keep the physical models that they were working with useful to the design decision "in hand," as they could be placed on tables in the office and moved around, pointed to, and modified, literally, by hands. The architect describes the physical models they used:

> chronologically there were 1:200 models on roof forms. There were 1:50 models of half the roof, once it became symmetrical, and mirrors; there were 1:20 beginnings of the abutment. There were 1:20 of the abutment and half the roof; there were 1:10, 1:5, 1:2, of the nodes. The 1:5 we often did in plasticine, so one model served for five or ten times. In total there, there must have been 300 models made. Then early on, with the 3D we went to... I think we went to [a university] to do a prototype, a wax prototype process. (architect)

Diverse media were enrolled to hold ideas stable or change them at different scales and on different temporal cycles. Thus, the first models at 1:200 brought the whole roof into view, while later models zoomed progressively in, to half the roof and the abutments, and then to individual nodes. The plasticine is particularly interesting, as it is the same lump of plasticine that is reused five to ten times to create a number of different models, being squashed together into a ball and then reworked by the architect and engineers into a new version of a node, which could then be considered architecturally, structurally, and in terms of fabrication and construction. The architect made a photographic record of all the different models that were made and destroyed in this design process.

The iteration between digital and physical representations was central to the design development. In referring to "the 3D" in the above quote *"Then early on, with the 3D,"* there is a reference to the digital model which was itself then output into a physical model, a wax prototype, which was used to check the digital design, in the same way that an author might print an article that had been written in the computer and check through it. Given that, at the level of the overall project, there was a strong focus on getting all professionals to work digitally and integrating all of this work, it is intriguing to unpack why in an area of the project that was seen to be successful there was so much iteration between digital and physical representations.

Digital modeling was primarily the role of the digital modeler, who joined the team in 2001 and came with skills using a solid modeling package that was became useful for considering the 3D shapes required for the abutments. The successful use of the digital model for these shapes led to his work modeling half a bay of the structure (the entire structure was duplicated out of this so there was no need to model it more than once). Employed to work alongside the two physical model-builders who were also employed by the architect, the digital modeler came to play a pivotal role in the team, not only in design coordination but also in the construction planning. Because of the nature of the design problems around nodes (landings for steel columns and junctions in the structure) and the background expertise he brought, he did most of the work in a solid modeling package which is more common in product design than architecture and construction. However, digital modeling was also used by the engineer who used a frame model in the structural analysis and wind-testing of the structure. From the 2D drawings provided by the engineers, the fabricator built a 3D model of the steel structure. While some cases where the data was reformatted or re-input might have been plausibly unnecessary inefficiencies in this process, many were related to active cognition—a rethinking and working out of the structure.

Physical models—whether in wax, cardboard, or plasticine—were also important, and this use of both the digital and the physical was mentioned by many of the professionals involved who found this roof project a particularly rewarding and innovative project in which to be involved. When asked what he had learned from the project, the fabricator said *"The importance of physical models—that was definitely a bonus."* These models made visible to the team the issues that were under discussion, and the decisions that had to be made to achieve the goal of a final building. As the engineer said:

> in understanding what the 3-D shape actually looks like and how it looks through angles and that kind of thing, very useful indeed, because there are things, especially as geometry has become complex, it's fairly difficult to see very much at a time on the screen, your perception of the complex shape is . . . is quite limited, you have to make the thing move around and then you can't quite remember what was on the other side, and it's all a jumble because all the on the tower . . . you see on our models of it, you see the framing on this side of the building and then you also see the framing on the other side of the building and it makes you feel there's something physical in front of you, you can . . . you can perceive it more, it's sort of—its there isn't it?—it's to do with the way we perceive things, so they were very useful. (Engineer)

Physical models were used not only in the design process but also in making decisions in construction planning. Many of the subassemblies were modeled and tested off-site. Half-size models of details of the abutments were made at the casting foundry at Burton-on-Trent for the team to inspect. The largest of these tests occurred after the design had been fixed, with half a bay built off-site in the abutment first-run study. Through these and other contexts, knowledge of the structure, its design, and its erection were developed through interactions across different types of models, each of which partially instantiated the characteristics of the final building.

The roof subproject is an intriguing context in which to examine the coordination work, as it was seen as successful, yet modeling work was not input into the overall project's single model environment until later in the process. While the team coordinated with the surrounding subprojects, it had also reduced the number of interfaces by designing the roof as a stand-alone structure. Much of the modeling work was thus focused on coordination within this multidisciplinary subproject team. The large number of evolving models used draws our attention to the way that models enroll particular actors and are malleable across different timescales. They also suggest the importance of switching between models in the generation of ideas. The models used have different data structures and affordances—some are homogeneous and easily pliable; others are more articulated and are at a scale at which one can walk around them. Digital models, which are viewed through the screen and rotated, are constructed in different formats, as solid models or as 2D vector frameworks. These objects are focal to the work and skills of different communities of actors, reorganizing social groups and creating different boundaries between practices.

The findings draw attention to the plasticity of objects in this setting, as work is coordinated through time toward the goal of a final building. By drawing from the Star and Greisemer study, as well as the literature on epistemic and trans-epistemic work, this analysis draws attention to how the production of boundary objects is part of the ongoing socio-material practices of coordination, where objects are developed, evolved, and discarded as a range of different actors coordinate their work to achieve the eventual goal. Some of these objects are more, and some less, malleable within the practice. As discussed below, in this context, the objects used are not neutral but rather are partisan in nature. To articulate how this is manifest, we will examine local design practices in which conflicts arose as the single model environment became salient as part of everyday practice.

The Partisan Nature of Objects in Socio-material Practice: Standard Development for the Single Model

The idea that objects are partisan comes out of the observations in the project of objects that are implicated in the conflicts around coordination. This also resonates strongly with previous work which has delineated the political intentionality that is imparted into objects as they are developed and shaped (e.g., Winner, 1985; Woolgar, 1991). It is for this reason that the debates about which objects to use for coordination can be so heated within organizations (Thomas, 1994; Henderson, 1999). People feel that there is a lot at stake in these debates, as knowledge is differently valued and given status according to what becomes core grounds for mutual intelligibility and what is seen as peripheral and subject to interpretative flexibility.

At the detailed design phase, there is little actual construction work happening on site, other than site preparation works; yet, there are negotiations over the set of standards for the production of the structural steel detail designs, in line with the single model environment. This is not a debate about design, but one about integrating the design, fabrication, and installation processes across time and disciplines. This integration through the "single model environment" offered a number of specific benefits, above and beyond the broader aim of digital coordination. By using diverse design, calculation, and production software tools that could nonetheless seamlessly transfer information between them, designs could be integrated with the fabrication of the steel work later in the construction process, avoiding the manual reworking of information which, as well as adding cost, potentially introduces dimensional inconsistencies. There was also a robust software package already available to enable this exchange (which was not always the case with other disciplines, such as building services engineering; see Harty, 2005). The task was therefore not one of selecting technical objects but rather about agreeing to standardized ways of producing CADs using these objects, which would be consistent across the various subprojects and across design and fabrication.

These negotiations involved structural steel engineers, drafters, project managers, and team leaders, and the information technology (IT) department who were the housekeepers of the IT system on the project. They centered on a particular document which would define the standardized way of producing the CAD models. The content of the document was largely concerned with the layering conventions to be used. Layering—the way that complex drawings and models are split into separate, overlaid sections—stops single models or drawings becoming too complicated and messy, and allows specific layers to be edited without affecting others. It is a way of representing extensive amounts of information while limiting problems of interdependence—for instance, where a specific layer remains an integral part of the model but can only be edited by certain users:

> layering is very important. Being the single model environment, using each others' files, we had to decide very early on what the layering was going to be... how we are going to differentiate between structural, architectural... we had to satisfy all disciplines. (Structural engineer)

The process of layering therefore is something that enforces boundaries between those groups and provides a mechanism for exchanging and coordinating information. In the case of the "single model environment," it was also a way to coordinate information exchange between the disciplines, a device for structuring information across these groups. Consistent layering methods are crucial to produce multiple and complex designs that can be understood, shared, and combined across different users and disciplines.

The document was circulated among the various relevant actors. Initially, there was skepticism over whether the single-model environment was either desirable or possible, given the work required to attain such an unprecedented level of technical and informational coordination within the "single model environment," and the significant disruption to existing ways of coordinating information it represented:

> I used to argue with him [one of the project managers] at the beginning—I say 'no you just can't do it—it won't work'—we'd have real argument sessions. (Structural engineer)

There was also some lack of understanding within the of drafting and designing domains about what the single model environment was and its implications for existing practices, as it was a significant shift away from usual practice:

> at the beginning I was shying away from 3D stuff—I couldn't see it working (CAD drafter). when I first joined... I wasn't sure what they meant [by 3D model]. (CAD drafter)

The layering document was passed back and forth between these different actors, who each redrafted and revised it in line with their own expectations and experience. Over successive iterations, it gradually became more fixed and stable. During this process, some project staff also became more convinced of the value of digital coordination and the single model environment, and initial skepticism was overturned to the point where existing practices and expectations were transformed to converge around a set of new standardized ways of producing CAD models.

> As the project has progressed you can see why we did it...the system is working...if you can understand the actual system it becomes a lot easier...it's all about how you set up the project at the beginning. (structural engineer)

> more and more you get into it—there's a certain logic there and it does make sense for everybody to be working in 3D.... (CAD drafter)

But other important influence on the outcome of these negotiations came from outside the project. Although project staff were colocated in shared offices, they could not forget or replace the expectations and the existing ways of working of their employing firm. In this case, one director of a large design practice who had a number of staff working at T5, but had no direct participation in the project, was not entirely convinced about the worth of the single model environment across the whole of the design and construction process.

However, specifically for structural steel detailing, it was considered that digital coordination would better connect design and fabrication, with significant cost savings:

> We're pushing it [the structural steel design] to be 3D—it's the easiest bloody way to draw it—it's in our interests to push it... All we are saying is, all we're boiling down to is that people should coordinate better—it's the biggest problem in the industry—people need to coordinate round something—is 3D the best way of doing that? I don't know... but in streamlining the process [we are] taking out the historical barriers—its nonsensical that we create things electronically, send them out on paper to the contractor whose got a little man in the corner typing them into a bending machine. (design practice director)

The standards document certainly acted as a boundary object as it moved across boundaries within and around the project. It served as a focus and a medium for negotiations over how to achieve the coordinated practices around structural steel design that the project's managers supported.

Once the document was stabilized after numerous journeys between the relevant actors, it also became part of a wider ecologies of objects and actors within the project—CAD packages, various disciplinary-based practices, etc., but in doing so did not increase the complexities of interactions between disciplines but provided a mutually understandable way of transacting across these boundaries—this is the "classic" version of what boundary objects do, supported by common methods. In line with Star and Griesemer's conceptions of boundary objects as deleting controversy, the document also drew clear boundaries around the layering conventions, shielding these debates from wider disagreements over the efficacy of the "single model environment." Issues over the broad intent behind coordinated design across the whole project were masked by a more focused set of engagements over particular layering practices.

Discussion

Objects are important in coordinating knowledge and practice, but perhaps even more so in their very constitution. Through our study, we observed the interactions between people and objects in achieving organizational goals. From our analysis, we can say that the idea of the "single model environment," for example, was fluid, uncertain, and contested. The roof subproject team largely achieved coordination with other subprojects through other means, with a range of media in their own coordinative work. In the other vignette of practice, the constituent parts of the single model environment did not function as instrumental objects; they could not enforce particular ways of using onto individuals; and they were themselves subject to much transformation as they were incorporated into hybrid practices. Various objects not only crossed boundaries but also constituted and reconfigured them. Hence it seems that, at different times and within different sets of interactions, objects can span across different boundaries and sets of actors as well as act as more fluid and evolving epistemic objects.

Star and Griesemer's work is richly suggestive and continues to inform recent work on objects in the organizational literature. The concept of an object that can span epistemological boundaries and can be simultaneously embedded into contrasting social worlds yet retaining some central, shareable identity is highly alluring, especially when discussing how diverse social groups might collaborate. But this is a considerable achievement for an object, and Star and Griesemer acknowledge the rigorous standards or methods within these domains that define what can be mutually understood and hence what can be unproblematically exchanged across their boundaries. At the same time, these methods, or practices, exclude the aspects of the boundary object that might not be mutually intelligible across these boundaries.

Their study emphasizes how volunteers coordinate their work through aspects of the work itself, with mutually negotiated objects of interest, while indicating how other forms of coordination may involve more managerial solutions, with nonmutual strategies being imposed onto the work. This may, for example, involve the imposition of particular representations as well as the coercion and silencing of stakeholders. In some instances, coordination may not be accomplished; instead, there would be a fragmentation and breakdown of an organizational practice. This is an important point: the concept of the boundary object is only one possible means of achieving coordination across disparate groups, and other techniques can be alternatively, or simultaneously, mobilized.

Our own empirical work, as well as our rereading of their study, suggests that diverse forms of coordination coexist. This is consistent with Star and Griesemer's discussion of coordination where consensus does not exist. It is in their articulation of methods that this comes through most strongly. They describe the "trick of translation" that required insistence on and success with standardized methods as a testament to skillful management of the complex, multiple translations involved in the natural history research museum. Cooperation itself may involve either consensus or conflict, but it differs from coercion in that it is negotiated. This is reflected in the case of the layering standards document—in the way it became structured to both limit the interdependence between different groups and their contributions to the single model environment, and also to draw a boundary between more pragmatic issues of layering conventions and broader (and more contested) debates around the efficacy of the single model environment as a whole.

The chapter contributes to the wide discussion of the concept of a "boundary object" (Bucciarelli, 1994; Henderson, 1999; Carlile, 2002, 2004; Bechky, 2003; Levina and Vaast, 2005; Oswick and Robertson, 2005; Ewenstein and Whyte, 2009; Panourgias et al., 2010) by highlighting the plasticity and partisan nature of objects in the setting we studied. In Star and Griesemer's work, objects are required to have interpretive flexibility (Bijker, 1992) to allow them to have meaning across epistemologically distinct groups, but also to remain the same object throughout these exchanges. The challenge for research is not to leave objects as neutral and boundary-spanning, oriented to producing mutually meaningful exchanges, but rather to examine how they are implicated in the setup of the social relations in particular contexts.

Hence the work suggests new directions for research to describe how the material and social are entangled and together become reconfigured in organizing practices. One area that is particularly in need of further theorizing is the notion of the "object" and its digital or material nature. The challenge is to articulate interactions without romanticizing them; so that the architect or model builder taking a ball of plasticine in their hands and work it into a shape that helps them to explore part of a building they are proposing is seen in the context of the wider digital system, but that this system is also seen in the context of the material practices it supports. These objects operate at different scales and span different types of boundaries between occupational groups.

While we have drawn particularly on the Star and Griesemer study in this chapter, later work by Star suggests a direction that researchers might take in approaching this issue. In later work, Bowker and Star (1999) introduce the idea of boundary infrastructures: an idea that has been picked up in later work on ecologies of practice and infrastructures (Edwards et al., 2007) and was returned to later by Star (2010). For example, Bowker and Star discuss:

> A hospital information systems... has to respond to the separate as well as combined agendas of nurses, records clerks, government agencies, doctors, epidemiologists, patients and so forth. To do so, it must bring into play stable regimes of boundary objects such that any given community of practice can interface with the information system and pull out the kinds of information objects it needs. (Bowker and Star, 1999: 313)

This concept of infrastructure seems a useful concept for organization and management researchers to return to and explore in understanding the institutionalization of particular socio-material practices, which involve both stable regimes of boundary objects and the production of new and evolving objects, as well as the interaction between people and objects in achieving organizational goals.

Acknowledgments

A previous version of this chapter was presented at the 2008 Academy of Management, Anaheim, USA, as part of the OMT/TIM Symposium: "The Role of Objects in Innovating across Boundaries: Venturing into the Unknown." Both authors gratefully acknowledge the feedback from participants at that workshop and discussions with the editors and participants at the Chicago workshop on Materiality and Organizing, which helped in the development of this work. They also gratefully acknowledge the support of the UK funders, EPSRC and ESRC, through grants EP/E001645/1, RES-331-27-0076, and EP/H02204X/1 at the University of Reading. The first author also thanks the EPSRC for previous support at the Imperial College Business School and colleagues Tim Brady, Catelijne Coopmans, Andrew Davies, and David Gann, with whom she worked in the first phase of her data collection.

References

Bechky, B. A. (2003). Object lessons: Workplace artifacts as representations of occupational jurisdiction. *American Journal of Sociology*, 109(3), 720–52.

Bijker, W. (1992). The social construction of fluorescent lighting, or How an artifact was invented in its diffusion shape. In W. Bijker and J. Law (eds.), *Shaping technology/building society: Studies in sociotechnological change* (pp. 75–104). Cambridge, MA: MIT Press.

Boland, R. J., Lyytinen, K., and Yoo, Y. (2007). Wakes of innovation in project networks: The case of digital 3-D representations in architecture, engineering, and construction. *Organization Science*, 18(4), 631–47.

Bourdieu, P. (1984). *Distinction: A social critique of the judgement of taste*. London: Routledge and Kegan Paul.

Bowker, G. and Star, S. L. (1999). *Sorting things out*. Cambridge, MA: MIT Press.

Bucciarelli, L. L. (1994). *Designing engineers*. Cambridge, MA: MIT Press.

Callon, M. (1986). Some elements of a sociology of translation: Domestification of the scallops and the fishermen of St Brieuc Bay. In J. Law (ed.), *Power, action and belief—A new sociology of knowledge*. London: Routledge and Kegan Paul.

Carlile, P. R. (2002). A pragmatic view of knowledge and boundaries: Boundary objects in new product development. *Organization Science*, 13(4), 442–55.

——(2004). Transferring, translating and transforming: An integrative relational approach to sharing and assessing knowledge across boundaries. *Organization Science*, 15(5), 555–68.

Contu, A. and Wilmott, H. (2003). Re-embedding situatedness. The importance of power relations in learning theory. *Organization Science*, 14(3), 283–96.

Edwards, P. N., Jackson, S. J., Bowker, G. C., and Knobel, C. P. (2007). *Understanding infrastructure: Dynamics, tensions, and design*. Ann Arbor, MI: DeepBlue.

Ewenstein, B. and Whyte, J. (2009). Knowledge practices in design: The role of visual representations as "epistemic objects." *Organization Studies*, 30(1), 7–30.

Flyvbjerg, B., Bruzelius, N., and Rothengatter, W. (2003). *Megaprojects and risk: An anatomy of ambition*. Cambridge: Cambridge University Press.

Gherardi, S. (2006). *Organizations knowledge: A practice-based approach to learning in the workplace*. Oxford: Blackwell.

Giddens, A. (1984). *The constitution of society: Outline of the theory of structuration*. Cambridge: Polity Press.

Goffman, E. (1971). *The presentation of self in everyday life*. Harmondsworth: Penguin.

Harty, C. (2005). Innovation in construction: A sociology of technology approach. *Building Research and Information*, 33(6), 512–22.

——Whyte, J. (2010). Emerging hybrid practices in construction design work: The role of mixed media. *Journal of Construction Engineering and Management*, 136(4), 468–76.

Henderson, K. (1999). *On line and on paper: Visual representations, visual culture and computer graphics in design engineering*. Cambridge, MA: MIT Press.

Hughes, T. P. (1983). *Networks of power: Electrification in western society, 1880–1930*. Baltimore and London: John Hopkins University Press.

Knorr Cetina, K. (1997). Sociality with objects: Social relations in postsocial knowledge societies. *Theory, Culture and Society*, 14(4), 1–30.

—— (1999). *Epistemic cultures: How the sciences make knowledge*. Cambridge, MA: Harvard University Press.

——(2001). Objectual practice. In T. R. Schatzki, K. K. Cetina, and E. von Savigny (eds.), *The practice turn in contemporary theory*, pp. 175–88, London: Routledge.

Lanzara, G. F. (2009). Reshaping practice across media: Material mediation, medium specificity and practical knowledge in judicial work. *Organization Studies*, 30(12), 1369–90.

Leonardi, P. M. and Bailey, D. E. (2008). Transformational technologies and the creation of new work practices: Making implicit knowledge explicit in task-based offshoring. *MIS Quarterly*, 32(2), 411–36.

——Barley, S. R. (2010). What's under construction here? Social action, materiality, and power in constructivist studies of technology and organizing. *The Academy of Management Annals*, 4, 1–51.

Levina, N. and Vaast, E. (2005). The emergence of boundary spanning competence in practice: Implications for the implementation and use of information systems. *MIS Quarterly*, 29(2), 335–63.

Nicolini, D. (2009). Zooming in and out: Studying practices by switching theoretical lenses and trailing connections. *Organization Studies*, 30(12), 1391–418.

Orlikowski, W. J. (2007). Sociomaterial practices: Exploring technology at work. *Organization Studies*, 28(9), 1435–48.

——(2010). The sociomateriality of organisational life: Considering technology in management research. *Cambridge Journal of Economics*, 34, 125–41.

Oswick, C. and Robertson, M. (2005). Boundary objects reconsidered: From bridges and anchors to barricades and mazes. In *European Group for Organizational Studies Conference (EGOS)*, Berlin.

——— (2009). Boundary objects reconsidered: From bridges and anchors to barricades and mazes. *Journal of Change Management*, 9(2), 179–94.

Panourgias, N. S., Nandhakumar, J., and Scarbrough, H. (2010). Material practices of coordination and innovation in the design and development of computer games. In *European Group for Organization Studies (EGOS)*, Vol. Sub-theme 18: Coordination in Action. Lisbon.

Pickering, A. (1995). *The mangle of practice: Time, science and agency*. Chicago: University of Chicago Press.

Rheinberger, H.-J. (1997). *Toward a history of epistemic things: Synthesizing proteins in the test tube*. Stanford: Stanford University Press.

Star, S. L. (2010). This is not a boundary object: Reflections on the origin of a concept. *Science, Technology & Human Values*, 35, 601–17.

——Griesemer, J. R. (1989). Institutional ecology, "translations," and boundary objects: Amateurs and professionals in Berkeley's museum of vertebrate zoology, 1907–1939. *Social Studies of Science*, 19(4), 387–420.

———(1999 [1989]). Institutional ecology, "translations," and boundary objects: Amateurs and professionals in Berkeley's museum of vertebrate zoology, 1907–1939. In M. Biagioli (ed.), *The science studies reader* (pp. 505–24). New York: Routledge.

Thomas, R. J. (1994). *What machine's can't do: Politics and technology in the industrial enterprise*. Berkeley and Los Angeles, CA: University of California Press.

Whyte, J. (2008). Reaching for the skies: Time and turbulence in digital technologies and practices. In International Symposium on Automation and Robotics in Construction (ISARC), Vilnius.

Winner, L. (1985). Do artefacts have politics? In D. Ma. J. Wajcman (ed.), *The social shaping of technology*. Milton Keynes: Open University Press.

Woolgar, S. (1991). The turn to technology in social studies of science. *Science, Technology and Human Values*, 16(1), 20–50.

Yanow, D. and Tsoukas, H. (2009). What is reflection-in-action? A phenomenological account. *Journal of Management Studies*, 46(8), 1339–64.

Zeiss, R. and Groenewegen, P. (2009). Engaging boundary objects in OMS and STS? Exploring the subtleties of layered engagement. *Organization*, 16(1), 81–100.

V
Materiality as Affordance

11

Theorizing Information Technology as a Material Artifact in Information Systems Research

Daniel Robey, Benoit Raymond, and Chad Anderson

Introduction

There is a long history of "computer impact" studies in the fields of computer–human interaction (CHI), organization studies, information systems (ISs), and related areas. These studies typically focus on an information technology (IT) artifact and use a variety of research methodologies to ascertain the effects of the artifact on dependent variables or outcomes of interest. The relevance of studying IT impacts continues as IT diffuses across an increasingly large scope of human activity. Organizations of all kinds are exposed to a seemingly endless progression of IT applications, fueling a keen interest in studying impacts at multiple levels of analysis. Like research in all fields, studies of IT impact are potentially hindered by a lack of clarity in the definition of concepts and the specification of relationships among concepts. Whether seeking to produce general explanations through a "grand" theory, or account for empirical observations through a locally "grounded" theory, researchers must apply or develop theoretical frameworks that correspond to the phenomena being studied. Without clear theoretical conceptualizations of IT artifacts, impact studies can only advance so far.

This chapter addresses this concern by focusing on theory about IT artifacts. We believe, along with others, that most theories currently used in impacts research do not engage adequately with the material properties of IT artifacts (Orlikowski and Iacono, 2001; Leonardi and Barley, 2008). We define material properties as features of IT, including hardware devices, software interfaces and applications, and communication services. In practice, it is difficult to separate hardware devices from applications and services (Jarvenpaa, Lang and Tuunainen, 2005; Cousins and Robey, 2012), so our definition focuses on the functions and capabilities that technology features make available in potential contexts of use. This definition preserves the distinction between material artifacts and social contexts of use and permits the analysis of the consequences of different technologies in different social settings. We believe the preservation of this distinction is important because, as Faulkner and Runde (2010) argue:

> ... people and technological objects, while no doubt often internally related and sometimes interpenetrating, are nevertheless generally distinct and different things, with their own intrinsic properties (e.g. while people are conscious beings, technological objects generally are not, while people have biological bodily functions, most technological objects do not, and so on). (p. 21)

This approach bears similarities to those taken in Chapter 2 by Leonardi and Chapter 5 by Pollock, in this volume, respectively but differs from the approaches taken in Chapters 6 and 19 by Scott and Orlikowski and Cooren et al., respectively.

Preserving the distinction between material artifacts and social contexts does not imply that technology determines social consequences. However, it does imply a functional relationship between them. As our discussion of technology affordances below affirms, material artifacts are relevant because they are associated with social purposes and needs, which indirectly serve to cause the existence of the artifacts. Stinchcombe (1968) argues that functional explanations of social phenomena involve a type of reverse causality wherein the consequences of human activity help to explain the presence of social structures and technologies that produce them. In the case of IT, it is clear that IT artifacts exist only because they are built with some functional purpose in mind. This does not mean that all features of material artifacts map directly to the outcomes imagined by their designers, or that users and designers share similar understandings of potential functions. Rather, specific features tend to be selected and used recurrently because of their usefulness, while other features may be disregarded completely. In other words, material artifacts are functional to the realization of social consequences, both desired and unintended, by "affording" those consequences. The "material agency" of artifacts becomes imbricated with human agency as people seek to realize their intentions through the artifact's material features (Leonardi, 2011).

Recently, scholars in information systems and organization studies have sought to restore materiality of IT in their work. Examples include Jonsson et al.'s (2009) study of the effects of remote diagnostic systems on boundaries between equipment suppliers and maintenance departments in client firms, and Pollock et al.'s (2009) analysis of online software support for customers of large enterprise applications. Jonsson et al. (2009) show that the material features of the technology are directly implicated in boundary-spanning, acquiring specific tasks in monitoring equipment, and intervening in preventive maintenance. Pollock et al. (2009) show that the virtualization of software support stimulates radically different work practices than those in locally situated software support. By engaging directly with the materiality of IT applications, these studies suggest revisions of established views on IT impacts.

We contribute to these efforts to elevate materiality to a more central theoretical role in the following ways. We begin by tracing the disappearance of materiality from the concept of technology in organization theory. Second, we address the challenge of carefully defining concepts and adapting them to a sociomaterial context. In Chapter 2 of this volume, Leonardi takes a similar approach in defining concepts such as materiality, sociomateriality, and so on. We advance the understanding by analyzing the case of technology affordance to reveal the concept's implicit functionality and its dependence on the relationship between material objects and human actors. Third, we address the challenge of including sociomaterial concepts in established theories commonly used in IS research. We propose two strategies for dealing with this challenge. In the first strategy, theories that nominally address technology can be extended by incorporating concepts that represent technology's material characteristics. Markus and Silver's (2008) extension of adaptive structuration theory and Raymond's (2012) extensions of organizational routines theory illustrate this first strategy. The second strategy is to extend theories that address substantive issues but neglect technology. Cousins and Robey's (2012) extension of work–life boundary management theory to include the affordances of mobile technologies illustrates this second strategy.

How Technology Became "Immaterial" in Organization Theory

In the 1950s and 1960s, organization theory was transformed from a largely prescriptive field espousing best practices for managers into a theory-based discipline focused on explaining variation in organization structures. The new "contingency theory" was exemplified by studies of organization structure (e.g., Burns and Stalker, 1961; Lawrence and Lorsch, 1967), which helped to identify the importance of fit between an organization and several contingencies. The most important contingencies at that time were organization size, environmental uncertainty, and technology (Robey and Sales, 1994). The basic argument was that organizations of different sizes operating in environments that varied in uncertainty and using different technologies were expected to exhibit different structural characteristics. Achieving a proper fit between these contingencies and structural design was expected to generate higher degrees of organizational effectiveness.

The technology contingency in organization studies was defined as "activities, equipment, and knowledge used to convert organizational inputs into desired outputs" (Johns, 1992: 8). This comprehensive definition combines social actions, material equipment, and abstract knowledge into a single theoretical concept. Retrospectively, such an all-inclusive definition can be seen as an early sign of conceptual difficulty, allowing scholars interested in technology to interpret its meaning in different ways. Chapters on "technology in management" textbooks typically review multiple approaches to technology in organizations without seeking to reconcile their fundamental differences (Robey and Sales, 1994).

One of the first research studies of technology in organization studies focused squarely on materiality. Woodward (1965) identified material differences among three types of manufacturing processes: unit and small batch production, large-batch and mass production, and continuous process production. These processes differ with respect to physical arrangements for transforming raw materials into finished products. By comparing the organizational structures of companies in each technology category, Woodward offered early evidence of the relevance of material aspects of technology to social organization. Each technology grouping adopted distinctive forms of organizing, which engendered consequences for organizational control, labor relations, and other outcomes. The most effective firms in each technology group conformed most closely to the modal structure for that group.

Despite this early focus on materiality, Woodward's approach to technology was limited to the study of manufacturing organizations. Organization theory, of course, encompasses a much wider range of organizations including those involved in rendering human services. Organizations such as schools, prisons, and government bureaucracies were also assumed to have technologies that allowed the inputs (clients) to be processed (served). In an influential article, Perrow (1967) adapted the concept of technology to include service organizations by defining technology in terms of the variety and difficulty of problems encountered during the transformation of inputs into outputs. This new definition allowed technology to apply to all organizations since it was no longer obliged to include the material aspects of an organization's technology. Thus, the technology of human service organizations such as schools was studied in terms of the variety of instructional and advisory programs in relation to the variety of student populations, not in terms of the material design of classrooms or instructional media.

Following this convenient extension in technology's definition, most scholars steered away from studying the material characteristics of organizations, so contemporary organization theory became mute on material aspects of organizations such as geographic location, architecture, office layout, and IT (Orlikowski, 2007). For example, Galbraith (1977) defined information systems in terms of decision frequency, scope of data base, degree of formalization, and capacity of decision mechanism. None of these categories represents material features or acknowledges the potential of software to enable or constrain human action. Hatch (2006), who includes a rare chapter on the physical structure of organizations, claims that the topic of physical structure was for many years a "theoretical backwater" in organization theory. With few exceptions (e.g., Fayard and Weeks, 2007), researchers largely avoid studying the material geography, office layout, décor, and technology of organizations. Those who pay attention to physical objects often treat them as cultural symbols, thus quickly shifting their focus away from the material artifact and toward social interpretation (Hatch, 2006).

The treatment of technology as "immaterial" in organization studies permits scholars to study technology without engaging with its material properties. As a more abstract concept, technology becomes amenable to measurement with survey questionnaires, thus enabling scholars to accumulate larger samples without worrying about the physical properties of organizations. Given the interest in producing empirical support for generalizable theory, it is understandable that the more abstract concepts and their related more convenient measures became unchallenged standards.

The material features of technology occasionally surface in the literature as a reminder of the physical artifacts that comprise organizations. Perrow (1981) produced insightful studies of material failures at the Three Mile Island nuclear power plant, which suffered a reactor core meltdown in the 1970s. Perrow (1983) also contributed an essay on the material consequences of human factors engineering for organizations. Perhaps to compensate for his contribution to the earlier dematerialization of technology in organization theory, Perrow writes:

> I hope I have suggested that organizational theorists pay attention to the way mere "things"—equipment, its layout, its ease of operation and maintenance—are shaped by organizational structure and top management interests, and in turn shape operator behavior. The early work on technology and structure, including my own, recognized a one-sided and general connection, but it failed to recognize how structure can affect technology and speculate about the large areas of choice involved in presumably narrow technological decisions. (Perrow, 1983: 540).

Our present aim is to respond, perhaps belatedly, to this call with a renewed focus on theories that accommodate materiality into explanations of technology impact. The following sections outline strategies for theorizing IT as a material artifact in IT impact studies.

Defining Concepts for Building Sociomaterial Theory

The primary need for theorizing the material aspects of IT is to define the theory's basic elements, that is, its conceptual building blocks. This fundamental requirement proves challenging because, as noted earlier, few theories about organizations address material aspects in any way. The first recourse, therefore, is for theorists to examine other disciplines that have addressed materiality. Perhaps the most obvious choice of disciplines would be the natural sciences, which began scientific inquiry by trying to explain the physical world. Unfortunately, the natural sciences offer little guidance to the researcher intent on theorizing how physical properties of technologies impinge on social systems such as organizations. A more productive first step would therefore be to examine the incorporation of materiality in related social sciences, such as sociology and psychology. Here, the concept of affordance has attracted attention because it focuses upon the possibilities for action that a material object offers to an actor. Like many other concepts in the social sciences, the concept of affordance fits with a functional explanation (Stinchcombe, 1968) that aims at understanding the functions that material objects play in human activities.

In sociology, Hutchby (2001) advocates using affordances as a way to counter the radical constructivist tendency to regard material reality as open to endless interpretation, which in turn tends to disregard the physical constraints imposed by material objects, including IT. Although Hutchby acknowledges the importance of social interpretation, as widely shown in social studies of technology, he uses affordances to draw attention to the material constraints on social action that cannot be removed through social interpretation.

In psychology, Norman (1988) uses the concept of affordance to focus attention on the design of computer interfaces. From a designer's perspective, making the possibilities of technology use visible to a computer user is as important as restricting attempts to use a computer for activities that it cannot support. Interestingly, Norman's work on affordances exhibits more emphasis on interpretation than on material objects. He even acknowledges at one point that: "I should have used the term 'perceived affordance,' for in design, we care much more about what the user perceives than what is actually true" (Norman, 2002).

Before rushing to incorporate the concept of affordance into studies of IT impacts on organizations, researchers should be aware of numerous issues surrounding the concept. Such awareness can be aided by a review of the concept's original development in ecological psychology. The term "affordance" originated with Gibson (1977, 1979) as a part of his theory of ecologically-based visual perception, which seeks to understand how an animal perceives and interacts with its environment.

> The affordances of the environment are what it offers the animal, what it provides or furnishes, either for good or ill . . . I mean by it something that refers to both the environment and the animal in a way that no existing term does. It implies the complementarity of the animal and the environment. (Gibson, 1979: 127)

In contrast to cognitive psychologists who believed that perception was entirely an internal mental process, Gibson theorized that information about affordances exists within the environment itself and can be directly perceived. For example, an earthbound animal can directly perceive that a horizontal, flat, rigid surface will afford it locomotion without having to first formulate a mental interpretation of the visual stimuli it takes in. Gibson further hypothesized that affordances are relative to an animal but exist independent of the animal's perception of them.

Although affordances became a foundational concept for ecological psychology, Gibson's writings sparked ongoing debate over the ontology of affordances. Stoffregen (2003) notes that the definitions of affordances within ecological psychology generally fall into one of two categories: (*a*) affordances are properties of the environment (Turvey, 1992; Reed, 1996; Michaels, 2000); or (*b*) affordances are relations between an animal and its environment (Kirlik, 1995; Chemero, 2003; Stoffregen, 2003). Representing the first category, Turvey (1992) formally defines affordances as dispositional properties of the environment that pose "real possibilities" for action. He equates a real possibility to a law, that is, "an invariant relation between or among substantial properties of things" (Turvey, 1992: 177). This definition allows for the prospective control of action, which Turvey considers essential to an accurate ontology of behavior.

The second category is represented by Stoffregen (2003), who argues that defining affordances as dispositional properties of the environment is problematic because affordances do not always result in the same actions. Theorizing affordances as invariant relations contradicts Gibson's characterization of affordances as representing what an animal can potentially do, not what an animal must do. Stoffregen further suggests that "affordances sometimes are described as entities that constrain behavior" (2003: 127), and yet properties of the environment alone cannot constrain behavior. He argues that these issues can be overcome by conceptualizing affordances as emergent properties of an animal–environment system. In this systems approach to affordances, the direct perception of affordances eliminates the need to consider properties of the environment as separate from properties of animals. This makes the animal–environment system the unit of analysis and allows affordances to be specified and detected prospectively, which preserves the notion of prospective control proposed by Turvey (1992). It also permits affordances to be seen as constraints on behavior because behavior may be constrained by the relationship between environmental and animal properties.

Stoffregen's arguments about the ontology of affordances parallels Orlikowski's (2007) assertion that we gain analytical insight by focusing on the ways in which the social and material intertwine in ongoing situated practice rather than treating them as distinct concepts. As such, the relational definition of affordance can be seen as a sociomaterial concept that encompasses both material objects and actors. IT practice research that treats imbrications as performative outcomes of a "sociomaterial nexus" (Introna and Hayes, 2011) appears to support the positioning of affordances as a property of actor systems rather than a property of the environment.

However, there is another way to conceive affordances as a relational concept that also maintains distinctions between human actors and material objects. Chemero (2003) suggests that affordances can be defined as "relations between the abilities of animals and features of the environment" (p. 189). This approach differs from Stoffregen's (2003) argument because Chemero defines an affordance as the relationship itself rather than an emergent system property.[1] Conceptualizing affordances as a relationship is similar to the economic concept of utility, which is the valuation of a commodity in relation to specific actors rather than some absolute value. In this relational view, affordances are seen as potentials for action that depend on both the material properties of objects and the ability of actors to perceive and use them. Material properties thus become necessary conditions for the existence of affordances, but material properties are not the affordances themselves (Markus and Silver, 2008). By defining affordances as a relationship rather than a property of the animal–environment system, the ontological distinction between actors and objects may be preserved.

Based on this review of debates in ecological psychology, it becomes apparent that adopting a concept such as affordance raises a number of issues directly relevant to IT impact research. If affordances are construed as properties of the environment, then they are assumed to exist independently of human actors. While this may seem sensible, pursuing this angle requires the specification of affordances associated with IT devices, software, and services. Taken to an extreme, IS researchers would need to develop descriptions of features at a highly detailed level, effectively assuming the "essentialist" position that specific technologies provide specific affordances and not others. By contrast, a more relativist stance is likely to be more useful to social

[1] Chemero's justification for defining affordances as relationships is based on Heft's (2001) argument that Gibson's ecological psychology is descended from William James's radical empiricism. According to radical empiricism, everything that is experienced is equally real. Among the things we experience are relations between things, so relations are real, with the same status as the things that stand in relations.

scientists who study technology and its social consequences. However, treating affordances as properties of the relationship between actors and their environments raises the level of abstraction and poses epistemological issues that might be phrased: "how do I recognize an affordance when I see it?" This epistemological issue is of course important if sociomaterial theory is to guide empirical research.

These issues regarding the concept of affordance (also see Faraj and Azad, this volume) hopefully alert IT researchers to the challenges of using new concepts in studies of the social consequences of IT's material properties. These challenges apply to other concepts such as material agency (Pickering, 1995), imbrication (Introna and Hayes, 2011; Leonardi, 2011), performativity (D'Adderio, 2011), ecologies of practice (Whyte and Harty, Chapter 10, this volume), and others. In our own research (Anderson, 2012; Cousins and Robey, 2012), we have engaged directly with the challenges of theorizing material artifacts in terms of their affordances. Researchers working with other concepts will also need to explain and justify their chosen definitions. In the following section, we turn to further challenges of including materiality into established theoretical frameworks.

Strategies for Theorizing Materiality

Extending Theories that Nominally Address IT

Producing novel theory is challenging. On one hand, theorists may be accused of borrowing too heavily from reference disciplines and not contributing original theory to their home disciplines. On the other hand, a novel theory may be rejected because it is too far removed from ongoing discourse. A middle ground through these challenges is to revise existing theories that nominally address IT but exclude or marginalize the material features of IT. Two examples of this middle ground strategy are illustrated here: adaptive structuration theory and organizational routines theory.

ADAPTIVE STRUCTURATION THEORY

Markus and Silver (2008) demonstrate this strategy by adding materiality to the well-known adaptive structuration theory (AST) developed by DeSanctis and Poole (1994). AST was developed with IT applications in mind, specifically group support systems, and posits that IT artifacts have "structural features" and a "spirit" that embodies the intentions for which IT applications were designed. For many years, AST has guided research studies showing how user appropriations of IT artifacts often differ from the artifact's intended spirit. These "unfaithful" appropriations support a growing interest in explaining the unintended impacts of IT artifacts.

Unfortunately, the concepts of structural features and spirit have generated confusion. Some critics object to the assumption that technology is capable of embedding structural features, given that Giddens' original version (1984) of structuration theory treats structure as virtual instead of tangible (Jones and Karsten, 2008). The use of vague terms such as "spirit" and "faithful" to describe design intentions and appropriations, respectively, further cloud AST's precision. For these reasons, Markus and Silver (2008) propose the addition of three concepts to AST: technical objects, functional affordances, and symbolic expressions. The first two of these relate directly to our interest in materiality.

Markus and Silver use the concept of technical objects to restore materiality to DeSanctis and Poole's concept of structural features. Technical objects include properties of IT artifacts, such as packaging, arrangement, and appearance, which exist apart from any functions or interpretations that users may bring to their relationship with IT. Markus and Silver view technical objects as separate and distinct from human users, and their causal potential depends on how the user appropriates them. The causal potential of IT artifacts is best understood by including the second concept, namely functional affordances, as part of the explanation for IT impacts.

Markus and Silver define functional affordances as "the possibilities for goal-oriented action afforded to specified user groups by technical objects" (Markus and Silver, 2008: 622). Thus, affordances are not equivalent to technical objects but rather depend on the relationship between technical objects and a perceptive and motivated user. Their treatment of affordance conforms to Chemero's argument (2003) that affordances are relations between the abilities of actors and features of the environment. Markus and Silver also demonstrate the inherently functional nature of affordances by linking technical objects to goal-oriented behavior.

In the case of AST, Markus and Silver's contribution is not the specification of new theory but rather an extension of a well-known theory. The extension specifically introduces the material aspects of IT artifacts, thus overcoming an acknowledged weakness in the formulation of the initial theory. With a revised version of AST, researchers may direct research attention to specific technical objects, such as portable smart phones, cloud computing services, or integrated enterprise systems. However, rather than treating technical features as determinants of user behavior, research using AST should emphasize the functions that such technical objects provide for prospective users. Thus, for example, users of cloud computing services may appropriate the technology in ways that fulfill certain needs and not others. Likewise, users of smart phones and enterprise systems may find specific technical features valuable because they afford communication and integrated material control, respectively.

ORGANIZATIONAL ROUTINES THEORY

The inclusion of material artifacts into existing theory is also illustrated by Raymond's (2012) and D'Adderio's (2008, 2011) extensions of the theory of organizational routines. Routines are a core component of any organization, so they comprise a natural focus for studies of IT impacts on organizations. Feldman and Pentland (2003) define organizational routines as repetitive and recognizable patterns of interdependent actions carried out by multiple actors. Although routines are generally associated with stability and inertia, they are also a source of flexibility and change. A routine may contribute to organizational inertia if it re-enacts an established pattern of action, but a routine may also generate new patterns of action (Feldman and Pentland, 2003).

Feldman and Pentland identify two aspects of organizational routines: the ostensive and the performative. The ostensive aspect embodies the abstract, generalized idea of the routine: the routine's ideal or schematic form, which may be codified as a standard operating procedure or exist as a taken-for-granted norm. The performative aspect consists of the actual performances of the routine and is inherently improvisational. Feldman and Pentland argue that neither aspect alone is sufficient to describe the properties of organizational routines. They argue that routines are "generative systems" capable of producing a wide variety of performances depending on the circumstances. Each performance of a routine is partly (re)-enacted from past experience, and partly improvised based on current circumstances and future goals (Pentland and Feldman, 2008).

Although IT artifacts are, in practice, often implicated directly in organizational routines (Volkoff, Strong, and Elmes, 2007; D'Adderio, 2008), Pentland and Feldman (2005, 2008) expressly exclude them from the definition of routines and thus marginalize the material aspects of artifacts. Although IT artifacts may embed a vision of organizational work processes and coordinate interdependent activities, they are theorized as external to routines. IT artifacts may have the capacity to influence and represent routines but do not have the capacity to become part of the generative system. Our preferred approach is to theorize IT artifacts as key aspects of the generative system of organizational routines (Raymond, 2012). This is consistent with D'Adderio's argument (2008, 2011) that artifactual representations of routines are important constituents of routines.

We develop two related arguments to support this claim. First, we argue that some (not all) IT artifacts can become embedded into organizational routines to the point where it becomes unlikely that the routine could be performed without the artifact's presence. Second, once they are embedded into routines, we argue that IT artifacts may acquire the status of material agents, thus becoming part of the generative system that Feldman and Pentland (2003) define as an organizational routine.

Artifacts become embedded into organizational routines through a process of adaptation and selection, similar to the way that IT applications are appropriated as described by AST (DeSanctis and Poole, 1994). This is consistent with D'Adderio's (2011) argument that artifacts can both influence the course of routines and evolve themselves as a consequence of their appropriation in specific contexts. D'Adderio notes that artifacts can produce various degrees of influence, from no influence to full influence. As performers of a routine are provided IT artifacts, and perhaps mandated to use them, artifacts may be discarded, ignored, merely used as accessories, or become embedded into the routine. This produces a range of possibilities by which artifacts can shape and be shaped by the performances of organizational routines. For example, actors performing an accounting routine may decide to use a calculator as a mere accessory, ignore it, or discard it altogether since the routine could be performed without its presence. In contrast, an accounting software package is more likely to become embedded into an accounting routine.

Embedded IT artifacts may acquire the status of material agents, interacting closely with human agents in the performance of organizational routines (Leonardi, 2011). Material agency can be defined as the capacity of artifacts to act independently of human action (Pickering, 1995). Many artifacts, as well as natural phenomena such as wind and water, have material properties that are capable of exerting agency on their own, rather than depending on the interpretations or interventions of social actors. For example, software applications embedded in countless work processes across thousands of organizations perform as material agents without oversight or external intervention. Rather than excluding such ubiquitous phenomena from the basic definition of routines, we seek to include them.

Raymond (2012) engages Feldman and Pentland's own arguments (2003) to explain how routines can be shaped by both IT artifacts and human agents. First, just like the ostensive aspect of organizational routines, embedded IT artifacts can play a *guiding* role for human action in organizational routines. By embedding organizational elements such as routines, data, and roles, IT artifacts serve as a template for behavior by influencing what actions ought to be taken (Volkoff et al., 2007). Second, embedded IT artifacts can play a *constraining* role for human action in organizational routines. Networked technologies such as workflow systems, for example, constrain individual human actions in part by constraining the actions of linked users. Although they can choose to bypass the software, their boycott holds consequences for their ability to have their work accepted by others in the organization (D'Adderio, 2008). Third, embedded IT artifacts may *monitor* the performance of organizational routines, identify aberrant actions that require correction, and enforce compliance to standard procedures. Software makes performance more visible across an organization, thereby enhancing control. Fourth, embedded IT artifacts can play a *legitimation* role for human action in organizational routines, for example by codifying the organization's values and norms within an IT artifact. This is consistent with D'Adderio's notion (2011) of the inscription of interests, intentions, assumptions, rationales, and logics into artifacts. Once embedded, artifacts can lend a sense of appropriateness to particular human actions and justify the routine.

Embedded IT artifacts may also act as material agents in the performative aspect of organizational routines, potentially altering the ostensive aspect. Consistent with a relational view of affordances, routines may be seen as possibilities for action that depend on both the material properties of objects and the ability of actors to perceive and use them. Because many IT artifacts take time to be fully understood, actors may perceive and use new affordances over time as they learn and experiment with them. Thus, embedded IT artifacts may contribute to the creation of new performances of organizational routines and, over time, help to establish more enduring changes in interdependent patterns of actions.

Acknowledging the material aspects of organizational routines helps to explain both their stability and flexibility without sole reference to human agents. By treating embedded material IT artifacts as part of the generative system of organizational routines, researchers may be less inclined to treat IT as an external or invariant influence on organizational routines. For example, by treating accounting software packages as part of a routine, researchers may be less likely to hypothesize the impact of IT on accounting practice and attend more directly to the sociomaterial practice of accounting. Integrating the materiality of IT artifacts within established theory thus promotes greater sensitivity to the close interdependence between material and human agencies.

Extending Theories that Ignore IT

The preceding section focused on theories, AST, and organizational routines theory in particular, which nominally address IT artifacts but do not theorize their materiality in their original formulations. A second situation commonly faced in IT impact research is engagement with a theory area that was established before the advent of IT artifacts that are commonly used today. Since much of social science was developed prior to the information age, neglect of IT artifacts in established theories is common. Researchers interested in studying the impacts of particular IT artifacts thus have an opportunity to extend these theories to include materiality. In the following section, we present one example from ongoing research on the role of mobile technologies in workers' management of work–life boundaries (Cousins and Robey, 2012).

THEORY OF WORK–LIFE BOUNDARY MANAGEMENT

The management of work–life boundaries addresses the social issues related to workers' ability to perform both work and nonwork roles effectively. Work and nonwork are conceived as domains within which certain activities and behaviors are deemed appropriate and others inappropriate (Clarke, 2000), and the borders between these domains are commonly referred to as work–life boundaries. Research in human resource management emphasizes the negative consequences of blurring work–life boundaries (Perlow, 1998; Ashforth et al., 2000; Clarke, 2000). For example, Perlow (1998) reports various managerial techniques to extend work boundaries beyond previous limits, thereby creating conflicts between workers' work and family domains. However, the literature also includes more positive views, such as Greenhaus and Powell's (2006) claim that work and family commitments do not necessarily conflict and that positive experiences in one role can reinforce positive experiences in the other.

With few recent exceptions, theories of work–life boundary management tend to exclude material features of mobile IT artifacts such as smart phones. Typically, researchers interested in the impacts of IT artifacts on work–life balance identify a sample of workers who use particular technologies, but they then exclude materiality in favor of the relationships between sample characteristics and outcomes. For example, Boswell and Olson-Buchanan (2007) studied users of cell phones, email, voicemail, personal digital assistants (PDAs), and pagers, finding that workers with higher ambition and job involvement were more likely to use those technologies after working hours. Clearly, this study lacks theoretical engagement with IT's materiality. Likewise, Kreiner et al. (2009) include "leveraging technology" (e.g., using voicemail, caller ID, and email) as a tactic to facilitate boundary management, but the material properties of these technologies are not central to the theory. To our knowledge, only Golden and Geisler (2007) and Hislop and Axtel (2011) grant IT artifacts a clear theoretical status in studies of work–life boundary management.

Mobile technologies may be seen, on one hand, as instruments that enable the spatial extension of work outside the office and the temporal extension of work beyond normal work hours (Cameron and Webster, 2005). Technologies may thus become implicated in the blurring of boundaries and disrupting of work–life balance (Middleton and Cukier, 2006). On the other hand, mobile technologies may be seen as tools that enable mobile workers to execute work–life transitions more frequently, with the promise of restoring work–life balance (Scheepers et al., 2006; Hislop and Axtel, 2011). Small, portable technologies may permit unobtrusive switching between work and family domains, thereby allowing urgent matters in either domain to be addressed in a timely manner.

Our specific research interest lies in the impact of mobile technologies on the work–life boundary management practices of mobile workers, whom we define as workers who use mobile technology at least 50 percent of the time (Cousins and Robey, 2005, 2012). These kinds of workers frequently find themselves in different locations. Although location can constrain the kinds of activities that can be performed, mobile workers can choose to transform places to become more amenable to work. For example, *cocooning* involves using mobile devices in ways that shelter users from active engagement with physical surroundings, and *encampment* involves constructing personal work spaces by bringing portable media to chosen public places such as cafés and libraries (Ito et al., 2009). Mobile technologies also allow users to construct novel self-representations using instant messaging and social networking services. Self-representations may be adjusted dynamically and become part of a boundary management strategy.

Because studies of mobile work practices treat mobile technology in a descriptive fashion, the impact of mobile technology on work–life boundary management remains undertheorized. To address this issue, we return to the concept of technology affordances, as discussed earlier. As a relational concept describing possibilities for action, the affordances of mobile technologies may become central in the explanation of how mobile workers manage their work–life boundaries. Based on empirical fieldwork (Cousins and Robey, 2012), we identified five affordances describing the relationship between mobile workers and the material characteristics of the technologies available to them. *Portability* is the potential for the user to transport a mobile device. *Connectability* is the potential to establish communications. *Openness* is the potential to use mobile technology to share information and data across various heterogeneous devices and applications. *Identifiability* is the potential to associate a mobile device or service with a single authorized individual, thus allowing the user to represent a unique identity. Finally, *configurability* is the potential to select mobile technology options and settings to match user's personal preferences or needs.

Each of these affordances is associated with different strategies and tactics for managing the social boundaries between work and nonwork domains. Some mobile workers pursue a segmentation strategy by using mobile technologies to strengthen the borders between work–life domains. Others use the same technologies to support a strategy of integrating work–life domains. Within these two strategies, mobile users employ IT artifacts to manage their use of space and time, and develop personalized practices for executing transitions between domains. Although managing boundaries could always be accomplished without material IT artifacts in common use today, understanding the affordances of mobile IT artifacts helps to explain contemporary boundary management practices more completely.

The extension of theories that were originally formulated without concern for or knowledge of IT is a strategy that can be applied widely. As IT has become ubiquitous in the workplace, it is no longer defensible to rely solely upon established theories that exclude IT. Several chapters in this volume illustrate the value of such extensions in different theoretical areas. However, the inclusion of materiality into existing theories invites critical response from scholars vested in traditional versions of theory, which in turn places higher demands on theorizing materiality. For this reason, we advocate great care in extending theories in ways that both preserve the theory's original strengths while strengthening its explanatory power. Our work on mobile IT and work–life boundary management thus seeks to extend the insights from traditional theory while questioning past assumptions about the negative effects of integrating activities across traditional work–life domains. Since the aim of social theory is to explain collective human behavior, theories that overlook salient aspects of human action (i.e., technology use) eventually become less useful than they could be.

Conclusion

We have addressed the issue of theorizing material IT artifacts in IT impact studies. Our aim is to guide the development of theoretical explanations in which IT's materiality plays a central role in explaining the impacts of IT on organizations. Theorizing IT artifacts in material terms should help to redress the neglect of material characteristics in IT impact studies. Our historical review of the treatment of technology in organization theory demonstrates how easily, and perhaps innocently, material aspects of organizations can disappear into the backwaters of theory development. Oddly, material properties of work settings have always been salient to those who work in them. Computing hardware and software are ubiquitous, yet so often their materiality is ignored by researchers who use more abstract concepts and convenient, self-report measures. Our aim is to restore materiality to studies of IT impact by guiding theory development. Our analysis of affordances emphasizes the array of issues that should be confronted when incorporating concepts borrowed from other disciplines to address sociomateriality. Clearly, scholars advocating materiality of IT as a research focus need to define material concepts and their relationships to social actors carefully.

The two strategies for extending theory are closely related. In the case of existing theories that nominally address IT artifacts, contributions may be relatively minor extensions or redefinitions of existing concepts. The example of AST, as revised by Markus and Silver (2008), shows how older theories can benefit from being updated rather than discarded. D'Adderio's and our own extensions of organizational routines theory to include material IT artifacts as part of generative systems are also beneficial. The second strategy of revising existing theories that ignore IT and its material properties may be more challenging. Often, established theories have received the bulk of their testing and refinement from scholars with no special interest in IT impacts. Our contribution to work–life balance theory tries to preserve the primary insights from existing theory while showing the relevance of mobile IT to boundary management strategies. Hopefully, these kinds of extensions can become part of the discourse in other established theory areas that are also affected by material IT artifacts.

Acknowledgments

We thank Karlene Cousins for her contributions to the theorizing of affordances of mobile technologies. We also appreciate the many suggestions from members of the Northwestern Workshop, especially Anne-Laure Fayard, Jennifer Whyte, Luciana D'Adderio, Wanda Orlikowski, Susan V. Scott, Neil Pollock, and Paul Leonardi.

References

Anderson, C. (2012). Information systems affordances: The materiality of information technology. Working Paper.

Ashforth, B., Kreiner, G., and Fugate, M. (2000). All in a day's work: Boundaries and micro role transitions. *Academy of Management Review*, 25, 472–91.

Boswell, W. and Olson-Buchanan, J. (2007). The use of communication technologies after hours: The role of work attitudes and work–family conflict. *Journal of Management*, 33, 592–610.

Burns, T. and Stalker, G. (1961). *The management of innovation*. London: Tavistock.

Cameron, A. F. and Webster, J. (2005). Unintended consequences of emerging communication technologies: Instant messaging in the workplace. *Computers in Human Behavior*, 21, 85–103.

Chemero, A. (2003). An outline of a theory of affordances. *Ecological Psychology*, 15, 181–95.

Clarke, S. (2000). Work/family border theory: A new theory of work/family balance. *Human Relations*, 53, 747–70.

Cousins, K. C. and Robey, D. (2005). Human agency in a wireless world: Patterns of technology use in nomadic computing environments. *Information and Organization*, 15, 151–80.

——— (2012). Managing work-life boundaries with mobile technologies: An interpretive study of mobile work practices. Working Paper.

D'Adderio, L. (2008). The performativity of routines: Theorising the influence of artefacts and distributed agencies on routines dynamics. *Research Policy*, 37, 769–89.

—— (2011). Artifacts at the centre of routines: Performing the material turn in routines theory. *Journal of Institutional Economics*, 7(2), 197–230.

DeSanctis, G. and Poole, M. S. (1994). Capturing the complexity in advanced technology use: Adaptive structuration theory. *Organization Science*, 5, 121–47.

Faulkner, P. and Runde, J. (2010). The social, the material, and the ontology of non-material technological objects. Working Paper.

Fayard, A. and Weeks, J. (2007). Photocopiers and water-coolers: The affordances of informal interaction. *Organization Studies*, 28, 605–34.

Feldman, M. and Pentland, B. (2003). Reconceptualizing organizational routines as a source of flexibility and change. *Administrative Science Quarterly*, 48, 94–118.

Galbraith, J. (1977). *Organization design*. Reading, MA: Addison-Wesley.

Gibson, J. (1977). The theory of affordances. In R. E. S. J. Bransford (ed.), *Perceiving, acting, and knowing: Toward an ecological psychology* (pp. 67–82). Hillsdale, NJ: Lawrence Erlbaum.

—— (1979). *The ecological approach to visual perception*. Boston: Houghton Mifflin.

Giddens, A. (1984). *The constitution of society*. Berkeley: University of California Press.

Golden, A. and Geisler, G. (2007). Work–life boundary management and the personal digital assistant. *Human Relations*, 60, 519–51.

Greenhaus, J. and Powell, G. (2006). When work and family are allies: A theory of work-family enrichment. *Academy of Management Review*, 31, 72–92.

Hatch, M. J. (2006). *Organization theory* (2nd edition). Oxford: Oxford University Press.

Heft, H. (2001). *Ecological psychology in context: James Gibson, Roger Barker, and the legacy of William James's radical empiricism*. Mahwah, NJ: Lawrence Erlbaum.

Hislop, D. and Axtel, C. (2011). Mobile phones during work and non-work time: A case study of mobile, non-managerial workers. *Information and Organization*, 21, 41–56.

Hutchby, I. (2001). Technologies, texts and affordances. *Sociology*, 35, 441–56.

Introna. L. and Hayes, N. (2011). On sociomaterial imbrications: What plagiarism detection systems reveal and why it matters. *Information and Organization*, 21, 107–22.

Ito, M., Okabe, D., and Anderson, K. (2009). Portable objects in three global cities: The personalization of urban places. In R. Ling and S. Campbell (eds.), *The reconstruction of time and space: Mobile communication practices*. New York: Transaction Publishers.

Jarvenpaa, S. L., Lang, K. R., and Tuunainen, V. K. (2005). Managing the paradoxes of mobile technology. *Information Systems Management*, 22, 7–23.

Johns, G. (1992). *Organizational behavior: Understanding life at work*. New York: Harper Collins.

Jones, M. and Karsten, H. (2008). Giddens's structuration theory and information systems research. *MIS Quarterly*, 32, 127–57.

Jonsson, K., Holmström, J., and Lyytinen, K. (2009). Turn to the material: Remote diagnostics systems and new forms of boundary-spanning. *Information and Organization*, 19, 233–52.

Kirlik, A. (1995). Requirements for psychological models to support design: Toward ecological task analysis. In J. Flach, P. Hancock, J. Caird, and K. J. Vicente (eds.),

Global perspectives on the ecology of human–machine systems. Hillsdale, NJ: Lawrence Erlbaum Associates, Inc.

Kreiner, G., Hollensbe, E., and Sheep, M. (2009). Balancing borders and bridges: Negotiating the work–home interface via boundary work tactics. *Academy of Management Journal*, 52, 704–30.

Lawrence, P. and Lorsch, J. W. (1967). *Organization and environment*. Boston: Harvard Business School.

Leonardi, P. M. (2011). When flexible routines meet flexible technologies: Affordance, constraint, and the imbrication of human and material agencies. *MIS Quarterly*, 35, 147–67.

——Barley, S. (2008). Materiality and change: Challenges to building better theory about technology and organizing. *Information and Organization*, 18, 159–76.

Markus, M. L. and Silver, M. (2008). A foundation for the study of IT effects: A new look at DeSanctis and Poole's concepts of structural features and spirit. *Journal of the Association of Information Systems*, 9, 609–32.

Michaels, C. (2000). Information, perception, and action: What should ecological psychologists learn from Milner and Goodale (1995)? *Ecological Psychology*, 12, 241–58.

Middleton, C. and Cukier, W. (2006). Is mobile email functional or dysfunctional? Two perspectives on mobile email usage. *European Journal of Information Systems*, 15, 252–60.

Norman, D. (1988). *The psychology of everyday things*. New York: Basic Books.

——(2002). Affordances and design. http://www.jnd.org/dn.mss/affordances-and-design.html

Orlikowski, W. (2007). Sociomaterial practices: Exploring technology at work. *Organization Studies*, 28, 1435–48.

——Iacono, S. (2001). Research commentary: Desperately seeking the "IT" in IT research: A call to theorizing the IT artifact. *Information Systems Research*, 12, 121–34.

Pentland, B. and Feldman, M. (2005). Organizational routines as a unit of analysis. *Industrial & Corporate Change*, 14, 793–815.

————(2008). Designing routines: On the folly of designing artifacts while hoping for patterns of action. *Information and Organization*, 18, 235–50.

Perlow, L. (1998). Boundary control: The social ordering of work and family time in a high-tech corporation. *Administrative Science Quarterly*, 43, 328–57.

Perrow, C. (1967). A framework for the comparative analysis of organizations. *American Sociological Review*, 32, 194–208.

——(1981). Normal accident at three mile island. *Society*, 18, 17–26.

——(1983). The organizational context of human factors engineering. *Administrative Science Quarterly*, 28, 521–41.

Pickering, A. (1995). *The mangle of practice: Time, agency, and science*. Chicago: University of Chicago Press.

Pollock, N., Williams, R., D'Adderio, L., and Grimm, C. (2009). Post local forms of repair: The (extended) situation of virtualized technical support. *Information and Organization*, 19, 253–76.

Raymond, B. (2012). How IT artifacts influence the design and performance of organizational routines: Extending organizational routines theory. Working Paper.

Reed, E. (1996). *Encountering the world: Toward an ecological psychology*. New York: Oxford University Press.

Robey, D. and Sales, C. (1994). *Designing organizations* (4th edition). Burr Ridge, IL: Irwin.

Scheepers, R., Scheepers, H., and Ngwenyama, O. (2006). Contextual influences on user satisfaction with mobile computing: Findings from two healthcare organizations. *European Journal of Information Systems*, 15, 277–84.

Stinchcombe, A. (1968). *Constructing social theories*. New York: Harcourt Brace & World.

Stoffregen, T. (2003). Affordances as properties of the animal-environment system. *Ecological Psychology*, 15, 115–34.

Turvey, M. (1992). Affordances and prospective control: An outline of the ontology. *Ecological Psychology*, 4, 173–87.

Volkoff, O., Strong, D., and Elmes, M. (2007). Technological embeddedness and organizational change. *Organization Science*, 18, 832–48.

Woodward, J. (1965). *Industrial organization: Theory and practice*. Oxford: Oxford University Press.

12

The Materiality of Technology: An Affordance Perspective

Samer Faraj and Bijan Azad

Introduction

In the last few years, there has been an increasing interest in redefining and reinventing the theoretical linkage between social organizing and technology. The popularity of sessions that attempt to bridge the social and the material at academic conferences such as the Academy of Management (AOM) and the International Conference on Information Systems (ICIS) indicates the increased interest in this topic. There are renewed calls to integrate the technical and the social to better understand related phenomena of information technology (IT)-driven social change (Barrett et al., 2006), IT-enabled changes in organizational form/function (Zammuto et al., 2007), IT's effects in organizations (Markus and Silver, 2008), the role of materiality in organizational change (Leonardi and Barley, 2008), and the binding of the social and the material (Orlikowski and Scott, 2008). These calls share three distinctive characteristics. First, they focus on the materiality aspect of technology in an attempt to highlight the objective, realist, and nonmentalist nature of technology. Second, they invariably opine that the current ways of representing and studying technology–organizational change, via lenses such as structuration, practices, or emergent views, have not offered sufficient theoretical depth or empirical richness regarding the technology appropriation process. Third, they identify technology *affordances* as a promising means of analyzing and researching the technology appropriation process—especially as a way to rectify the shortcomings of the earlier approaches.

The difficulty in developing integrated or holistic formulations that overcome the social–material divide links back to deep-rooted philosophical divides that go back a century or more. These divides include the subject–object dualism that has encouraged psychological investigations of technology as a reflection of the reality "out there" by the cognizing subject. Another is the structure–agency split about the sources and constraints of human action which has ultimately been regarded as limiting our understanding of practices around technology (Orlikowski, 2000). Another well-known divide that many researchers have investigated is the determinism versus voluntarism dichotomy where technology is either a driving force that reshapes society and organizations, or, alternatively, is a force whose development and utilization is under the control of managers and social groups (Winner, 1978; Smith and Marx, 1994).

Traditionally, these fundamental dichotomies in viewing the world have been addressed through duality-of-technology, technology structuration, and technology-in-practice framings, all of which can be seen as attempts to develop frames of analysis that recognize the complex interplay between social structure and the agency of the actors and recognize that technology is both represented in someone's mind but also existing out in the world. However, recent assessments of these approaches indicate that we have veered from the pole of determinism to one that emphasizes voluntarism (Leonardi and Barley, 2010). Further, in spite of the increased presence of technology in organizational life, the vast majority of organizational studies, including those focused on practices, do not engage the materiality of organizational life and its obvious reliance on the technology infrastructure (Zammuto et al., 2007; Orlikowski and Scott, 2008). In spite of this growing awareness of the need to better theorize technology in general and to deal with its material aspect, organization scholars are still not clear on how to represent technology–organization change so that theoretically it is neither determinist nor voluntarist, nor how to translate this theoretical stance to a practical approach to research that allows a deeper study of technology–organization change.

In this chapter, we suggest that researchers have too often accepted vendor-based categories of technology, reduced them to features, and avoided facing differences in the same technology over time. We propose the technology affordance lens as a way to bring materiality back in and as a theoretical lens for studying the sociomaterial nexus. *Technology affordances* are action possibilities and opportunities that emerge from actors engaging with a focal technology. Affordances are rooted in a relational ontology which gives equal play to the material as well as the social. Several researchers have highlighted the importance of developing an alternative conceptual apparatus for the study of the technology appropriation process and have advocated the affordance perspective as a promising approach to overcome the subject–object and agency–structure dichotomies that have stifled much of the research at the intersection of technology and organizations (Zammuto et al., 2007; Leonardi and Barley, 2008; Markus and Silver, 2008).

Our development of technology affordances recognizes the challenge faced when introducing new concepts that explicitly address the material aspects of technology (see Robey et al., Chapter 11). As several authors in this volume remind us, technology's materiality is more than the physical and needs to reflect the form, function, symbol, and its imbrication with the social (see, e.g., the Chapters 2, 4, 6, 9, 14, and 15). We offer the technology affordance concept as one of the ways to take materiality and sociomateriality seriously when studying work and social interactions involving technology.

In the rest of this chapter we proceed as follows: First, we summarize deep problems in the dominant organizational conceptualizations of technology as commensurate bundles of features and products. Our analysis suggests that in the push to develop/theorize the construct of technology, extant research has cast aside crucial issues about material aspects of technology in use. Next we explore the affordance perspective as a promising avenue to bring materiality back into organizational studies of technology and evaluate how it has been used in the organizational literature. Finally, we develop conjectures of how technology researchers can elaborate the affordance lens to build a technology-in-use perspective that more effectively integrates materiality with human action.

Viewing Technology as a Bundles of Features and Product Classes

In the eyes of most technologists and computer scientists, information technologies consist of bundles of features,[1] analogous to the features and functions that are used to characterize the objects of the physical world in a typical Cartesian perspective. Cognitivists and psychologists still view the world as an external reality to be characterized and recognized by cognitive processes. For example, a traditional cognitivist views an encounter with a chair based on information processing around features such as color, size, apparent solidity, and where the new chair is compared and classified (e.g., chair has no back) based on one's memory traces of previously encountered chairs. In a similar vein, most analysts view technology as an object in the world defined by the functionality it provides (e.g., Arthur, 2009). It is, above all, an object separate from the subject and thus needs to be distinguished and described.

Among IT researchers, there is wide acceptance of nomenclature developed by commercial vendors or the specialized press that categorizes hardware and software into product classes and defines lists of criteria and specific

[1] In this chapter, we use the terms "feature" and "functionality" interchangeably.

functionalities to evaluate these products and compare vendor offerings. For example, there is wide acceptance of IT products or classes of computer systems, for example, word processing, spreadsheets, browsers, accounting systems, human resource management systems, customer relationship management (CRM), enterprise resource planning (ERP), etc., over the technologies-in-use. Unfortunately, these product classes have become accepted as core technology frames and act as the primary objects of analysis for IT researchers. This is due to three distinctive historical and commercial trends. First, the technology categories constitute convenient labels and a shorthand description to refer to a set of recognized features. Second, they represent market categories that the IT industry employs and legitimates in its marketing efforts and interaction with organizations. Finally, each product class enshrines a shared worldview among its adherents in regards to what the feature/product class is supposed to do and is good for. As a result, the product classes (market categories) and categories are generally accepted by researchers as pointing to important differentiation in technology. However, it is our contention that this nominal differentiation is problematic for researchers who are concerned about technology's materiality—we refer to this as the *product conflation* issue.

Indeed, many IT researchers remain unwittingly accepting of this product conflation issue because they accept the technology-as-designed perspective where their research role is to study the acceptance and diffusion of a said technology (e.g., Azad et al., 2010)—where partial use and different use are seen as surprising and especially worthy of note. In other words, when organizational actors use a technology, let us say email, in their daily work, researchers tend to presume that users have computing needs that are being met by the product category. Little attention is paid regarding what the technology means to the user or how it fits with their activities. Thus, users are assumed to "browse," "word process," or utilize the "ERP system" without much attention being given to the actual activities that affect the fit between the technology-in-use and the product category. The presumption is that the technology provides a well-defined set of features that meet a user's individual or organizational need. Thus, any adaptation, modification, unplanned change, or workaround that are introduced are seen as surprising and worthy of deeper investigation. A more phenomenological viewpoint, one that is steeped in the social constructivist perspective, would take this noncanonical use of technology as the starting point rather than as an unexpected finding (Zammuto et al., 2007; Orlikowski and Scott, 2008; Leonardi and Barley, 2010).

Unfortunately, features and product classes have become accepted and the primary objects of analysis for researchers because of three distinctive historical and commercial trends. First, they constitute convenient labels to group products held together by similarity of features. Second, they represent market categories that the IT industry employs and legitimates in its marketing efforts. Finally, each feature/product class enshrines a shared world view among its adherents in regards to what the feature/product class is supposed to do and is good for. As a result, the features/product classes (market categories) and IT research highly coincident and similar.

Why has the vast majority of IT researchers adopted with little questioning this feature bundles/product class industry perspective on technology? Is this because the perspective provides a compelling formulation for social science researchers? Or is it because feature and product classes have become a taken-for-granted part of social landscape of the computer systems market and have imposed themselves as the only legitimate categorical lenses through which we observe phenomena? Whatever the answer, it is important to realize that applying this traditional cognitivist lens leaves researchers unable to analytically distinguish between taken-for-granted features (and product classes) and the actual categories-in-use that emerge when actual users encounter IT in a situated work context. This grip of the so-called feature-centric approach to research in information systems and organizations has been noted and criticized, so far with little effect (Orlikowski and Iacono, 2001; Orlikowski and Scott, 2008; Leonardi and Barley, 2010). There are at least three problems with this category and feature-centric approach to IT research. One problem is the conflation of product categories and the technology in use; another is the decomposition to features problem; and, finally, there is the incommensurability over time.

Conflation of Features and Product Category and Technology-in-Use

The conflation of product category and technology-in-use refers to the often ad hoc superimposition of vendor-defined categories as representing the technology being appropriated. Taking email as an example, we can start cataloging the variety of user conceptions of what is email to gauge how prevalent the above assumptions are in practice. One morning, the spouse of one of the authors points to her laptop saying that "the email is not working," and that is why she cannot send a needed report to her colleague. On closer inspection, it turns out that the file could not be uploaded and sent because the router (wireless access point) was disconnected. On a separate occasion, in checking his email from home, a teenager tells his parent "there is something wrong with email." Upon inspection, the problem turned out to be related to exceeding the ISP's monthly download quota. In a third instance, this time at the office, one gets a call from a colleague to report that he cannot download an MS-Word attachment "because his email is acting up." A closer investigation indicates nothing wrong with the email functionality but that the colleague is still using MS-Word 2003 without the MS-Word 2007 upgrade module so when he opens the later version files, it shows up as "garbage."

An additional scenario emerges through an email communication that is sent by the president of the university announcing a new university logo as a part of a rebranding campaign. A debate ensues. A professor attempts to send an email objecting to the new logo to other faculty members. The critical email does not get distributed. The author complains that his email was "blocked" and "censored" by the Vice-President of Communications. An email storm involving faculty and administration follows regarding the appropriateness of blockage/censoring behavior and whether it is allowed under university by-laws. Finally, the provost sends out an email explaining that the university has two sets of email address lists: moderated and nonmoderated. It turns out that the professor had mistakenly responded to the president's announcement that had gone out on the moderated "administration" list. He should have instead used the "all-faculty" list which is nonmoderated.

All of these episodes have two things in common. First, the notion of email as a specific feature/product is largely not in alignment with email-in-use. In all four examples of email not functioning, the problem did not occur with the core email technology (e.g., the SMTP protocol) but from related technologies: the router being disconnected, the cap on DSL data transfer, and the incompatibility between Word versions. However, from a user standpoint, the email technology, broadly defined, did not work: the ISP and physical connectivity problems had become entangled in the mind of the user with his inability to perform his emailing activity. This reflects the more accurate view that email is seen as a technological blackbox rather than a clearly delimited bundle of functionalities (Latour, 2005; Zammuto et al., 2007). Second, the users' employment of email has less to do with email's specific feature/product categories, and more to do with their action strategies in regard to email as a capabilities bundle. As indicated by the university email example, a breakdown in transmittal may have less to do with the core technology and more to do with the social and organizational context it is entangled with. Indeed, the actual breach in delivery was in the use of the moderated address list as opposed to the unmoderated one. The lists included the same respondents but, in the social context of institutional life, they represented completely different communication tools: one "official" for use by the administration for formal communication; and the other identical but unofficial list and thus devoid of the same social, legal, and institutional symbolism.

Feature-Centricity and the Infinite Regress Problem

Let us consider a word processing package as an illustrative example. When researchers refer to word processors, they have in mind a distinctive grouping of capabilities that can serve many different purposes. On one hand, they can mean a software package, along with the personal computer and the printer, which is probably a secretary's perspective view of a features bundle. He/she is interested in doing office management, and for that purpose the word processor is basically viewed as a bundle of office management functions and features. This would include typing memoranda and storing them in an orderly manner. On the other hand, word processors can mean an application software that provides text editing and manipulation capabilities for authors of academic papers. These often can integrate functions that allow bibliographic importing of references in an automated manner. For academic authors, the orderly manipulation of references and their compilation are of utmost importance. This is largely derived from the importance of impact factors and citations to academic careers. Alternatively, if a user is largely interested in doing brochure and newsletter design, then the primary features/functions of interests will be those that allow for text placement, layout, and font-kerning (a process used to produce smoother lettering). The brochure designer's features of interest lie in his professional concern with making sure that one can fit pieces of text within the allotted space of a brochure/newsletter and also the aesthetic font qualities of a brochure.

Effectively, all of the above users are using a word processor but each has a distinctive emphasis and a group of focal features that he would accent in his process of software use. This process of accenting technology-in-practice can be significantly varied based on the use of the same software in an "infinite" number of ways and perceiving the software as providing a distinctive bundle of capabilities. If this accurately represents the manner in which users appropriate and use technologies, then it is not clear at all that a "word processor" can be represented as a core feature set universally. Should it be a software application as a whole including peripherals such as a printer and a PC (e.g., based on the secretary's role)? Should it be the bundle associated with the academic writer's perceived needed capabilities of a word processor, that is, bibliographic integration (e.g., based on the professor's role)? Or should it be the software feature collection that is associated with brochure design, that is, text layout and placement as well as font-kerning (e.g., based on the brochure designer's role)? In other words, based on specific roles, the focal list of features and functions will probably change. This "infinitely" malleable reference to what constitutes a software's core features as an accurate representation of its design and material properties is a thorny research problem that has been recognized by researchers.

Indeed, some researchers have long recognized the shortcomings of the feature-based research approach: the subjective and highly variable nature of what constitutes a feature. One problem with features is that they may not focus attention on what is truly important in the technology-in-use from the actor's perspective. For example, according to Gutek et al. (1984) (cited in DeSanctis and Poole, 1994: 126): "Most systems are really 'sets of loosely bundled capabilities and can be implemented in many different ways'." Furthermore, DeSanctis and Poole (1994: 333) postulate that computer "systems vary so much in the presentation of their features that information based on features alone makes it virtually impossible to compare systems or versions of systems." The potential variation in packaging, development, and user interface makes it difficult for information system (IS) researchers to decide the features and the level of details that are to be investigated. According to DeSanctis and Poole, the problem is the "repeating decomposition problem: there are features within features... So how far must the analysis go to bring consistent, meaningful results?" (DeSanctis and Poole, 1994: 124). Griffith (1999) has concurred with this problematization: "The concept of a feature remains somewhat elusive. It is possible to examine some technology features at increasingly smaller (or larger) units of analysis. For example, the personal digital assistant may take input from a stylus, the stylus may be plastic or metal, the plastic may be hard or soft, ad infinitum." Ironically, despite their criticism of the repeated decomposition problem, these perceptive authors offer no alternative to the use of features as an essential element in their theoretical formulations. We refer to this as the "infinite regress" perspective, consistent with Markus and Silver (2008).

Erasure of Qualitative Differences of Technology over Time

Let us go back to the "word processor" example, this time to illustrate another issue, namely, the presumed "constancy" of technology over time. This can undermine the theorization of technology's materiality or context within extant research. To highlight what we mean by the neglect of change over time, in Table 12.1 we provide a summary of four different eras in "word processing" technology: Wang-style era, WordStar- and WordPerfect-style era, MS-Word 1995 era, and MS-Word 2007 era. All these systems are often captured via the label "word processor." Nevertheless, even the highly superficial presentation in Table 12.1 points to significant differences whereby potential variation in fundamental design and material properties can be glossed over via the term "word processor."

First, the Wang-style machine represents the quintessential first-generation word processor which came on the scene in the late 1970s and dominated the landscape for some time. These machines used the latest microprocessor technology at the time to design and deliver computing power for handling a significant office function, namely, electronic handling of text and documents. In other words, although text editors had been employed for programming purposes, that technology had proved too cumbersome to be employed

Table 12.1 How traditional technology description neglects the time dimension

Prevalent "word processors"	Screen look and feel	Key design/material properties
Wang, etc. (ca. 1980)		Character-based screen, dedicated hardware for screen, computer and printing
WordStar and WordPerfect (ca. 1980–92)		Character-based screen, with formatting based on standard CTRL-key keyboard menus, can work on most PCs, works with printers whose drivers are available by the corresponding vendors
MS-Word Windows 1995 (ca. 1995–2003)		Graphic user interface screen, a la MS Windows, with drop-down menus, what-you-see-is-what-you-get formatting using mouse-based commands, and use of "styles"
MS-Word Windows 2007 (ca. 2007+)		Inaugurates the dominance of "ribbons" as the fundamental text handing process in MS-Word, thorough change in menu structure (most users continue to install the pre-2007 menu structure as a macro not finding functions in the new configuration!)

for office document handling. A key characteristic of the first-generation word processors was their dedicated and proprietary nature. A system typically included, as in the graphics in the Table 12.1, an "integrated terminal" of keyboard/screen/CPU in one artifact. To produce hard-copy documents, the word processor could only be connected to a printing machine manufactured by the same maker and worked only with it.

Second, the WordStar and WordPerfect applications software (prevalent in the 1980s) represented a significant shift in design philosophy, capabilities, and material configuration. Commensurate with the rise of the PC as the primary platform of end-user computing, the technology of handling documents for office use (i.e., other than programming) became available on a wide scale. It included similar capabilities to the Wang-style machine but relied on the standard keyboard functions of the PC. It also marked a shift to the concept of limited menus that were available on the screen and could be invoked with a few keystrokes. More importantly, the application software worked on any "IBM-compatible" PC. Finally, the hard-copy output could be produced on any manufacturer's printer as long as that maker provided a printer driver software for the application program.

Third, the MS-Word 1995 version offered a more mature graphical user interface (GUI) by Microsoft to the market. This application software for "word processing" included sophisticated use of what-you-see-is-what-you-get (WYSIWYG) interface and menus. The GUI-based manipulation of characters, words, sentences, and paragraphs was made much easier via the use of a mouse interface. In addition, the use of "styles" was also introduced. This capability visually combined a series of formatting functions "under a single name" which then could be applied (as mouse clicks) to a number of paragraphs or to the whole document. Also, with each release of the software, manufacturers worked with Microsoft to make available printer drivers. This made the process of interfacing with these devices easier and more transparent for the end users. These developments together constituted significant design and material changes to the "word processing" technology over the earlier configurations (e.g., WordStar and WordPerfect).

Finally, the release of MS-Word 2007 inaugurated the much touted "ribbon" approach to menus and functionality handling within this popular application software for text manipulation. In GUI-based application software, a "ribbon" is a grouping of functionalities within an interface where a "set of toolbars are placed on tabs in a tab bar." The "ribbon contains tabs to expose different sets of control elements, eliminating the need for many different icon-based tool bars." As a technology-in-practice, most users who would engage in migrating from prior versions find it difficult to trade off the ribbon capabilities with the ease of recall of prior functions they are used to. Indeed, it is now a stylized fact that the "screen real estate" (i.e., how an aspect of an application software's materiality is manifested in practice to users) of a system compared with the previous versions may only superficially "shield" new fundamental design and material properties from the users.

A key takeaway from reviewing these instances of change in the underlying technology (e.g., Wang-style to WordStar/WordPerfect style, to MS-Word 1995, to MS-Word 2007) is that the label word processor hides the fundamental design and material changes occurring over time. The context of the early dedicated word processors reflected dominant designs of integrated hardware and software within a machine. The steep learning curve and high cost limited the primary users of this technology to offices. The second-generation character-based menus and associated implementation of systems such as WordStar and WordPerfect on a PC meant that the price along with the design and material properties made the technology available for use by large numbers of individuals with access to personal computers. The third era of systems represented by MS-Word represented a significant change from the past since they incorporated GUI and WYSIWYG formatting and easy printing. The contextual changes in the increased computing power of CPUs, the Windows operating system, as well as the support for printer drivers by manufacturers were critical to the rise in popularity of this version of the word processor. Finally, in the era of the MS-Word 2007 systems, it is a telling sign that a giant software manufacturer such as Microsoft would take for granted that its existing users would actually go along with such a massive shift in the "screen real estate" a la ribbons. To do so would simply mean that twelve to fifteen years of investment by users are easily traded in because of the alleged "simplicity" of the new approach. Thus, word processing as technology-in-use was challenged when the vendor chose to alter its design "to make it simpler" by making access to software functions "follow a ribbon based scheme." To a large number of users invested in previous material practices, access to their software's capabilities was deappropriated as technologies-in-use. In that context, the materiality of a change to ribbons from traditional menus can be simply too much to fathom. It is a pity that since the original studies of "word processing" by Gutek et al. (1984), Gerson and Star (1986), Gasser (1986), Rice et al. (1986), little work has been done to investigate the implications of the materiality of changes in technology from one era to the next. The core question that needs to be addressed is whether all these examples of word processors are commensurate given the focus on the technology-in-use. In other words, is word processing on WordStar equivalent to word processing in MS-Word 2007?

To Address Materiality a Fresh Perspective on Context/Technology is Needed

Table 12.2 summarizes the above discussion of how the acontextuality of the extant approaches to technologies-in-practice has been manifested as follows: conflation of product categories over technology-in-use; infinite regress or arbitrary selection of features as a technology-in-use; and neglect of change over time in technology-in-use. It is our considered view that for researchers to be able to address materiality in a more theoretically rigorous manner, elements of context need to be specified in greater precision.

Several researchers have favored broader conceptualizations of the technology–organization nexus as a way to avoid some of the problems discussed in the preceding section. Some have favored a resolutely human agency prioritization of how we study the appropriation of technology because technology-induced change is, in the final analysis, at the discretion of the human agents (Boudreau and Robey, 2005). Recognizing that the human agency perspective leaves out the materiality of the technology, Orlikowski and colleagues have suggested the elimination of the dichotomy between the social and the material and the development of a unified conceptualization that incorporates the social and the material under the label of

Table 12.2 How the materiality of technology is typically neglected

How technology is acontextualized	Materiality aspect	Example
Conflation of product/category with technology-in-use	Superimposition of product and vendor categories over technologies-in-use leads to the blackboxing of key material aspects	The "product" category is almost always superimposed on the technology-in-use as synonymous. However, their correspondence is questionable at best. For example, the email software (e.g., Outlook or Thunderbird) is sold/licensed on the market under a single product category according to commercial conventions. But, from a user perspective, it may be better described as a broad communication and networking *technology-in-use*, which provides possibilities to inform, misinform, advertise, campaign, and collaborate
Feature-centricity and the infinite regress problem	Referring to an arbitrary selection of technology features as representing a focal technology-in-use	"Word processing" as a bundle of capabilities can refer to: a software package and an accompanying printer that render it useful to a secretary for office management (e.g., typing memos and reports); an application software that provides the ability to integrate bibliographic data; a collection of features which provide for text layout/placement/font kerning. Equating any arbitrary and selective technology functionality set and technology-in-use is not likely to hold
Erasure of differences among technologies-in-use over time	Neglect of material differences in technologies-in-use at different times by referring to them via the same generic technology label	"Word processing" at different points in time can be qualitatively different though such variations recede into the background and the same label is used to refer to very different bundles of capabilities and material properties. The significant qualitative differences between (*a*) the Wang machine in 1979, (*b*) WordStar and WordPerfect in 1985, (*c*) the GUI MS-Word in 1997, and (*d*) MS-Word in 2007 are examples of such qualitative differences which are hidden under the rubric of "word processor"

sociomateriality (Orlikowski, 2007, 2010; Orlikowski and Scott, 2008). Similarly, through the notion of imbrication, Leonardi (2011) has attempted to show that each instance of technology is interwoven with organizing processes so as to render a distinction between the technological and the social difficult to untangle theoretically. While most researchers agree on the need to take materiality into account, differences exist on how best to do so. A narrow point of debate focuses on whether the social and the material need to be disentangled analytically and unraveled empirically in order identify their constituent contribution, or, to the contrary, analyzed together via a relational ontology focused on constitutive entanglement and a focus on technology enactment in practice (cf. Leonardi and Barley, 2010; Orlikowski, 2010).

Thus, alternative perspectives that do not take feature/product categories for granted and that balance between the material and the social are needed. Affordances are one way by which the relationality of technology in practice can be represented. In the rest of this chapter, we review the notion of affordances, offer a modified conceptualization, and suggest how affordances can provide promising grounds for a relational perspective on the social and the material.

Affordances as a Way to Bring in Sociomateriality

The notion of affordances is an influential ecological psychology theory originally conceptualized by Gibson (1979). It has been deepened since by a number of ecological psychologists (Turvey, 1992; Sanders, 1997; Chemero, 2003; Jones, 2003) and sociologists (e.g., Hutchby, 2001). Gibson (1979: 129–30) states his basic relational proposition as follows: "An affordance is neither an objective property nor a subjective property; or it is both if you like... It is equally a fact of the environment [artifact] and a fact of behavior [action]. It is both physical and psychical [social]... An affordance points both ways, to the environment [artifact] and to the observer."

Gibson (1979) developed affordance as a bridge concept to escape the confines of cognitivism when explaining perception and how species orient themselves to the objects in their environment in terms of possibilities for action. For example, the same object (e.g., a tree) affords different possibilities for action to a monkey (as a refuge from predators) compared to a giraffe (which sees it primarily as a source of food). Gibson developed the term to emphasize a reciprocal and immediate relation between the environment and an organism compared to previous theories of perception that emphasized an organism-centric processing model of cognition. This emphasis on the actionable properties of an object, seen as an invariant bundle of features, has been accepted in ecological psychology and is now the focus of an intense program of research to understand how direct perception of actionable objects is activated (You and Chen, 2007).

Inspired by Gibson, ecological psychology researchers have invested significant energies investigating affordances vis-à-vis physical–structural properties of visual perception. For example, Warren (1984) conducted a series of pioneering stair-climbing experiments seeking to understand affordance-based perceptions via body-scaled metrics. Specifically, he showed that actors perceive their environment in terms of intrinsic or body-scaled metrics, not in absolute or global dimensions. That is, judgment of whether one can climb a stair step is not determined by the height of the stair step, but by its ratio to one's leg-length. In general, this stream of Gibson-inspired research has been focused on physical–structural affordances. However, there have been numerous calls in the ecological psychology literature for affordance researchers to pay greater attention to more clearly account for the mutuality relations and to better include the social dimensions (see E. J. Gibson, 2000; Heft, 2003; Jones, 2003).

One point of contention about affordances is their location. Are affordances properties of the environment (or the object) or are they specific to the species/actor? Most early research took the view that affordances are species-relative properties of the environment that are perceived to afford action (Warren, 1984; Michaels, 2000). Turvey (1992) builds on the Gibsonian early work to offer a definition of affordance as a property of the environment framing it as a dispositional property—in the sense of property of a thing that is latent or that manifests itself under specific circumstances. In turn, he concludes that there is a need for a complement to affordance from the actor's side, a property he calls effectivity. The key aspect of such definitions is to frame affordances as dispositions or possibilities that are not yet actualized and thus clarify the opportunities for action. Such a definition emphasizes a components view of affordances, locates the affordance squarely in the material side of the actor/environment divide, and necessarily entails the presence for a matching concept from the actor's side. Both properties are needed for the affordance to be actualized.

Several researchers prefer to view affordances more relationally. Stoffregen (2003: 124) defines affordances as emergent "properties of the actor–environment system that determine what can be done." Thus, affordances are opportunities for action rather than properties of the environment per the earlier definition. Chemero (2003, 2009) affordances in relation to both the actor and the environment. Taking the example of stair-climbing, he suggests that it may not be useful to talk about a specific height or even a "golden" ratio of the stair height to the climber's leg length as the climbing affordance for humans. Rather, it is important to realize that the ratio will vary depending upon the age of the participant (older adults are less flexible). Thus, one has to get more specific in terms of affordances: moving from a language of species–environment fit to one of a relation between particular environment feature and the actor's specific abilities or circumstances. This means that an object in the environment will offer different possibilities of action depending upon the actor's abilities, something that can be quite differentiated among humans based on age, training, cognition, motivation, and expertise—a more rigorous, precise, and nuanced view context for human action.

Appropriation of the Affordance Concept in Technology Studies

Norman is primarily responsible for popularizing the affordance angle to technology design. In a series of well-received books, Norman uses the term affordance to nudge technology designers toward intuitive design (Norman, 1988, 2007, 2011). Whereas Gibson had emphasized that affordances were invariant characteristics of the relationship between an object and an actor, Norman reformulated the term to emphasize an implicit communication of meaning. Perceived affordances refer to design choices that serve as a more direct communication between the designer and the user. When users interact with a designed object, they perceive affordances that imply or invite a certain way of interacting with the object. In contrast to the Gibsonian view that affordances exist irrespective of whether they are immediately perceived, Norman's formulation (2007: 68) emphasizes that affordances need to be perceived to be useful: "if you didn't know that an affordance existed . . . then it was worthless, at least in the moment." Thus, the use and interaction possibility of objects is based on the user's mental representation of what the object performs. Each user would have a different affordance with an object depending on specific conditions and experience. The idea of perceived affordances is a step away from Gibson's emphasis that perception is not based on information processing but is direct and action oriented.

Furthermore, the "Norman perspective" is deeply ingrained in the affordance-is-in-the-object view, in spite of using the qualifier "perceived" regarding affordances. Designers typically work with a typical user or a class of users in mind. They often do not see a wide variability among users and emphasize the goal of "good" as opposed to "bad" design. A good technology design is one that makes the majority of users effortlessly aware of the usability of the device in question. Both designers and users have parts to play: designers must tame and conceal the complexity of the technology based on good design principles. In turn, the users must accept the perspective of the designer and learn to engage with it (Norman, 2011). This perspective underlies the success of Apple Inc. designed computers and devices where the products seem to function seamlessly and intuitively as long as users follow the preestablished use pattern. We suggest that the affordance perspective is less about intuitive design and more about recognizing the unexpected, situated, and emergent actions that actors may want to engage in with their devices.

In recent years, several IT researchers interested in new theorizing of the technology-organizing nexus have become interested in affordances. Unfortunately, much of the work relies on a cursory use of the concept of affordances and a reliance on the original Gibsonian definition now over three decades old. In a review piece on the status of IT research in *Organization Science*, Zammuto et al. (2007: 752) offered that: "An affordance perspective recognizes how the materiality of an object favors, shapes, or invites, and at the same time constrains, a set of specific uses." The authors then go on to offer affordances for organizing as a bridging concept emerging from the intertwining of IT and organizational features. Five major technological organizing possibilities are then described using language that considers simultaneously the features of both IT and organizations. Yet, the language is still close to the idea that technology (an external force for most organizations) affords new ways of organizing, something akin to the concept of affordance as a feature of the environment in ecological psychology.

Other researchers adopting an affordance angle face the difficulty of describing the relationship between technology and organizing. In an attempt to extend the theory of adaptive structuration, Markus and Silver (2008) define affordances as the relation between a technology object and a group of users. They then offer a definition of technology affordance that emphasizes the environment, in this case the material aspect of the technology: "functional affordances are defined as the possibilities for goal-oriented action afforded to specified user groups by technical objects" (p. 622). They suggest that functional affordances clarify the goal-oriented actions provided by the technology and simultaneously match the concept with the concept of symbolic expression to refer to the communicative aspect of the technology. Through these twin concepts, the interaction of users with specific technological objects is explained.

In an innovative field study of how routines and technology affect each other in automotive design, Leonardi (2011) uses affordances and constraints to illustrate the imbrication between human and material agencies. He uses the twin notions of affordances and constraints to represent the twin dimensions of human and material agencies and suggests the image of imbrication to recognize the weave between the two intertwined yet distinctive agencies. Developing a common representation and a unified treatment of material and social agencies has long been called for but has been difficult to achieve in practice (Pickering, 1995; Latour, 2005; Barad, 2007). In spite of adopting a relational perspective on affordance, Leonardi ends up having to analytically separate the material and human agencies to explain how they get constituted and evolve as routines and technologies.

Taken together, all three studies show the difficulty of using affordances in a truly relational manner. One source of the difficulty is often the external nature of the technology, that is, it is generally developed outside the focal organization. Then, it is brought into a context, giving the technology salience as an external object and thus leading to language such as "affordances for organizing" (Zammuto et al., 2007), "functional affordances" (Markus and Silver, 2008), or "affordances and constraints on action" (Leonardi, 2011). Another challenge is how to conceive of the materiality of the technology. Is there more to the affordance lens than a different and possibly more sophisticated vocabulary but that nevertheless falls back on the traditional view of "technology-out-there?" In other words, can the concept of affordance offer a qualitatively different representation than a product/features in use? We believe so, but the conceptual difficulty that has led Latour, Orlikowski, and others to call for a unified treatment of the technology–social imbroglio remains.

Expanding the Concept of Technology Affordance

One way to expand the concept of affordances to be of greater use in exploring the sociomaterial nexus is to clarify and extend some boundary conditions that may be of relevance to a context different from ecological psychology. For humans acting in a social setting, the concept of affordance needs to expand to take into account the relational constraints that affect most human interaction. This relationality manifests itself in two ways. First, for humans, much more than for other species, the identity of the specific actor matters a great deal. More than members of other species, natural capabilities vary a great deal. These abilities can be nurtured by training, education, and years of working. For example, a charcoal pencil and a piece of paper afford different affordances to an artist compared to an untrained human.

Second, as noted by Hutchby (2001), human action is governed by a web of social constraints and rules. Thus, for a student, a computer game may afford some escape and fun but the affordance may not be available due to a variety of social reasons such as the inappropriateness of playing games in class. In that fashion, affordances are both functional (artifact has a material presence) in the sense of enabling and constraining action with the technology, and relational (differs from one person to another and based on context of use). Thus, we can conclude that, within organizations, the affordance of artifacts is not simply based on their materiality but also on the relational properties that arise because of the symbolic and social nature of the setup.

Thus, a truly relational perspective on affordance would need to use the concept at a level more specific than that of the generic "user" so dear to technology designers. We suggest that an affordance is a *multifaceted relational structure*, not just a single attribute or property or functionality of the technology artifact or the actor. That is, affordance is often realized via the enactment of several mutuality relations between the technology artifact and the actor, with, i.e., mutuality is the existence of reciprocal relations among *role, line-of-action, practice/routine*, and *artifact material/functional bundle*). Because these mutuality relations are situated and emergent in practice, there is a potential for the existence of multiple affordances of the same artifact depending on the focal context. That is, because of the mutuality of the relation between an artifact and actor, the same technology can be enacted vis-à-vis various actor groups and thus exhibit different affordances.

Elaborating the Affordance Lens in Technology Research

Technology affordances come about from the confluence between an actor's line of action and the generative action possibilities in the technology. An affordance is thus a bridging concept that conceptually links the design and use of technology. Depending on the viewpoint of the user, her context, and the narrative she is following, the same object can have meanings independent of the material aspect of the object. Thus, an object may have different possible uses and meanings depending on the actor's viewpoint and the narrative they are espousing (Harre, 2002). If features are defined as technical attributes and ways of working inscribed in the artifact by technology designers, then affordances represent possibilities of using select features or combinations of features in a way meaningful to the user's goals, abilities, and lines of action. In that sense, affordances are constitutive of and instantiated within materially bound practices (Orlikowski, 2007).

Our interest in the notion of affordances diverges from the purely physical structural properties of actor–artifact relations and focuses on the promise of an embedded and socially aware view of technology affordance. Specifically, we see potential in the notion of affordances as a relational construct linking the capabilities afforded by technology artifacts to the actors' purposes. We recognize that there is a need to significantly evolve the original affordance construct and go beyond the ecological and perceptual focus of the physical affordances stream of research. This is in line with the various calls by technology–organization change researchers highlighting the affordance lens as a fruitful avenue of research to pursue (Leonardi and Barley, 2008; Markus and Silver, 2008; Orlikowski and Scott, 2008).

One useful start is to recognize that the term "affordances," and "technology affordances" in particular, can have many but related meanings that need to be recognized and clarified. However, it may be that attempts to identify "one true meaning" are likely to be counterproductive. This *modus operandi* is quite common since fields of study often superimpose their specific conceptualization atop prior usage within sister fields. For example, concurrent with the "practice turn" in the social sciences (e.g., Schatzki et al., 2001), terms such as *practice, in-practice,* and *situated* have enriched the lexicon of technology researchers. Indeed, theoretical developments surrounding IT as *technology-in-practice* (Orlikowski, 2000) have offered one influential way—technology appropriation as technology in practice—to view technology from the practice lens. Similarly, we would also highlight the need to further develop, in as precise a way as possible, the assigned and distinctive focal meaning(s) for the term *technology affordances* so that the concept becomes useful for understanding technology–organization change.

Conclusion: Affordance as a Translation of Sociomateriality

If we accept that prior approaches provide only partial accounts of the technology appropriation process or of the technology-in-use activity, then the affordance lens can significantly enrich our theoretical insight by giving to the artifact's materiality equal theoretical importance vis-à-vis social factors. Second, and more importantly, the technology's *affordances* are an intuitively appealing operationalization of a relational approach (artifact–actor) to understand activities involving the social and the material. The affordance lens provides a means of interweaving the specific elements from each of the prior perspectives into a framework that integrates, enriches, and extends the insights of earlier approaches.

Affordances can be a significant element in developing a sociomaterial explanation to the technology–social nexus where constructivist explanations that do not downplay the material have been difficult to come by (Leonardi and Barley, 2010; Orlikowski, 2010). The implication of accepting a relational view of affordance is to abandon the talk of generic user or to think of technology as bundles of features. Context such as user intent, abilities, social, environment, as well as the specifics of the situation will matter even more. We will also need to abandon the view that affordances are about a technology or an object. They are about actions in the world that involve technology. Thus, the theoretical focus shifts away from the actor or the object towards an expanded view of the interaction with the object. What becomes ontologically important in this context is how the specific action unfolds in that unique moment and situation, whom and what it enrolls, and how it affects the world.

In conclusion, technology affordance, in spite of its label, is not about the artifact itself. Owing to the influence of Descartian thinking and psychological training, technology has been represented as an external object that is processed via the human perceptional system, acted upon based on an individual's plans for action, and evaluated according to utilitarian, symbolic, or social schemes. Affordances as a relational concept that allows us to function in the world has suffered from consistent attempts to reduce it to a characteristic of the material object. Accepting such a way of viewing the world brings us back to dualistic thinking and to a Cartesian representation of the world. Much potential progress is to be gained by embracing a relational view of technology affordances.

Acknowledgments

We are grateful to the editors of this volume for a generative and enriching collaborative process. We especially thank Paul Leonardi for insightful suggestions and constructive feedback that helped improve our arguments. Feedback on how to theorize affordances was provided by Steve Barley, Beth Bechky, Terri Griffith, Pam Hinds, Ann Majchrzak, and Brian Pentland. Their generous comments helped shape our exploration and refine our thinking. Both authors contributed equally to the chapter.

References

Arthur, W. B. (2009). *The nature of technology*. New York: Free Press.

Azad, B., Faraj, S., Goh, J. M., and Feghali, T. (2010). What shapes global diffusion of e-government: Comparing the influence of national governance institutions. *Journal of Global Information Management*, 18(2), 85–104.

Barad, K. M. (2007). *Meeting the universe halfway: Quantum physics and the entanglement of matter and meaning*. New York: Duke University Press Books.

Barrett, M., Grant, D., and Wailes, N. (2006). ICT and organizational change. *The Journal of Applied Behavioral Science*, 42(1), 6–22.

Boudreau, M. C. and Robey, D. (2005). Enacting integrated information technology: A human agency perspective. *Organization Science*, 16(1), 3–18.

Chemero, A. (2003). An outline of a theory of affordances. *Ecological Psychology*, 15(2), 181–95.

——(2009). *Radical embodied cognitive science*. Cambridge, MA: MIT Press.

Gibson, J. J. (1979). *The ecological approach to visual perception*. Reading, MA: Houghton Mifflin.

Gibson, E. J. (2000). Where is the information for affordances? *Ecological Psychology*, 12 (1), 53–6.

Griffith, T. L. (1999). Technology features as triggers for sensemaking. *Academy of Management Review*, 24(3), 472–88.
Gutek, B. A., Bikson, T. K., and Mankin, D. (1984). Individual and organizational consequences of computer-based office information technology. *Applied social psychology annual*, 5, 231–54.
Harre, R. (2002). Material objects in social worlds. *Theory, Culture & Society*, 19(5–6), 23–33.
Heft, H. (2003). Affordances, dynamic experience, and the challenge of reification. *Ecological Psychology*, 15(2), 149–80.
Hutchby, I. (2001). Technologies, texts and affordances. *Sociology*, 35(2), 441–56.
Jones, K. S. (2003). What is an affordance? *Ecological Psychology*, 15(2), 107–14.
Latour, B. (2005). *Reassembling the social: An introduction to actor-network-theory*. Oxford: Oxford University Press.
Leonardi, P. M. (2011). When flexible routines meet flexible technologies: Affordance, constraint, and the imbrication of human and material agencies. *MIS Quarterly*, 35(1), 147–67.
——Barley, S. R. (2008). Materiality and change: Challenges to building better theory about technology and organizing. *Information and Organization*, 18(3), 159–76.
————(2010). What's under construction here? Social action, materiality, and power in constructivist studies of technology and organizing. *The Academy of Management Annals*, 4(1), 1–51.
Markus, M. L. and Silver, M. S. (2008). A foundation for the study of IT effects: A new look at DeSanctis and Poole's concepts of structural features and spirit. *Journal of the Association for Information Systems*, 9(10/11), 609–32.
Michaels, C. F. (2000). Information, perception, and action: What should ecological psychologists learn from Milner and Goodale (1995)? *Ecological Psychology*, 12(3), 241–58.
Norman, D. A. (1988). *The design of everyday things*. New York: Doubleday.
——(2007). *The design of future things*. New York: Basic Books.
——(2011). *Living with complexity*. Cambridge, MA: MIT Press.
Orlikowski, W. J. (2000). Using technology and constituting structures: A practice lens for studying technology in organizations. *Organization Science*, 11(4), 404–28.
——(2007). Sociomaterial practices: Exploring technology at work. *Organization Studies*, 28(9), 1435–48.
——(2010). The sociomateriality of organisational life: Considering technology in management research. *Cambridge Journal of Economics*, 34(1), 125–41.
——Iacono, C. S. (2001). Research commentary: Desperately seeking the "IT" in IT research—A call to theorizing the IT artifact. *Information Systems research*, 12(2), 121–34.
——Scott, S. V. (2008). Sociomateriality: Challenging the separation of technology, work and organization. *The Academy of Management Annals*, 2(1), 433–74.
Pickering, A. (1995). *The mangle of practice: Time, agency, and science*. Chicago: University of Chicago Press.
Sanders, J. T. (1997). An ontology of affordances. *Ecological Psychology*, 9(1), 97–112.

Smith, M. R. and Marx, L. (1994). *Does technology drive history? The dilemma of technological determinism*. Cambridge, MA: MIT Press.
Stoffregen, T. A. (2003). Affordances as properties of the animal-environment system. *Ecological Psychology*, 15(2), 115–34.
Turvey, M. T. (1992). Affordances and prospective control: An outline of the ontology. *Ecological Psychology*, 4(3), 173–87.
Warren, W. H. (1984). Perceiving affordances: Visual guidance of stair climbing. *Journal of Experimental Psychology: Human Persception and Performance*, 10(5), 683–703.
Winner, L. (1978). *Autonomous technology: Technics-out-of-control as a theme in political thought*. Cambridge, MA: MIT Press.
You, H-C. and Chen, K. (2007). Application of affordances and semantics in product design. *Design Studies*, 28(1), 23–38.
Zammuto, R. F., Griffith, T. L., Majchrzak, A., Dougherty, D. J., and Faraj, S. (2007). Information technology and the changing fabric of organization. *Organization Science*, 18(5), 1–14.

13

Pencils, Legos, and Guns: A Study of Artifacts Used in Architecture

Carole Groleau and Christiane Demers

In this chapter, we describe our work on situated practice considering its material and social dimensions, as well as the embeddedness of everyday situations in spatially and historically extended contexts. Drawing on activity theory (Engeström, 1987) and on our own research dealing with technology (Groleau, 2008), more specifically with technology in architecture practice (Groleau et al., 2012), we present a comparative analysis involving the use of artifacts in concept development in eight architecture firms. As Leonardi and Barley (2008: 166) have suggested for technology, we believe that we will learn more about materiality by comparing the use of different tools in similar contexts. In this chapter, we explore how different sociomaterial configurations emerge in one professional context, namely, architecture.

Exploring Sociomateriality Using Activity Theory

In recent years, numerous authors have argued that materiality has not been taken seriously in social and organizational studies (Latour, 1993; Heath et al., 2000), even among those who study technology (Orlikowski, 2007, 2010).

Some have criticized extant literature for adopting either a technological determinism perspective which neglects agency (Barley, 1988; Suchman, 1994), or a social constructionist view that is too "human-centric" (Latour, 1993; Pickering, 1993). With calls to explore sociomateriality, the integration of the material and the social in organizations have become more and more common (Orlikowski, 2007, 2010; Suchman, 2007; Leonardi and Barley, 2008). As a way to overcome the limitations of previous research, practice studies characterized by a conception of organizations emphasizing the situated nature of action (Orlikowski, 2000), the importance of materiality (Nicolini, 2009), and the connection between situated and extended social practices have been brought to the forefront (Corradi et al., 2010).

Among the numerous theoretical approaches that have been labeled "practice studies" (Nicolini et al., 2003), actor network theory (ANT), and particularly its notion of symmetry between humans and nonhumans (Latour, 1991), has been proposed by some scholars a way to take into account the "constitutive entanglement" of the social and the material (Orlikowski, 2007). As such, referring to this relational ontology that has become increasingly popular recently (Pickering, 1993; Barad, 2003; Suchman, 2007), Orlikowski (2010: 135) suggests that "technological artifacts should be treated symmetrically to the humans, and as equivalent participants in a network of humans and nonhumans that temporarily align to achieve particular effects."

However, while ANT has the merit of focusing attention on materiality and its effects, and of suggesting an interesting avenue for integrating the material with the social, it is not without its critics. For instance, Rose et al. (2005), while finding ANT's proposition interesting, criticize ANT studies for not taking symmetry seriously. They argue that, in fact, nonhumans' (or actants') doings are mostly analyzed in terms of human interests, in terms of how they impede or advance human agency. Pickering (1993), on his part, while agreeing with the idea of material agency, rejects the idea of perfect symmetry between humans and nonhumans, which is based on the argument that humans have intentions while nonhumans do not.

Like Kaptelinin and Nardi (2006: 237), we agree that ANT has contributed to the debate on sociomateriality by calling attention to the "agency of things," but believe that activity theory provides an alternative that circumvents some of the problems associated with the principle of perfect symmetry.

Like Leonardi (this volume), we think Kaptelinin and Nardi's conceptualization (2006) of different forms of agency derived from activity theory is particularly interesting. While maintaining a separation between humans and nonhumans, they argue that agency is not a monolithic property "that is either present or absent in any given case" (Kaptelinin and Nardi, 2006: 247) and distinguish three different forms of agency: conditional, need-based, and delegated. Furthermore, they assert that these forms of agency can only be assessed in context: "producing effects, acting, and realizing intentions, while potentialities of certain kinds of agents, vary within the enactment of a specific activity" (Kaptelinin and Nardi, 2006: 247).

More generally, we think that the central concepts of activity theory provide us with a way to articulate the material and the social that is very promising and has not been sufficiently developed so far. Activity theory is an analytical framework that allows us to apprehend organizations as arenas where multiple strands of socio-historical contexts manifest themselves in the conduct of everyday activities (Kaptelinin and Nardi, 2006). It is through the manipulation of a number of tools and different forms of interaction that socio-historical constructs actualize themselves in situated practice.

Activity Theory

Mediation is a central concept in activity theory, which was first developed by Vygotsky (1978) to argue that situated practices are embedded in wider historical contexts (Kaptelinin and Nardi, 2006). The concept of mediation refers to the manipulation and use of tools through which individuals are exposed to social-historical means and methods from which they draw to orient the conduct of their daily activities. Engeström (1987: 59) argues that "[t]he tool's function is to serve as the conductor of human influence on the object of activity." Activity theorists depict the ways subjects rely on tools to act upon their environment. In organizational contexts, for example, these tools can be instruments, such as technologies, or more abstract entities such as signs, languages, or codes, both sustaining the enactment of work practices.

Engeström also recognizes that activities require the involvement of various parties. More specifically, he differentiates between two types of human participation in activity systems, namely, as subject and as part of a community. Subjects are those directly involved in the conduct of the activity, such as employees and managers. Other individuals sharing an interest and influencing the unfolding activity, for example, clients, suppliers, competitors, governmental agencies, pressure groups, etc., are viewed as part of the community. The way individuals, either as subjects or members of the community, come together to take part in the activity is captured through two other concepts: rules and division of labor. Rules, such as codes of conduct or management policies, are constituted through a set of relations, explicit or not, involving subjects and community and are sustained through their patterns of interaction. Division of labor, including job descriptions, methods, and routines, refers to the organizing process through which multiple parties strive to meet the object of the activity system. Like the tools referred to before, rules and division of labor are also mediators, resting on socio-historically derived constructs brought into the actual context of the unfolding activity to be reproduced or contested (Groleau, 2006).

Within organizations, socio-historical constructs guiding practices may originate from different sources, such as professional codes, governmental regulations, and organizational traditions, just to name a few.

The system, formed by subjects interacting with the community drawing on tools, rules, and division of labor to perform activities, is the basic unit of analysis of this framework. Activity theory explores the conduct of collective practices which are guided by a common orientation captured through the concept of object. The object is conceived as a project under construction, something that is given and anticipated. The object is different from a motive because it is tentative and collective, and can be renegotiated as the activity unfolds. Furthermore, it differs from individual goals, as these goals are subordinated to the broader collective orientation captured in the definition of the object. It can be applied to explain how activities are conducted in a variety of organizational settings. For example, in a supermarket the object is providing food, while in a hospital it is providing health services.

We want to build on activity theory's framework by raising a series of questions pertaining to its concepts that have remained unexplored and that can contribute, in our opinion, to the extant literature on sociomateriality. More particularly, we want to investigate how historically grounded means and methods, mediated in material tools, are actually deployed by organizational members. To make the specificity of various tools explicit, our field research will contrast the use of different tools used within various organizations coexisting in the same industry. The variety of tools used in the conduct of activity will allow us to explore different forms of agency. Delegated agency (Kaptelinin and Nardi, 2006: 248) is inscribed in activity theory's definition of tools as mediators of means and methods, but we will investigate how this delegation takes form in practice while remaining attentive to other forms of agency associated with materiality. Our focus will extend beyond the study of particular tools to consider how they are combined with the means and methods supporting division of labor. As such, we strive to grasp particular configurations of mediations in which material and social dimensions of practice mutually inform one another.

The resulting analysis will help us to further reflect on the way material and social dimensions are intertwined in practice. We will thus seek to respond to Leonardi and Barley (2008: 164), who, in their call "to integrate materiality with a more voluntaristic stance (which) requires that researchers attend directly to the specific ways in which the features of particular artifacts become entangled in the social practices of people's work," note that "students of computer-supported cooperative work whose work is rooted in ethnomethodology and activity theory have made the most progress on this score."

Methodology

We decided to conduct our comparative study of organizations using a variety of tools within the architecture industry. The importance of materiality within that profession has already been discussed by authors such as Henderson (1991, 1995, 1999) and Ewenstein and Whyte (2007, 2009). These researchers have explored the materialization of ideas through drawings. We want to extend their work by examining how various tools, among them drawings, give concrete form to concepts as architects come together to develop a building project. Our exploratory study focuses on the design process through which architecture projects are conceptualized using activity theory concepts. We investigate how this process, within different firms, rests on various material and social configurations and whether it perpetuates, or not, socio-historical constructs in that profession. By choosing the concept development process, we overlook other tasks as well as the output of architecture practice as it materializes itself in a building.

This chapter is based on case studies drawn from the work of Elke Krasny (2008) who, with collaborators, visited twenty architecture studios to examine how they use various tools to generate ideas giving form to their building project. Her quest to understand "the relationships and constellations between architects, their tools and their work spaces... in the process of designing" (Krasny, 2008: 5) led her to produce a museum exhibition, as well as a book documenting her exhibition, using data collected through interviews and observations. We also gathered data through public documents, articles, and web sites of some of the firms we present in the following section.

We analyzed our data by drawing on central concepts of activity theory, particularly those of tools, division of labor, and object, to try to explain different sociomaterial configurations that characterize the ways architecture firms conduct their practice. Our analysis draws primarily on eight firms for which we had sufficient and relevant data with regard to the important elements of our analytical framework. A comparative analysis of these cases allowed us to identify three categories of firms based, primarily, on their choice of tools.

In the next section, we will present each group, emphasizing the use of tools, as well as the patterns of interaction that characterize the firms that are part of it. In the following section, we will carry out a comparative analysis of the three categories of firms stressing, first, the differences in the tools, their material properties, and the way they mediate between sociocultural and situated practices. Then, we will investigate the various patterns of interaction that are actualized as forms of division of labor, each with its own mediation of sociocultural means and methods, that intersect with tools. Finally, we will explore the link between tools, division of labor, and orientation toward the object.

Drawing, Modeling, and Inventing

Drawing to Pursue an Artistic Quest

Architects in the first group of firms all emphasize the use of drawings to develop the concepts guiding their projects. We will investigate how they use drawing as a tool, as well as how the use of this tool is inscribed in practice by focusing on similarities and differences among the firms.

The practice of Rudolf Olgiati, a Swiss architect, rests on the use of a series of drawings he produces at different steps of the concept development process. One of his collaborators Alfred Candrian says: "Rudolf Olgiati's most important tool he always had on him was a B6 pencil" (Hausegger, 2008*a*: 85). Regarding his sketches, he adds: "These were sketches into which everything flowed, from the most fundamental considerations to Greek architecture, to Graubünden architecture, down to color and textiles" (Hausegger, 2008*a*: 86). Olgiati's sketches were produced on thin tracing paper. As the concept evolved, he laid new sheets of tracing paper on the existing sketches to further develop his drawing. The concept took form through evolving sketches but also in a series of iterations during which Olgiati talked to his employees and isolated himself to further reflect on the project. He drew inspiration from other traditions, among them the work of Le Corbusier and Greek architecture. Olgiati also collected small artifacts from past buildings which he integrated in his work. As one of his collaborators put it: "He included many of the pieces in his buildings in a fragmentary way, as a kind of aesthetic sensation, a contrast, and a link to old building traditions" (Hausegger, 2008*a*: 85).

Steve Holl, head architect of a firm named after him, also uses drawing in the form of small watercolor illustrations to punctuate the concept development process. As he put it: "The process starts always on a 5 × 7 inches (piece of paper). Sometimes it is just a painting, it is not necessarily a built one. Might be a building later...." (Hausegger, 2008*b*: 69). These watercolor drawings, created on 5 inch × 7 inch pieces of paper, are drawn, redrawn, scanned, and shared through the development of the building project. They are the medium through which Holl expresses his ideas, and reworks them, but also guide other members of the team regarding the orientation he wants to give to the project. "Steven's watercolors are kind of a conceptual guide for us, to guide us through the development of a project," explains one of his partners (Hausegger, 2008*b*: 69). Holl shares his take on the evolving concept with the project's architects and also the project's team. To solve a problem, he may even extend the consultation to the entire office or other professionals close to him. It is clear that his illustrations orient the work of the project teams, even though models may be produced by other organizational members. Regarding the impressive number of watercolors produced, Hausegger observed: "The shelves are full of books and grey boxes in which he keeps his chronologically ordered sketchbooks" (Hausegger, 2008*b*: 69). He himself describes his office in the following terms: "My office is messy, messy like an artist studio.... I see something in the mess that becomes part of my creativity" (Hausegger, 2008*b*: 70). While he sees the future of architecture as moving away from nineteenth and twentieth century architecture, Holl argues that "once we reach the 21st century there is a freedom and I believe that the source and the connective link could be a piece of music, could be a scientific principle, could be a relations to morphology that is the ground of a particular place" (Hausegger, 2008*b*: 71). In presenting the future of architecture along the lines of arts, science, and morphology, Holl remains within the traditional disciplines constituting this form of professional practice.

Finally, Karl Schwanzer, directing an atelier bearing his name, also worked with sketches to test and develop ideas as architecture projects took form. This case differs from the two others because even though drawings are the medium through which concepts are elaborated and developed, they are not produced by Schwanzer himself. Schwanzer drew little; rather, he asked his firm's architects to prepare drawings presenting various concepts from which he selected the ones that were closest to his vision of a project. Consequently, teams of architects within the firm competed with one another in an effort to present a drawing that would be as faithful as possible to the concept imagined by Schwanzer. They worked by trial and error, going through numerous attempts in drawing a concept before one was finally selected. As one of his collaborators commented: "[Architects drew] many different variations on large amounts of butter paper.[1] Schwanzer came along, saw, and filled the wastepaper basket" (Krasny, 2008: 107). Schwanzer recognized the stress inherent to this creative process: "The idea of reducing stress is inconceivable in a creative profession... My staff have shown great patience with me in bearing with my temperament and my restlessness" (Krasny, 2008: 108, 110). He drew his inspiration from travel and art, including discussions with painters and sculptors. Finally, this creative process, even though Schwanzer was not drawing himself, led to a style that bears his signature. According to Feuerstein who studied his work: "Almost every design [was a] 'Schwanzer'" (Krasny, 2008: 108).

In all three firms, drawings, resting on various supports such as tracing paper or sheets taken from sketchbooks, or relying on different tools whether a B6 pencil or watercolors, play a central role in the design process by giving form to a concept and allowing it to be developed through subsequent drawings. As such, the design process is punctuated by numerous drawings, some of which might be a painting that will eventually reveal itself as an inspiration for a concept, while others are more faithful visual representations of future projects.

Furthermore, architects using drawings all see themselves within a particular tradition of architecture inspired by fine arts. For some of them, it manifests itself in their drawings, while, for others, it surfaces as they describe their

[1] Butter paper is a thin yellowish tracing paper.

inspiration and their practice. In the case of Holl, this fine arts tradition supports the production of his watercolors, but also appears when he refers to his work area as an artist's studio. Furthermore, his view of architecture of the twenty-first century draws on disciplines associated with the arts. Schwanzer also associates himself with this tradition when he describes his sources of inspiration. Finally, Olgiati's inspiration is described as taking its root in Greek cubism. One of his colleagues also points to his interest in a particular form of aesthetic in the way he uses past fragments of architecture in actual projects. We see here a common orientation toward a particular view of architecture focusing on aesthetics, style, and art, which reflects itself, beyond the use of drawing, in the way the architects make sense of, and situate, their professional practice.

Finally, our comparative analysis revealed different patterns of interaction supporting concept development. The first pattern of interaction, found in the firms founded by Olgiati and Holl, puts the lead architect in the center of the project development process because he creates the drawings that, first, constitute the concept. Furthermore, within these two firms, the lead architects take a directive role in the design process. While they consult other members of the organization, they use this feedback to subsequently continue, on their own, the exploration of the architectural concept. This process differs greatly from the one used by Schwanzer, requiring architects working for him to draw until they succeed in giving form to the concept he has imagined. Instead of working in a top-down logic, as Olgiati and Holl did, Schwanzer worked in a bottom-up logic. Still, in both interactional patterns, the lead architect remains the one who orients the selection and elaboration of the concept. The staff helps him give form to his concept. As a result, it is not surprising that these firms are all designated by the name of the lead architect who is the key player in identifying and developing the concept.

Models to Maximize Efficiency

The firms in the second group, for their part, rely mostly on models to develop their architectural concepts. As we will investigate in this section, these models have particular characteristics and are used in specific ways by these firms.

The architects of Atelier Bow-Wow produce a great number of models to support the concept development of their architecture projects, often creating up to sixty models for a project. These models are made to scale with exact proportions in order to explore, among other aspects, volumes and space. As Momoyo Kaijima explains, "With the model we check the design process spatially" (Krasny, 2008: 29). He describes the particular constraints under which they have to work: "The clients want a lot of things. So many functions and so little place" (Krasny, 2008: 29). Consequently, Krasny describes the firm in the following terms: "Extreme precision and a sophisticated handling of proportion and scale characterize the way in which Atelier Bow-Wow works" (Krasny, 2008: 29).

Edge Design Institute also uses models to explore concepts during the designing phase. As Gary Chang, the lead architect, describes it, their use of models serves different purposes: "In terms of speed, Lego blocks are a very good tool to work with. I can do a model in three minutes. If you are brainstorming and you want to explain the idea you don't have to focus on the scale so much. They are extremely good for conceptual models" (Krasny, 2008: 47). Within their fast-track design methodology, spatial economy is central. Krasny notes: "The acceleration of work process is just as important as the convertibility of space" (Krasny, 2008: 47). Or, as they define themselves on their web site: "a studio for innovative multidisciplinary design with the core focus on space."

Finally, SOM (Skidmore, Owings & Merill) has a different approach to modeling using a software package developed within the firm to maximize structural calculations.[2] As such, their models made with digital tools allow them to push the limits of technical possibilities pertaining to space, proportions, volumes, and shapes. For them, modeling is essential to building and distinguishes itself from other means, such as drawing. As one of the senior architects of the firm put it: "Your daily task is that you are no longer drawing, you are building, physically constructing a building... A partner will say: Build me a model. If you can't build a model, you can't build a building" (Krasny, 2008: 116–17). They are particularly interested in digital modeling because these tools inform new spaces and allow them to innovate both in engineering as well as in architecture.

Among the firms relying on models, we see a common preoccupation with space. Naturally, space is a concern for all architects, but these three firms distinguish themselves from others presented by Krasny because of their focus on efficiency. The group of architects presented in the previous section, even though they also raised issues pertaining to space, did so in terms of aesthetics following the fine arts tradition.

In this group, space is approached in terms of efficiency. From the data, we noted that efficiency is defined using a variety of criteria. First, efficiency as spatial economy is central to concept development among architects working either for Atelier Bow-Wow or Edge Design Institute. Second, at SOM, efficiency is described in technical terms, as the digital modeling tool allows architects to maximize engineering efficiency in the development of the architecture concept.

[2] In our study, we define materiality like Leonardi (Chapter 2, this volume), recognizing the materiality of both physical and virtual models.

Efficiency also surfaces as a major concern in the process through which concepts are developed. Here, again, we see different criteria to approach efficiency. For the architects of Atelier Bow-Wow, efficiency is expressed through a constant preoccupation with preciseness, expressed by Yoshiharu Tsukamoto in the following terms: "All our models are to scale and scientific and precise. We are as precise as possible" (Krasny, 2008: 31). Within Edge Design Institute, they describe their working method as "fast-track design," emphasizing efficiency through time. Finally, SOM sees their digital software as an efficient technical tool that supports concept development, but also exchanges among the various parties taking part in the project. As noted by Krasny: "The work method based on the collaboration of a number of specialist disciplines was directly expressed in the software developed by SOM, as it could be used by different specialist planners at the same time and facilitated communication between them" (Krasny, 2008: 115).

Finally, this collective development process, through which concepts are created and evolve, is a common thread in the firms using modeling. It is interesting to note that none of the firms described within this section is identified through the name of a specific architect. Regarding SOM, Krasny comments: " . . . the acronym SOM precisely embodies the original idea of the firm's founder, not to attribute the work of many to a single creator" (Krasny, 2008: 113). More importantly, in the three firms, many organizational members take part in brainstorming sessions during which concepts are created and subsequently elaborated. The profile of individuals taking part in interactions supporting concept development may vary from one firm to the other. In the case of Atelier Bow-Wow, architects also integrate clients in this process. SOM brings together workers from various occupational profiles, such as engineers, urban planners, and interior designers, to take part in the development of a concept. In all cases, they see the use of models as a particularly relevant tool to render their ideas visible to one another whether the individuals taking part in the process are architects, other professionals, or even clients.

Reinventing Tools to Experiment

The third group of firms is totally different from the two previous firms, as their aim is to completely redefine architecture. The two firms which form this category, each in their own way, distinguish themselves by creating their own tools to develop concepts that question the nature of architecture itself. As we will see, not only do they use materials and tools that are not ordinarily part of architecture's repertoire but they also continually redefine the process through which they create projects that transcend the boundaries of architecture.

At Diller Scofidio + Renfro, the three partners who give their name to the firm all participate equally in the concept development process, as well as all members of the project teams who are all architects. The design process itself is reinvented each time. For instance, Diller noted: "I think more than almost any other studio the projects are so extremely varied that we kind of customize methodologies to each project" (Hausegger, 2008*c*: 43).

This customization involves a long research phase during which new tools are invented. As Elizabeth Diller explains: "There are always inventions that are created for our projects... Like with the Blur Building for the Swiss Expo 2000, we wanted to make a building out of water. But is there something like a specialist for fog, a fog engineer? And a lot of our work has to do with producing effects, so we often need to transform existing technologies and come up with materials that have never been used in these ways" (Hausegger, 2008*c*: 44). She added: "We see almost anything as a potential tool to help us think. The real challenge lies in the circumstances themselves, because sometimes the circumstances produce the need to invent a new tool. Sometimes the fact that there is a tool allows you to think about things you have never thought about before" (Hausegger, 2008*c*: 44).

Furthermore, theirs is a collaborative process in which, apart from the three partners and the members of the project team, artists from very different disciplines and other professionals are invited to participate, according to the need of the project. For instance, Diller commenting on a particular project noted: "At the moment there is a seamster (tailor) working in the model shop who is helping us evolve the shaping of a particular bag, because we are developing a light design for Swarovski" (Hausegger, 2008*c*: 43).

Hausegger describes their work in these terms: "... this team has transcended the boundaries between the disciplines of architecture, visual and performing arts, has carried out dance projects, multi-media art installations and investigated innovative media" (Hausegger, 2008*c*: 43). Diller said: "The quality of our work mostly comes from the desire to think through something that has not existed before." She added: "We have spent our careers in trying to break down the disciplinary boundaries. We can say what we do, but we cannot say that architecture is one thing or another" (Hausegger, 2008*c*: 44).

The second firm, R&Sie(n), a name drawn from Lacan's 1972 "Colloque réel, symbolique et imaginaire" and pronounced "hérésie," is an agency founded by François Roche, an architect, and Stéphanie Lavaux, an artist. They cite Deleuze, Artaud, and Bachelard, among others, as sources of inspiration. Writing about their design process, Krasny (2008: 95) said: "Each beginning differs in the conscious use of the illusion of uniqueness." As Roche explained: "It is a postmodern way of producing... " (Krasny, 2008: 95). He added: "The starting point of each project is the notion of situationism, where we could extract from a situation the ambiguity of its own transformation" (Krasny, 2008: 95–6).

R&Sie(n) uses a wide variety of tools to create an architecture that evolves in interaction with the nature surrounding it. For example, the development of a project called "He shot me down," for a demilitarized zone between North and South Korea, involves the use of a gun to shoot clay blocks full of holes, the cutting up of the clay blocks, their scanning and their reinterpretation in 3D parametric design, as well as the creation of a robot. In an article discussing their work, Fabian Neuhaus (2010) explains that they use technologies such as genetic algorithms, scripts, and modeling to explore such concepts as self-organization, adaptation, contingency, and indeterminacy in relationship to architecture. Roche notes: "The tooling reproduces the unique attitude" (Krasny, 2008: 96).

As Krasny (2008: 98) argues, for them: "The issue is not the proof offered by the building, by the ability to build, [but] the emphasis is on what points beyond building, the network of stories, the narratives that are spun and can be spun further by means of the project." Roche added: "We are really dreaming of an architecture without architects... We try to never dominate our subjec... We are weak or a slave, in servitude to the story itself" (Krasny, 2008: 99). That is why, according to Roche, they create as a collective: "If... we produce as a group, between architects and artists (among others), it's mainly to 'eviscerate' the work, the project, from its presumed 'author,' and 'make fuzzy' who is the one speaking" (Olivier, 2003: 2). This collective evolves constantly, bringing together for each project a different group of experts, including, among others, neuroscientists, robot designers, and artists. They often change their name, because, as Roche says in an interview: "When you become a brand, you have to repeat yourself... To change the name is to recognize that our office mutates" (Inaba and Clouette, 2008: 2).

As we see, while Diller Scofidio + Renfro seek to redefine architecture as part of a multidisciplinary project that is "about defining space in relationship to engagement, sites and activities" (Hausegger, 2008*c*: 44), R&Sie(n), inspired by postmodern thinkers, reflect critically on architecture and have the ambition to create a radical architecture that is "non-macho." Each in their own way experiment with tools coming from the arts, new media, and the sciences to question the very foundations of architecture.

Comparing Work Practices among the Three Categories of Firms

Using activity theory's main concepts, we will now compare the three categories of architecture firms identified in the last section. To help us get a better grasp of sociomateriality, we will analyze their use of tools, their division of labor, and their object, as well as discuss the interaction between them. We also refer to relevant sources to document the socio-historical traditions that become manifest in situated practices.

Different Tools and Various Uses

When we consider their use of tools, firms focusing on drawings and models can be distinguished from firms that reinvent the tools of architecture.

Drawings and models are traditionally associated with the architecture profession. As described by Henderson (1995), drawings are tools that are socio-historically grounded in architecture as part of its fine arts heritage. They have been used by architects through the centuries to design building projects. Models, while also commonly used in architecture, are inscribed in a different tradition. Throughout history, this tool has been used, like drawings, to represent the building project for patrons, but it also bears a more technical function serving as a support for craftsmen who build the project (Arpak, 2008). This other use of models, as a technical tool for different parties involved in the building project, highlights a different architectural tradition associated with technical efficiency, rather than the one linked to the fine arts tradition's focus on aesthetics illustrated by architects using drawings. As such, while both these tools are historically associated with architecture, they refer to alternative dimensions of its practice.

The use of these drawings and models is also governed by historically grounded conventions and codes to which architects generally conform when they manipulate them. More specifically, Latour (1986) discusses, along these lines, the use of perspective, while Henderson (1995) explains the use of fine arts conventions in perspective drawings. In this regard, architects share a form of visual and spatial literacy, specific to their profession, which allows them to make sense of ideas as they materialize them through drawings and models. Architects in our study reiterated and reactualized these traditions even though, in some cases, they seemed to play with them. For example, while Edge Design's architects used Legos to construct their models, they still followed shared assumptions regarding spatial literacy in the production of these models.

Drawings and models both support the design process by materializing the architecture concept. Through drawings or models, the design in an architect's mind takes a material form, allowing them to further work on it through iterations between what they see and what they think. Because of the tangible form these ideas take, either by sketching on a paper or constructing a model, ideas not only become concrete to the designer elaborating the concept but also became visible to other parties who can take part in the elaboration of the concept of the future building project. We will see in the next section that the possibility to integrate various parties in the design process leads to various interactional patterns through which concepts are negotiated and evolve. But regardless of these various forms of participation, both drawings within the artistic tradition and models within the technical logic represent the central medium through which design takes form, evolves, and is finally fixed.

While drawings and models both support the design process, the material properties of these artifacts offer different possibilities that distinguish them from one another. For instance, drawings and sketches are flexible and can be reworked easily. They make the design process open and reversible due to the small investment needed in the production of one more drawing. On the other hand, models need more investment in terms of time and resources to produce them. Therefore, they are not as malleable. The use of Lego blocks, as described by architects working for Edge Institute, allows them to overcome some of these limitations, but the shapes that can be created remain more restricted than in drawings. Thus, we observed that, as design tools, drawings and models, because of their material characteristics, offer different levels of malleability.

While this first grouping brings together firms relying on tools traditionally associated with architecture, this is not the case for the remaining firms that tend to move away from methods and means associated with architecture. Instead, architects from Diller Scofidio + Renfro and R&Sie(n) import tools associated with activities other than architecture to support their design process. For them, almost everything can constitute a tool to guide this process. The starting point of the building project, which they frame as the circumstances, the location, or the narrative characterizing the building site, leads them to choose a particular tool that is related to the specificity of the project. Consequently, each project is customized both in terms of the choice of tools as well as the method used in the design process. As we saw in our examples, a shotgun, a piece of fabric, or a fog machine becomes the driving tool through which the concept evolves and takes form.

In their use of tools, these architects aim to confront traditional architecture. Unlike architects using drawings and models, because they work with tools they are not familiar with and are not traditionally viewed as compatible with the object (i.e., a building project), they must find a way to integrate them in their practice. This experimental (i.e., trial and error) process is repeated for each of their projects as they are confronted by different circumstances and situations leading them to choose different tools.

We want to emphasize that our analysis is based on the tools that the architects themselves regard as being fundamental in the concept development phase. We do not want to suggest that they do not use other tools; rather our vision is based on the architects' view of how they use tools in their practice.

It is interesting to note that these tools intervene differently in the design process. For example, drawings and models allow architects to give a material form to their ideas, making them accessible to further develop them individually or collectively. On the other hand, guns and fog machines are experimental tools to generate ideas. In the first case, the tools play both a constitutive (generative) and a representational function. Being part of architecture tradition, their use is guided by well-known conventions based on certain forms of visual literacy. In the second case, architects are driven by the potential of the tool, without a priori conventions, because its use is not part of architecture's tradition. The experimentation thus leads to successive temporary materializations of ideas that remain exploratory until a choice is finally made.

Finally, the way tools punctuate the design process varies in firms we studied. More specifically, firms relying on drawings and models more or less follow the same procedure each time they go through the design process, even though projects vary and may even be dramatically different in terms of output. In the case of architects reinventing tools, the choice of artifacts differs from one project to another. Consequently, they are constantly adapting and manipulating new tools and must reinvent the process through which the project takes form each time.

To conclude, these two groups of architects highlight two different ways of doing architecture: one more representational and inspired by conventions, the other more generative and experimental. This is not to say that one group is more creative or original than the other; rather it points to divergent visions of architecture that we will discuss further in our comparative analysis of the object guiding these activity systems.

Division of Labor: Similarities and Differences in Patterns of Social Interaction

Within the three categories of firms described in our analysis, we also distinguish two different forms of division of labor that characterize the way architects interact. But here we see similar interactional patterns among the firms that rely on modeling and those that reinvent tools in the development of the architecture project. In both cases, collective participation in concept development is emphasized. In the case of firms relying mainly on drawings, we observe a different interactional pattern resting primarily on the logic of apprenticeship. We will first characterize this logic and subsequently explore the similarities and differences between the two other categories of firms who use various collective forms to conduct their activity.

In the first category of firms, an architect leads the concept development process while being helped by other architects in the firm to give form to his ideas. We see in this interactional pattern similarities with the apprenticeship logic, a tradition in the architecture profession going back to the guilds of the Middle Ages (Cuff, 1991). Apprentice logic, which is reproduced through internships mandated in university programs and, in some countries, by professional associations, creates a hierarchical relationship between the master and the intern, which is the basis of the pattern of interaction.

In each case, we see a renowned architect, the firm bearing his name, leading the concept development process in interaction with more or less experienced architects and interns who choose to devote themselves to the articulation of the concept proposed by the lead architect. Within this logic resting on an authoritative relationship between lead architect and the other architects, we see two different patterns of interaction around the production of drawings supporting concept development among the studied firms. In the cases of Olgiati and Holl, the lead architects produce visuals supporting concept development. They do so through iterations between exchanging ideas with coworkers and working alone. Ultimately, they choose to integrate or not suggestions made by their colleagues. In the case of Schwanzer, we see a different way of conducting work resting on a similar authoritative relationship. While Schwanzer does not draw, the architects working for him draw sketches trying to give form to the concept he is imagining. Schwanzer exercises his authority through the choices he makes, either opting to keep some sketches to be further explored or rejecting them in the hope of having new ones closer to his vision. In these three firms, drawings, produced either by the lead architects or his subordinates, are the medium through which concepts circulate to support a particular social dynamic based on an authoritative figure distinguishing himself through his reputation and experience.

In the two other categories of firms, we see a more collaborative approach around concept development which is representative of the consensual logic found in professional partnerships. Traditionally, professional service firms, such as architecture practices, are characterized by collegial decision-making among partners (Winch and Schneider, 1993; Pinnington and Morris, 2002), as well as their reliance on networks (Brock, 2006). Cuff (1991) also emphasizes the relational quality of the architecture profession arguing that it requires the skills to interact with multiple parties having different interests.

Within the vignettes presenting the two other categories of firms, architects describe the design process as collective, involving a variety of different actors. While the apprenticeship logic is also present, it is downplayed in favor of a more egalitarian view of the division of labor based on a consensual and collaborative concept development process. However, the constitution of the collective supporting the design process differs greatly between these two groups. Within the category composed of firms using models, architects work together with engineers, town planners, and interior designers. All of these occupations revolve around a common interest tied to building. In that sense, they differ from the more eclectic collectives constituted by the third group of firms, which can involve neuropsychologists, tailors, or dancers, just to name of few. While both categories of firms rely on collaboration, they adopt very different approaches. The first group creates a collective converging toward building issues and uses means and methods traditionally associated with architecture, in this case modeling. The second group creates an eclectic collective through which architecture borrows from a variety of disciplines, with experts using their own various means and methods not traditionally associated with building projects. This division of labor involves parties that are constantly changing. The collaborative logic of these groups can seem similar to the one supporting collectives such as SOM, for example. But, the fact that new experts, not linked to architecture, are integrated in the design team each time a new project is conceived requires architects to work out varying forms of cooperation with the new team members. As such, different forms of collaboration are constantly renegotiated within these projects. Finally, in this case, the collective is by definition ephemeral and continuously reconstituted, while in the first group one can expect more stability of the network.

To conclude, both stable collectives bringing together building experts and organizations working within the apprenticeship logic refer to socio-historical constructs within architecture. On the other hand, the emergent patterns of interaction integrating experts from various fields construct a division of labor that goes beyond architectural traditions.

Common Object, Different Dynamics

As described earlier, in activity theory, the object represents the orientation of collective practices. The object is not necessarily a goal, it is rather something that is collectively constructed that can be given, anticipated, but also renegotiated as the activity unfolds. In the case of architecture firms, the object consists of conceptualizing and managing building projects. This object is shared by all the architecture firms we studied. But regardless of this common orientation, our comparative analysis shows that the way tools and division of labor come together to meet this object differs greatly from one group of firm to the other.

The way tools and division of labor support the object illustrates two different dynamics. First, architects using tools such as drawings and models supported by established collectives in which roles are well defined strive to meet their object by relying on a relatively stable process through which they seek to be original within recognized architectural practices. Second, in the case of architects reinventing tools and dealing with eclectic and constantly renewed collectives, the object rests on a process that is constantly changing, which requires them to deal with a high degree of indeterminacy. Beyond its repercussion on the process, this distinction is also a way to contest the way traditional architecture is conducted. Thus, as we see, the choice of tools and division of labor in a situated practice can either reenact or challenge socio-historical constructs characterizing architectural practice.

For instance, the use of architectural convention, through the manipulation of drawings and models as well as in the way the division of labor is organized, reenacts existing socio-historical constructs supporting that specific professional practice. Even though, for example, the division of labor is not the same among architects working in the apprenticeship logic or in a collective bringing together numerous building experts, they both coexist as traditions within that profession. While the situated practice of these firms reenacts architecture as a practice resting on a series of existing socio-historical constructs, it is interesting to note that different sociomaterial configurations can emerge from existing conventions.

On the other hand, architects describing their tools as guns and fog machines confront architecture's traditions. They redefine the boundaries of architecture by opting for tools that are unusual. They do the same thing in the choice of collaborators they rely upon in their building projects. Thus, their situated practices challenge architectural traditions. Challenges occur when newly created patterns of material and social interactions are deemed incompatible with existing socio-historically constituted practices (Groleau et al., forthcoming).

Discussion and Conclusion

Our analysis of the case studies illustrates how the social and the material mutually constitute one another in practice. In this section, we will demonstrate how using an activity theory framework contributes to existing literature by offering a way to explain the entanglement between the material and the social that does not conflate time and space like other approaches which focus on performativity. Concentrating on situated practice, we will first discuss how the material characteristics of tools interacting with the embodied abilities of subjects mediate the means and methods that are actualized in practice. Second, we will show how various forms of social organization of work inform different readings of tools and their use. Finally, we will extend our unit of analysis beyond particular mediated means and methods to explore how they come together in situated practice to reproduce or challenge historically grounded traditions.

Apart from discussing our contribution to the debate on sociomateriality, we will also highlight our contributions to the literature pertaining more specifically to architecture practice and tools.

Material Enablements and Constraints of Tools in the Conduct of Human Activity

We abide by activity theory's definition of tools as mediators of historically constituted means and methods, but the material characteristics through which these social mediations are rendered accessible, or not, are rarely discussed, even in activity studies. Our data led us to question enablements and constraints specific to the materiality of tools and the consequences they had on the way material and social dimensions of human practice come together as activity unfolds. We will now focus on material enablements and constraints of tools to discuss some or our contributions.

Materiality in the context of architecture is important, as work practices are actualized through the body and the senses of architects as they interact with their physical work environment (Ewenstein and Whyte, 2007; Styhre, 2010). In our empirical data, some architects were explicit in describing their aesthetic experience as they felt the tip of a particular pencil running over paper. The materialization of ideas through drawings or models also became, for some of them, symbols of their profession. As such, the choice of tools, for architects, rests on criteria pertaining to relationship between their bodies and their material environment, considering the particular aesthetic and emotional experience provided by these tools. Beyond the aesthetic and emotional dimensions of architects' experience, our study focuses mostly on the functional dimension of these artifacts (Kallinikos, Chapter 4, this volume).

Organizational members need to master the particular means and methods necessary to physically manipulate these tools and make sense of their output, which has consequences for social configurations supporting architecture as an activity. For example, a specific set of skills is necessary for architects to produce visuals that allow them to give form to their ideas. While it can be relatively easy to acquire the visual literacy necessary to grasp the essence of an architectural drawing, using visuals as a tool to materialize concepts requires the mastery of complex skills. Consequently, we see that tools are not enabling or constraining per se, but through the means and methods they mediate they have a potential to structure practice. However, this potential can be more or less accessible and can always be questioned in practice.

For instance, our case studies propose a dynamic between the means and tools that expands our understanding of this particular dynamic. In the case where architects use tools outside the realm of architecture to generate ideas, material enablements and constraints are different. Even though architects can learn to use some tools through trial and error, to make a building out of a fog machine requires a technical expertise that is very specific and rarely found in architects. Here, means and methods developed through time are not readily accessible, and to use imported tools architects must either invent a new usage (i.e., using a gun to shoot clay blocks) or work with other experts to develop their building project.

The question of the necessary expertise to manipulate tools is particularly important because it implies different social configurations. In the case of concept development using a fog machine, collaboration between architects and a "fog" engineer is absolutely necessary to generate ideas through tool experimentation. When using drawings to create a concept, architects taking part in the process need to be able to produce these drawings as well as to read them.[3] It is not surprising, in this last case, that architects work among themselves. Thus, the physical attributes of the tool, the more or less specific expertise it requires, lead to various social configurations supporting the generation of ideas among the different categories of studied firms.

Our examples demonstrating the use of various tools, some of which are in the architecture tradition while others are not, led us to reflect on the tool–object relationship. The use of tools such as drawings, models, and fog machine are not exclusively determined by their physical characteristics or affordances, but also by the object to which they are contributing. For example, an artistic rendering of a building will follow different conventions if it is drawn in an art class rather than drawn by an architect. In an architectural drawing, the ability to visualize the projected space might be a priority, while in the case of an artistic drawing the choice of color might prevail. The use of a fog machine for a rock concert or for generating ideas to conceptualize a building project will not be the same, although it remains the same material entity. Consequently, the material characteristics of tools do matter in thinking about how they integrate into human practice, but they must be considered contextually, in light of the specific practice they are serving.

Social Enablements and Constraints Defining the Use of Tools in the Conduct of Activity

In our section analyzing the division of labor, we identified three interactional patterns characterizing the studied architecture firms. We will continue to explore the material and social entanglement of practice by examining how

[3] In the case of Schwantzer, his position of authority allowed him to have others draw for him, but to become an architect one must still develop the requisite skills. This applies to drawing, modeling and, now, computer-assisted design.

these different patterns contributed to define in various ways the tools used to materialize ideas during the design phase.

Like material tools mediating particular means and methods, the three interactional patterns we identified are associated with different methods that frame the place of material artifacts supporting the design process in different ways.

First, within apprenticeship logic, adopted by three of the firms, drawings are used as a guiding tool. The materialization of ideas through drawings orients the design process as senior architects pick the central concept which will be further developed through iterations between individual and collective work sessions.

Second, within the collaborative mode, involving building experts, models are used as a sharing tool. The concept, proposed by architects, is rendered materially accessible to others through models, and becomes negotiable among members of the work team who may propose various alternatives, even questioning the central concept proposed by the head architect.

As in the case of drawings, models are thinking tools materializing ideas to facilitate their development. But the negotiation process through which concepts are developed varies in both cases.

Third, within another type of collaboration involving experts in various disciplines, material artifacts are used as experimental tools. In this last category of firms, tools are taken outside the architecture tradition to explore, through experimentation, how various types of material artifacts can provide the design concept on which the building project will rest.

Interestingly, we see here divergent interaction frames that define material tools. The distinction is particularly surprising in the way drawings and models are approached, since both are tools that allow architects to materialize and exchange ideas. Consequently, these two material artifacts provide similar functions within the design process. Even though these functions are identical, the way they are used diverges. This distinction contributes to the literature on thinking tools developed by Henderson (1995) in her study of drawings, by extending its application to other tools, and by suggesting that there are different types of thinking tools, such as guiding tools, sharing tools, and experimenting tools.

To conclude, material artifacts, beyond their physical characteristics, are defined in relationship with other social constructs that make further means and methods accessible to the activity system. These mediated means and methods mutually influence one another as they come together to constitute human practice around an object.

Bringing Socio-historical Practices in Situated Context: Reenactement or Challenge

Until now our discussion focused on particular mediated means and methods and how the interplay among them can help us better develop a new understanding of material tools within human practice. In this last section, we contribute to the study of sociomateriality by adopting a broader level of analysis to investigate how situated practices resting on particular configurations of means and methods contribute to reproduce or challenge socio-historically grounded traditions in architecture. In this, we respond to Kallinikos' call (this volume) to transcend situated practice and take into account a more historical perspective.

Our empirical data revealed that, in some architecture firms, organizational members drew upon means and methods which were socio-historical constructs characterizing the architecture profession through the choice of tools as well as the social organization leading to the actualization of their building projects. As our analysis showed, architectural tradition is broad and integrates different tools and various forms of division of labor. By selecting these tools and these labor configurations, organizational members reenact socio-historically grounded traditions associated with architecture.

The reenactment of these traditions contrasts with the position adopted in other firms relying on other tools, outside the architecture realm, to design their building projects. In the choice of their tools, as well as in the selection of their collaborators, the means and methods supporting the profession were challenged.[4] Challenges manifest themselves when tools and division of labor historically associated with this particular practice are rejected to be replaced by others outside the realm of that profession.

The choice to reenact or question traditions is not individual but collective. Thus, the evolution of various traditions over time rests on particular social dynamics that also inform our view of sociomateriality.

Furthermore, in recognizing that organizational members reenact or challenge socio-historical constructs in situated practices, we escape framing tools using a deterministic logic in which the means and methods impose themselves to organizational members. Instead, we see organizational members as reflective actors, collectively, choosing to replicate or contest socio-historical constructs made available to them through material entities. As we saw in our case studies, the social dynamics of each firm rests on collectives configured differently and in which power relationships also vary.

[4] In that sense, architects challenging traditions can be considered as contributing to what Kallinikos (Chapter 4) has called technological development.

Finally, our analysis allows us to extend the discussion on various forms of agency characterizing tools and humans (Kaptelinin and Nardi, 2006). As discussed in the activity theory section of this chapter, delegated agency is compatible with the conceptualization of tools as mediators of historically rooted means and methods. But the tools integrating various means and methods compete with one another as architects are faced with tool selection. Some of the means and methods reenact architectural tradition while others challenge it. Consequently, the question of delegated agency needs to be examined beyond the immediate use of one tool to see how different means and methods combine themselves to reinforce or question socio-historically grounded traditions.

Moreover, delegated agency supposes that the means and methods historically integrated in material form are actually deployed as they were intended in a particular practice. Our data show that tools are sometimes used in an alternative way to serve an object that has not necessarily been conceived for, such as using guns to generate design concepts for a building project. Thus, delegated agency becomes challenged in context as tools are used in ways they were not originally intended for. Beyond challenging delegated agency, our data show a form of conditional agency as tools, such as fog machines and guns, produce unexpected effects which guided architects in the development of their design.

By comparing different sociomaterial configurations that link situated and socio-historical practices, we contribute to the debate on sociomateriality by exploring the entanglement between the material and the social without conflating time and space like authors abiding by a relational ontology tend to do (for a similar critique see Faulkner and Runde, Chapter 3, this volume). Furthermore, our discussion suggests a promising avenue to expand the understanding of different combinations of material and human agencies. Finally, our study also contributes to the activity theory literature by providing an empirical analysis comparing different configurations of mediations supporting a similar professional practice in three different contexts. Our investigation also allowed us to examine how a common object could be realized by either reenacting or challenging socio-historical constructs as situated practice unfolded. Finally, our research continues the work of authors such as Kaptelinin and Nardi (2006; Nardi, 2005, 1996), Bardram (1998), Bodker (1991, 1997; Bodker and Anderson, 2005), and Blackler (1993; Blackler et al., 1999), who explore the analytical and explanatory power of activity theory, by investigating issues pertaining to materiality and organizing.

Acknowledgments

We would like to thank Bonnie Nardi for her helpful comments. We also gratefully acknowledge the financial support of HEC Montreal and of the Social Sciences and Humanities Research Council of Canada (CRSH 410-2011-0375). A prior version of this chapter was presented at the Third International Symposium on Process Organizational Studies held in Corfu in June 2011.

References

Arpak, A. (2008). Physical and virtual: Transformation of the architecture model. Master's thesis, Middle East Technical University.

Barad, K. (2003). Posthumanist performativity: Toward an understanding of how matter comes to matter, *Signs*, 28(3), 801–31.

Bardram, J. (1998). Designing for the dynamics of cooperative work activities. *CSCW*, 98, 89–98.

Barley, S. (1988). Technology, power, and the social organization of work. *Research in the Sociology of Organizations*, 6, 33–80.

Blackler, F. (1993). Knowledge and the theory of organizations: Organizations as activity systems and the reframing of management. *Journal of Management Studies*, 30(6), 863–84.

—— Crump, N., and McDonald, S. (1999). Managing experts and competing through innovation: An activity theoretical analysis. *Organization*, 6(1), 5–31.

Bodker, S. (1991). *Through the interface—A human activity approach to user interface design*. Hillsdale: Erlbaum.

—— (1997). Computers in mediated human activity. *Mind, Culture and Activity*, 4(3), 149–58.

—— Anderson, P. B. (2005). Complex mediation. *Human–Computer Interaction*, 20(4), 353–402.

Brock, D. M. (2006). The changing professional organization: A review of competing archetypes. *International Journal of Management Reviews*, 8(3), 157–74.

Corradi, J., Gherardi, S., and Verzelloni, L. (2010). Through the practice lens: Where is the bandwagon of practice-based studies heading? *Management Learning*, 41(3), 265–83.

Cuff, D. (1991). *Architecture: The story of practice*. Cambridge, MA: MIT Press.

Engeström, Y. (1987). *Learning by expanding: An activity-theoretical approach to developmental research*. Helsinki: Orienta-Konsultit Oy.

—— (2000). From individual action to collective activity and back: Developmental work research as an interventionist eethodology. In P. Luff, J. Hindmarsh, and C. Heath (ed.), *Workplace studies: Recovering work practice and information system design* (pp. 150–66). Cambridge: Cambridge University Press.

Ewenstein, B. and Whyte, J. (2007). Beyond words: Aesthetic knowledge and knowing in organizations. *Organization Studies*, 28(5), 689–708.

—— —— (2009). Knowledge practices in design: The role of visual representations as epistemic objects. *Organization Studies*, 30(1), 7–30.

Groleau, C. (2006). One phenomenon, two lenses: Understanding collective action from the perspectives of coorientation and activity theory. In F. Cooren, J. Taylor, and E. Van Every (eds.), *Communication as organizing: Empirical and theoretical approaches to the dynamic of text and conversation* (pp. 157–77). Mahwah, NJ: Lawrence Erlbaum.

—— (2008). Integrative technologies in the workplace: Using distributed cognition to frame associated with their implementation. In B. Grabot, A. Mayère, and I. Bazet (eds.), *ERP systems and organisational change: A socio-technical insight* (pp. 27–46). London: Springer-Verlag.

—— Demers, C., Lalancette, M., and Barros, M. (2012). From hand drawings to computer visuals: Confronting situated and institutionalized practices in an architecture firm. *Organization Science*, 23(3), 651–71.

Hausegger, G. (2008*a*). Rudolf Olgiati. In E. Krasny (Ed.), *The force is in the mind: The making of architecture* (pp. 84–7). Basel: Birkhauser.

—— (2008*b*). Steven Holl architects. In E. Krasny (Ed.), *The force is in the mind: The making of architecture* (pp. 68–71). Basel: Birkhauser.

—— (2008*c*). Diller Scofidio + Renfro. In E. Krasny (Ed.), *The force is in the mind: The making of architecture* (pp. 42–5). Basel: Birkhauser.

Heath, C., Knoblauch, H., and Luff, P. (2000). Technology and social interaction: The emergence of "workplace studies." *British Journal of Sociology*, 51(2), 299–320.

Henderson, K. (1991). Flexible sketches and inflexible data bases: Visual communication, conscription devices and boundary objects in design engineering. *Science Technology and Human Values*, 16, 448–73.

—— (1995). The visual culture of engineers. In S. L. Star (ed.), *The cultures of computing* (pp. 196–218). Oxford: Blackwell Publishers.

—— (1999). *On line and on paper: Visual representations, visual culture and computer graphics in design engineering*. Cambridge, MA: MIT Press.

Inaba, J. and Clouette, B. (2008). Interview at C-Lab, November. http://www.new-territories.com/Columbia%20interview.htm accessed on April 29, 2011.

Kaptelinin, V. and Nardi, B. (2006). *Acting with technology: Activity theory and interaction design*. Cambridge, MA: MIT Press.

Krasny, E. (2008). *The force is in the mind: The making of architecture* (188 pp.). Basel: Birkhauser.

Latour, B. (1986). Visualization and cognition: Thinking with eyes and hands. *Knowledge and Society: Studies in the Sociology of Culture Past and Present*, 6, 1–40.

—— (1991). *Nous n'avons jamais été modernes*. Paris: Éditions La Découverte.

—— (1993). *Aramis ou l'amour des techniques*. Paris: Éditions La Découverte.

Leonardi, P. M. and Barley, S. R. (2008). Materiality and change: Challenges to building better theory about technology and organizing. *Information and Organization*, 18, 159–76.

Nardi, B. A. (1996). *Context and consciousness, activity theory and human-computer interaction*. Cambridge, MA: MIT Press.

—— (2005). Objects of desire: Power and passion in collaborative activity. *Mind, Culture and Activity*, 12(1), 37–51.

Neuhaus, F. (2010). R&Sie(n): The question of morpho-ecological architecture and engineering. *SustainableCitiesCollective* (online journal). http://sustainablecitiescollective.com/urbantickurbantick/7598/sien-question-of-morpho-ecological.html (accessed April 29, 2011).

Nicolini, D. (2009). Zooming in and out: Studying practices by switching theoretical lenses and trailing connections. *Organization Studies*, 30(12), 1391–418.

—— Gherardi, S., and Yanow, D. (eds.) (2003). *Knowing in organizations: A practice-based approach*. Armonk, NY: M.E. Sharpe.

Orlikowski, W. J. (2000). Using technology and constituting structures: A practical lens for studying technology in organizations. *Organization Science*, 11(4), 404–28.

—— (2007). Sociomaterial practices: Exploring technology at work. *Organization Studies*, 28(9), 1435–48.

—— (2010). The sociomateriality of organisational life: Considering technology in management research. *Cambridge Journal of Economics*, 34, 125–41.

—— Scott, S. V. (2008). Sociomateriality: Challenging the separation of technology, work and organization. *Annals of the Academy of Management*, 2(1), 433–74.

Olivier, M. (2003). François Roche: R&Si, agence d'architecture. *Le journal des arts*, No. 183, décembre 19.

Pickering, A. (1993). The mangle of practice: Agency and emergence in the sociology of science. *American Journal of Sociology*, 99(3), 559–89.

Pinnington, A. and Morris, T. (2002). Transforming the architect: Ownership form and archetype change. *Organization Studies*, 23(2), 189–210.

Rose, J., Jones, M., and Truex, D. (2005). Socio-theoretic accounts of IS: The problem of agency. *Scandinavian Journal of Information Systems*, 17(1), 133–52.

Styhre, A. (2010). *Visual culture in organizations: Theory and cases*. New York: Routledge.

Suchman, L. (1994). Do categories have politics? The language action perspective reconsidered. *Computer Supported Cooperative Work*, 2, 177–90.

—— (2007). *Human-machine reconfigurations: Plans and situated actions*. Cambridge: Cambridge University Press.

Vygotsky, L. S. (1978). *Mind in society: The development of higher psychological processes*. Cambridge, MA: Harvard University Press.

Winch, G. and Schneider, E. (1993). Managing the knowledge-based organization: The case of architectural practice. *Journal of Management Studies*, 30(6), 923–37.

VI
Materiality as Consequence

14

Materiality: What are the Consequences?

Brian T. Pentland and Harminder Singh

"In my opinion, the financial statements referred to above present fairly, *in all material respects*, the financial position of XYZ Corporation."

(Annual Audit Report for XYZ Corporation)

Lately, social scientists have been struggling to bring materiality back into our work. Of course, *materiality* is a subtle and fluid idea that applies to all manner of things (Borgmann, this volume). While the influence of any particular material artifact is often obvious, theorizing about materiality in general has been extremely problematic. In this chapter, we offer an approach to materiality that is grounded in pragmatism (James, 1910; Mead, 1929; Rorty, 1982), a philosophical perspective that emphasizes context and consequences, rather than intrinsic properties. Our argument entails a reversal of figure and ground from the conventional perspective on materiality. Rather than focusing on different kinds of artifacts, agents, actors, or actants, we focus on *actions*. We define actions simply as the things that actants do.

Our analysis is inspired by the concept of materiality as it is used, in practice, by financial and information systems auditors (Bernstein, 1967). Auditors assess materiality in terms of the impact (or consequences) of an error in a given organizational context. In practice, *materiality* depends entirely on the context and consequences. If there is an error in an account, or a security breach in a network, how bad would that be? Who might be affected, and by how much? In short, auditors use criteria for materiality that are entirely pragmatic. Here, we consider the implications of adopting a similar approach to the idea of materiality in social science research.

Materiality in Theory

Materiality has posed an ongoing challenge for social scientists for a variety of reasons. When we try to mix humans and nonhumans in our theorizing, the following kinds of problems seem to crop up.

The Pendulum of Determinism

The problematic quality of causality in socio-technical systems has been discussed quite extensively (Berg, 1998; Leonardi and Barley, 2008). Robey, Raymond, and Anderson (this volume), for example, describe a move away from materiality and a trend back toward it. At times, the social seems to determine the technical. Other times, the technical seems to determine the social.

Plausible examples can be offered in each direction. For example, in a typical building, the configuration of hallways and doors determines, or at least constrains, the movement of people within the structure. In contrast, on most university campuses, one can find well-worn shortcuts through the grass that have been formed by students who prefer to take the shortest route. The world of information technology is rife with examples of systems that have been appropriated by users (Descanctis and Poole, 1994).

It has been difficult to resolve this issue one way or the other. We have developed some useful vocabulary to work around the lack of determinacy. For example, actor network theory (ANT) has introduced notions such as translation and enrollment, and has helped us move away from actors and artifacts to the more general category of "actants." Still, the indeterminacy of determinism remains.

Agency Stew

When we mix humans and nonhumans in the same theory, there seems to be an ongoing question about where to locate agency and intentionality. Agency is closely related to the question of determinism, of course, but from the opposite perspective. For example, do artifacts reflect the intentions of their designers? If so, what is the relative influence of the designers versus the users? To help sort this out, we distinguish between human and material agency (Leonardi, Chapter 2, this volume).

The question of where to locate agency has been particularly salient in research on information systems. Technologies such as workflow systems offer a complex mix of human and material agency, because some actions are taken by humans and some are taken by the computer system itself

(Pentland et al., 2010). When implementing new systems, Boudreau and Robey (2005) argue that managers need to overcome resistance or unfaithful use (Desanctis and Poole, 1994) while tolerating or even encouraging reinvention.

As with determinism, there have been attempts to sort this out in theory (e.g., Leonardi, Chapter 2, this volume; Robey et al., Chapter 11, this volume). But in any practical situation, we always have an agency stew. It is hard to tell where the agency is: some of it is in the carrots (people), some of it is in the potatoes (things), and some is in the sauce (protocols, languages, etc.). You can pick it apart, it but it would not be the same dish.

Immaterial Artifacts

The difficulty of isolating the role of material artifacts (e.g., computers) is compounded by the presence of artifacts that have no material substance (e.g., computer programs). The world of computation and information technology offers many prominent examples (Leonardi, 2010; Kallinkos, Chapter 4; Yoo, Chapter 7). For example, the World Wide Web (WWW) is made possible by a collection of standards such as TCP/IP and HTTP, which can be classified as nonmaterial, symbolic artifacts. Unlike the door and hallways of a building (or even the dirt pathways across campus), these artifacts have no physical presence. They are not "material" in any normal sense of the word.

Nonmaterial artifacts play an active role outside the world of computers. This is illustrated vividly by the role of financial derivatives in the recent economic upheaval. A financial derivative has no material substance, nor does the mathematical formula used to compute its value: the Black–Scholes–Merton option pricing theory (Black and Scholes, 1973; Merton, 1973). Donald Mackenzie has argued, following Callon's work on the performativity of economics (Callon, 1998), that this formula made it possible to have markets in derivatives and facilitated their rapid growth (MacKenzie and Millo, 2003; MacKenzie, 2006).

Nonmaterial artifacts have had enormous practical consequences, but there is no-*thing* there. At the very least, these examples challenge our intuition about ideas such as materiality and artifact.

Materiality in Practice

Theorizing about materiality has been problematic, so perhaps we can try a different approach: let us consider how materiality is handled by practitioners who confront the issue every day. We consider two related examples, both of which have well-established practices concerning materiality: financial accounting and information systems. We realize, of course, that "materiality" takes on a peculiar sense when used in each of these communities. In terms of Cooren et al.'s etymological analysis (this volume), auditors use the word "material" to mean, roughly speaking, "something that matters." This meaning may (or may not) be somewhat peculiar to the world of auditors, but this practical sense of the term "materiality" is what motivates our analysis.

Financial auditors have a concept of materiality that is based on the level of risk that could be associated with a particular account or transaction. When conducting an audit, they set a level of materiality and use this to guide their attention during an audit. This is an important issue for them, because auditors have traditionally argued that it is impossible, and unnecessary, to provide complete assurance that an organization's financial statements are accurate in every detail (Bernstein, 1967). Rather, auditors should allocate their limited time and resources on "matters of importance and substance," not on "an item that is of no consequence" (Bernstein, 1967: 88).

Auditors use a variety of heuristics to determine the threshold at which something is considered material. The key insight is that, for some audits, transactions below $100,000 would be immaterial, but in other audits such transactions would be highly material. It depends on whether errors in recording and reporting those transactions might influence the judgment of a reasonable person who is using the financial statement to make a decision. Materiality depends not only on the size of the item "but also (on) the nature of uncorrected misstatements, and the particular circumstances of their occurrence" (ISA 320, 2009: 315, para. 6). More specifically, labeling an item or error in the financial records as material requires an understanding of the context in which it was found: for example, which element/s of the financial statements it pertains to, whether it is an item that users focus on, the nature of the entity being audited, the entity's ownership and financing structure, and so on (ISA 320, 2009: 318, para. A3).

Materiality is not just a question of magnitude—it is also a question of classification. Consider a scenario where an office equipment company sells a valuable piece of real estate. The auditor has to decide if the revenue should be considered as regular income or reported separately as extraordinary gains/losses. Since the sale of the real estate does not occur within the course of normal operations (selling office equipment), it should not be considered part of the firm's regular income. The categorization matters—it is *material*—because, if it is incorrect, users of the financial statements might misinterpret the transaction as a sudden growth in the firm's regular business. Of course, if a real estate developer sold the same piece of property, it would be classified as regular income, because it would be part of their normal business.

Setting the level of materiality is always somewhat subjective; it depends on the auditor's professional judgment, and it can be revised at any time during an audit, based on new information (ISA 320, 2009: 314–16). This poses somewhat of a dilemma for a profession that emphasizes the importance of objectivity (Bernstein, 1967) and has been the object of research in the accounting literature (Carpenter and Dirsmith, 1992; Messier et al., 2005). For our purposes, the important point is simply this: the materiality of a transaction is never an intrinsic property of the transaction itself. It depends on the context and consequences of that transaction.

Information systems (IS) auditors also use the concept of materiality.[1] Like their financial counterparts, IS auditors cannot provide 100 percent assurance that an organization's IS resources and processes are correctly configured and managed. Thus, to conduct an audit, they focus on systems and risks that they consider to be the most "material." In particular, they consider the likelihood and severity of events that might occur if a particular system is not adequately secured, or not properly backed up, and so on. The consequences from these events could be financial losses, data theft, privacy infringements, sub-par service levels, and so on. IS auditors are not concerned with the material properties of the systems (which, after all, contain layer upon layer of immaterial artifacts). Rather, they focus on the potential consequences of actions or that could occur in that context.

Pragmatism: Consequences, Values, and Context

From these simple examples, we make a small leap: auditors use the term *materiality* in a pragmatic sense. There are many variations of pragmatism (Thayer, 1968; Murphy, 1990), but they share some common themes.

First, pragmatism emphasizes consequences rather than intrinsic properties. A statement or an idea can be recognized as "true" if it works, if it is useful, or if it helps solve a problem (James, 1910; Mead, 1929). This clearly resonates with Burrell's analysis (this volume) of the effects of rumor.

Second, pragmatism emphasizes context. Context is critical, because what works in one context may not work in another. The materiality and classification of our fictional real estate transaction depends on who engages in the transaction. In fact, the *same* transaction might be material for the buyer and immaterial for the seller. The importance of context can be seen in Pollock's analysis (Chapter 5) of the pragmatics of making meaningful distinctions in ranking systems.

[1] By "information systems auditors," we refer to individuals who hold the Certified Information Systems Auditor (CISA) certification. The CISA is the dominant professional qualification in this field, and is awarded by the Information Systems Audit and Control Association (ISACA).

Third, pragmatism acknowledges values. So it is not just consequences, but consequences for someone with specific values or interests. Financial auditors care about money; IT auditors care about risk (which usually translates into money). Leonardi (Chapter 2) argues that artifacts become material when they influence outcomes that matter to a particular social group. A similar theme can be found in many of the chapters in this volume. So if we adopt a pragmatic point of view, perhaps we can say this: *Something is material insofar as it has consequences we value in a particular context.*

Actions have Consequences

Our brief examination of audit practice leads us to another insight about materiality: if materiality is defined in terms of consequences, it also depends on events or actions. A consequence flows from a series of events or a course of action. Without action, there can be no consequences.

In practice, IT auditors ask, "what could go wrong?" For example, if bad guys get your password, what could happen? If the backup system goes down, what could go wrong? Implicitly or explicitly, they use the "five 'w's":[2] who, what, when, where, and why? Who is affected? What could happen? When could it happen? Where could it happen? These questions help determine the scope and severity of the potential problem. By asking these questions, auditors try to detect and prevent events or actions with undesirable consequences.

IT auditing is filled with examples of things that can go wrong. For example, if someone gets the password to the account of a regular user, there are a variety of possible problems. However, if someone gets the password to a *system administrator* account, then the possibilities are multiplied. The severity of the consequences depends on the context of the action: same action, different consequences. In these examples, notice that the password is itself has no material properties. Even if it did (e.g., if it was a passkey), it has no consequences in and of itself. It is the act of losing (or stealing) the password that has consequences.

So perhaps we can take one step farther and say: *Only actions have materiality*.

Suddenly, we have arrived at a proposition that seems entirely counterintuitive. In the common sense view, *things* have materiality. As Cooren et al. (Chapter 15) point out, "the question of materiality is often posed in terms of physicality or corporeality" (see also, Ekbia and Nardi, Chapter 8; Robey, Raymond, and Anderson, Chapter 11; Borgmann, Chapter 17). Here,

[2] This heuristic was supplied by a senior manager in the IT audit practice at a large public accounting firm.

we are suggesting that we reverse figure and ground: shift attention away from the objects or artifacts entirely and focus directly on the actions.

Reversing Figure and Ground

The stereotypical, object-centered approach to the study of materiality is based on the intrinsic properties of artifacts. In such an approach, we might use the properties of the artifacts to construct taxonomies based on physical characteristics (e.g., material vs. immaterial), technical features, constraints, affordances, and so on. This kind of taxonomy could be integrated with our familiar units of analysis (individuals, groups, organizations, and so on) to create the foundation for a more inclusive, sociomaterial theory.

Of course, such taxonomies already exist in research on information systems: we have personal computers and personal digital assistants (individual level); groupware and collaborative systems (group level); enterprise systems (organization level); and social networking systems (such as Facebook) (societal level). We have included material objects in social science analysis by augmenting our traditional categories of human actors with a parallel set of technological objects. This approach tends to direct theoretical imagination and empirical inquiry toward understanding how the objects in the expanded, heterogeneous taxonomy operate to generate the phenomena we study. As mentioned above, ANT has offered many insights into phenomena involving heterogeneous collections of actants. More recently, we find nonhuman actants being incorporated into traditional "social" networks (Contractor et al., 2011). Indeed, many of the chapters in this volume strive to bring artifacts back in, or examine connections between different categories of actants.

We do not mean to seem dismissive or minimize the value of these lines of inquiry. In many cases, the relationship between people and their things is exactly what needs to be understood. But whenever the emphasis is on the actants themselves, distinctions between different categories of actants tend to be maintained. Ironically, ideas that are intended to signal a blurring of categories (such as intertwining, entanglement, and imbrication) also serve to reinforce the underlying distinctions. Further, approaching the phenomenon as a mixture of two different kinds of things leads to the problems we mentioned above: indeterminate determinism, immaterial materials, and agency stew.

To avoid these problems (or at least to bracket them off temporarily), we suggest a different strategy that entails a reversal of figure and ground. Rather than focusing on things (people and technology), why not focus on *actions*? This idea is not entirely new, of course. For example, it resonates with the framework described by Cooren et al. (Chapter 15) on Communicative

Constitution of the Organization (CCO) (Putnam and Nicotera, 2009). CCO is agnostic about the nature of actants, and stays focused on interaction. Still, most of our theories and methods are built around measuring the characteristics of objects and relating them to the characteristics of other objects. It is beyond the scope of this chapter to offer a full-blown alternative, but to the extent that we focus on actions, we may be able to analyze "what matters" without needing to unravel any intertwined, entangled, imbricated networks of heterogeneous actants. Auditors seem to do this every day, so perhaps we could give it a try?

Conclusion

Our interest in materiality has led us to the least material of phenomena: actions, consequences, and values. This chapter is short because our point is simple: materiality is a pragmatic concept. Materiality is not about artifacts, people, ideas, or any *thing*. Or rather, it is about all of them, but they only become *material* when they influence a particular course of actions or events that we value. Materiality is all about actions, values, and consequences in context.

References

Berg, M. (1998). The politics of technology: On bringing social theory into technological design. *Science Technology and Human Values*, 23, 456–90.

Bernstein, L. A. (1967). The concept of materiality. *The Accounting Review*, 42, 86–95.

Black, F. and Scholes, M. (1973). The pricing of options and corporate liabilities. *Journal of Political Economy*, 81, 637–54.

Boudreau, M.-C. and Robey, D. (2005). Enacting integrated information technology: A human agency perspective. *Organization Science*, 16, 3–18.

Callon, M. (ed.) (1998). *The laws of the markets*. Oxford: Blackwell.

Carpenter, B. and Dirsmith, M. (1992). Early debt extinguishment transactions and auditor materiality judgments: A bounded rationality perspective. *Accounting, Organization and Society*, 17, 709–39.

Contractor, N. S., Monge, P. R., and Leonardi, P. M. (2011). Multidimensional networks and the dynamics of sociomateriality. *International Journal of Communication*, 5, 682–720.

DeSanctis, G. and Poole, M. S. (1994). Capturing the complexity in advanced technology use: Adaptive structuration theory. *Organization Science*, 5, 121–47.

International Auditing and Assurance Standards Board (IAASB) (2009). *International Standard on Auditing (ISA) 320—Materiality in planning and performing an audit*. New

York: IAASB. http://www.ifac.org/sites/default/files/downloads/a018-2010-iaasb-handbook-isa-320.pdf accessed on October 10, 2011.

James, W. (1910/1963). *Pragmatism and other essays*. New York: Washington Square Press.

Leonardi, P. M. (2010). Digital materiality? How artifacts without matter, matter. *First Monday*, 15(6). http://firstmonday.org/htbin/cgiwrap/bin/ojs/index.php/fm/article/view/3036/2567 accessed on October 10, 2011.

—— Barley, S. R. (2008). Materiality and change: Challenges to building better theory about technology and organizing. *Information & Organization*, 18, 159–76.

MacKenzie, D. (2006). *An engine, not a camera: How financial models shape markets*. Cambridge, MA: MIT Press.

—— Millo, Y. (2003). Constructing a market, performing theory: The historical sociology of a financial derivatives exchange. *American Journal of Sociology*, 109, 107–45.

Mead, G. H. (1929/1964). *A pragmatic theory of truth, selected writings*. New York: Bobbs-Merrill.

Messier, W., Martinov-Bennie, N., and Eilifsen, A. (2005). A review and integration of empirical research on materiality: Two decades later. *Auditing: A Journal of Practice & Theory*, 24, 153–87.

Murphy, J. P. (1990). *Pragmatism: From Peirce to Davidson*. Boulder, CO: Westview Press.

Putnam, L. L. and Nicotera, A. M. (eds.) (2009). *The communicative constitution of organization: Centering organizational communication*. Mahwah, NJ: Lawrence Erlbaum Associates.

Rorty, R. (1982). *Consequences of pragmatism*. Minneapolis: University of Minnesota Press.

Thayer, H. S. (1968). *Meaning and action: A critical history of pragmatism*. New York: Bobbs-Merrill.

15

Why Matter Always Matters in (Organizational) Communication

François Cooren, Gail T. Fairhurst, and Romain Huët

The so-called "linguistic turn" purportedly has allowed scholars to demonstrate why it seems so important to focus on language, discourse, and social interaction when studying organizational phenomena (Deetz, 2003). However, it could be argued that it also led them to neglect some key aspects of the role *material agency* plays in organizational processes (Pickering, 1995; Ashcraft et al., 2009), a negligence that the more recent "material turn" could be said to be addressing (D'Adderio, 2011). This chapter proposes to show, both theoretically and empirically, that analysts do not actually need to keep turning in one direction or another, that is, choose between materiality and discourse, so to speak, but that they should rather focus on the multiple ways by which various forms of reality (more or less material) come to *do things* and even *express themselves* in a given interaction.

In order to acknowledge the variety of beings or entities that participate in the performance and definition of what is taking place in an (organizational) interaction, we first contend, following Orlikowski (2007), that, "*every* organizational practice is *always* bound with materiality" (p. 1436, original emphasis; see also Orlikowski and Scott, 2008; as well as Robey, Raymond, and Anderson, Chapter 11). We illustrate this position by analyzing an excerpt taken from fieldwork completed through the three-day video-shadowing of a building manager working in a sixty-story skyscraper in New York City. Drawing from this excerpt, we show that acknowledging the variety of these (textual, architectural, artifactual, technological, and human) forms of agency implies that we *decenter* our analyses by not systematically taking what people are doing as the only point of departure of our inquiry (Cooren et al., 2006; Cooren and Fairhurst, 2009; Fairhurst and Cooren, 2010; D'Adderio, 2011).

According to such an approach, agency thus has to be understood as *relational* (Law and Mol, 1995; Bechky, 2003; Robichaud, 2006), which is another way to say that action should be considered as always *shared* and/or *distributed* among a variety of agents with variable ontologies (Hutchins, 1995; Latour, 1996; D'Aderio, 2011). For instance, what *matters* or *counts* in a given situation is certainly something that human interactants negotiate and co-define through discursive means, but such human actions do not mean that objects, technologies, and documents are the simple means or products of these interactions and discourses (Fayard and Weeks, 2007). As we will show, a relational approach to agency demands that we properly situate the discursive and symbolic with the material and technological (Barad, 2007; Leonardi and Barley, 2008; Scott and Orlikowski, Chapter 6). Latour's notion (1996) of *hybrid agency* is our touchstone here.

As demonstrated in his oft-cited example on the NRA slogan, "Guns don't kill. People kill people," the human being is a different agent by virtue of the gun she holds—along with its enablements and constraints (such as close- or long-range shooting capabilities). The gun, too, is a different agent by virtue of this human being, her intent, and specific manipulation of it (as guns have other uses such as historical artifact or trophy). If both humans and objects have the power to transform the other, we propose studying their relationalities.

Following Latour (1996) then, our goal is to avoid reducing the world to terms overly favorable to one or the other—either a discursive/symbolic *or* material/technological determinism. Such a view enjoins us to pay attention not only to (*a*) what people are doing in interaction but also to (*b*) *what* leads them to do what they are doing, that is, what *animates* them in a specific situation or in their daily activities, as well as (*c*) what speaks or acts *through* them, that is, what *constitutes* them and what they constitute as social or organizational agents (Putnam and Nicotera, 2009).

What or who is acting or active in a given situation is therefore always an open question that can be solved only by studying the details of organizational—or, shall we say, "organizing"—interaction given our performative ontology. As we will show in our analyses, acting or *speaking as* a building manager means, for instance, that specific *preoccupations/concerns/worries/reasons* are supposed to iteratively or repetitively animate the person who enacts this role or function. Figuratively speaking, it is these preoccupations/concerns/worries/reasons that consequently express themselves and partly define what issues/problems/questions this person is *speaking to* in his daily activities. Furthermore, acting as a building manager also means that one is acting or *speaking for* a variety of principals—not only the company he is officially in charge of managing the building but also the clients this person is supposed to serve—whose interests are supposed to be represented, embodied, envisioned, and defended by this person in his daily activities.

Echoing ethnomethodological reflections on the incarnated/embodied/material aspects of social and organizational life (Garfinkel, 1967, 2002; Heritage, 1984), we will show that this reflection on various forms of speaking—speaking as, speaking to, and speaking for—leads us to problematize the questions of organizational interaction and discourse in terms of *im/materiality*. (Organizational) communication is, to a certain extent, material because any interaction always has to be incarnated/embodied/materialized in specific circumstances, artifacts, bodies, texts, architectural elements, facial expressions, intonations, etc. However, it is *also*, to another extent, immaterial (and discursive) precisely because these activities of incarnation or embodiment make present what appeared to be initially absent, disincarnated, or disembodied in rather specific ways (on the question of materiality and immateriality, see also Yoo, Chapter 7).

Organizational communication and discourse thus constantly navigate between these effects of presence/absence where certain questions, issues, beings, and realities come to materialize themselves in specific circumstances. A preoccupation, for instance, is material to the extent that it takes the form of or expresses itself through specific activities, discussions, and artifacts. However, it also immaterial because these actualizations and incarnations also point to *what* is supposed to be expressed through them, something that will never be exhausted by these materializations, that is, the preoccupation itself, so to speak, as experienced by the persons whose minds are putatively occupied by it.

This reflection on im/materiality thus enjoins us, as analysts, to focus not only on what appears to *count* or *matter* to interactants in a given situation but also what does *not* count or matter to them. When something or someone happens to count or matter, it/he/she materializes itself in their discussion or activities, making it/him/her spectrally or hauntingly present in what is said or done. Interestingly enough, this means that something whose materiality appears a priori unproblematic (an artifact, for instance) can lose a part of its materiality—etymologically, the substance from which it is made—because it does not appear to ma(t)ter(ialize itself) or count in what is said or done. Additionally, this means that discourse and interaction *also* support or stand under (i.e., substantiates) what exists or not.

Echoing Étienne Souriau (1956) whose work has been recently rediscovered (Cooren and Bencherki, 2010; Bencherki and Cooren, 2011; Latour, 2011), we could then say that the question of in/existence and im/materiality should never be posed in absolute, but always in relative and transactional terms. In keeping with our relational and performative ontology, we will thus demonstrate that what matters or does not matter has to do with what materializes (or does not materialize) in an interaction, whether through the form, for instance, of preoccupations, discussions, or activities. As we will show in our analyses, (organizational) interactions are literally and figuratively filled with these "matters of concern" (Latour, 2005) whose articulation triggers various form of organizing.

Discourse and Materiality: A CCO perspective

For almost twenty-five years now, several organizational communication scholars have proposed to reverse the traditional way of conceiving of their own object of study: Instead of studying how communication takes place in organization, they decided that it would be much more productive to study how organization takes place in communication (Taylor, 1988, 1993; Taylor et al., 1996), a paradoxical stance that could be traced back to James Dewey's pragmatic perspective (1916/1944, see Taylor and Van Every, 2000). Originally inspired by the interpretive movement (Putnam and Pacanowsky, 1983), whose main focus was sensemaking (Weick, 1979; Eisenberg, 2006), this approach took on a more actional turn some fifteen years ago by studying how organizations are *performed into being* through interactions (Fairhurst, 1993; Boden, 1994; Cooren, 2000, 2010; McPhee and Zaug, 2000; Taylor and Van Every, 2000, 2011; Kuhn, 2008).

This "grounded in action" perspective, as Fairhurst and Putnam (2004) call it, was recently coined the CCO approach (for Communicative Constitution of Organization; see Ashcraft et al., 2009; Putnam and Nicotera, 2009), which can be summarized by the following premises, proposed by Cooren et al. (2011). According to these authors, CCO scholarship (*a*) studies communicational events, (*b*) is as inclusive as possible about what is meant by (organizational) communication, (*c*) acknowledges the co-constructed or co-oriented nature of (organizational) communication, (*d*) holds that who or what is acting always is an open question, (*e*) never leaves the realm of communicational events, and (*f*) favors neither organizing nor organization.

Concretely, this means that one of the challenges of this approach consists of never leaving the *terra firma* of interaction (premises *a* and *c*, see Cooren, 2006) while extending what the term "interaction" or "communication" means or refers to (premise *b*). Since we cannot leave the realm of action and communication, proponents of this approach have tried to illustrate to what extent multiple forms of agency can be said to express themselves or do things in any given situation (Latour, 1996, 2005). For instance, positioning oneself or being positioned as speaking in the name of an organization means that, for all practical purposes (even legal ones), it is this organization that can *also* be deemed as acting or saying something, that is, communicating, in this situation (Cooren, 2006).

As exemplified in this illustration, recognizing that multiple forms of agency express themselves in an interaction does not mean that human beings lose their own agency, that is, their own capacity to make a difference. On the contrary, human beings actively participate in these effects by implicitly or explicitly, unconsciously or consciously staging these various forms of agency in their conduct or talk. Furthermore, these agencies are never an a priori given, but need to be co-constructed in interaction by the human participants (premise *c*), meaning that speaking in the name of an organization is something that ought to be acknowledged by the interlocutor(s) for the organization to be recognized as speaking and acting.

As mentioned in premise (*d*), who or what is acting or active always is an open question for CCO scholars because it is not only the analysts but also and especially *the human participants themselves* who co-construct and co-recognize the various forms of agency that express themselves in any given interaction. For instance, when someone positions herself as speaking in the name of a contract previously signed between two parties, the fact that this contract might make a difference, that is, do something, in a given situation is not something that can be determined a priori by the analyst, but results from how the interaction actually unfolds (Ashcraft et al., 2009). If the person happens to say, "Our contract stipulates that this task be completed by April 1," it is not a given that what this contract is supposedly doing will be recognized by her interlocutors, which means that some discussion about what the contract actually says and dictates might take place.

Finally, and in keeping with premise (*f*), CCO scholars do not have to choose between organization and organizing precisely because the mode of being and acting of any collective form has to be found *in interaction*. Coordinating actions is something that can result from the mobilization of several forms of agency (contracts, titles, procedures, protocols, rules, etc.), each of these agencies contributing to the (re)configuration of a collective, that is, to its organizing/organization.

If we agree with these six premises, then materiality should not be considered as lying outside of discourse in interaction. On the contrary, it actively *participates* in the production of communication, much as Latour (1996) tried to capture in his writings on hybrid agency (see also Whyte and Harty, Chapter 10, as well as Pollock, Chapter 5). As long as human interactants are deemed the only ones doing things with words (Austin, 1962), which, for a time, was the misguided legacy of the linguistic turn, the question of materiality cannot be integrated in the communicational equation, so to speak (Cooren and Matte, 2010). However, if the question of who or what is acting becomes an open question, we can then start to identify how materiality *constantly* invites itself in people's conversations while still acknowledging (read, not overemphasizing) the discursivity required for the interpretation of materiality's impact on human actors.

In order to achieve this stance, we first need to examine what we usually mean by the terms "matter" or "materiality," a difficult question that tends to be avoided by the literature, even when the question of material agency is introduced (but see Barad, 2007, as well as Scott and Orlikowski, Chapter 6). Both materiality and matter etymologically come from the Latin word *materia*, which means "the substance from which something is made" or the "grounds, reason or cause for something" (etymonline.com). When we wonder about the matter or materiality of something (whatever it is), we thus tend to question what it is made of, although we can also question the grounds or reasons for what is taking place.

It is interesting that substance etymologically means "what stands [*stare*] under [*sub*]," a meaning that can be implicitly recognized when we speak, for instance, about what *substantiates* a given position or argument (Chaput et al., 2011). Whether we speak of materiality or substance, we then implicitly refer to what appears to *stand under* something, that is, what might explain its existence or mode of being. For example, asking someone, "What is the matter with you?" amounts to literally questioning this person about what might be the cause of her present conduct or state of mind.[1]

While the question of materiality is often posed in terms of physicality or corporeality (and we, of course, have to acknowledge these dimensions of its meaning, see Ekbia and Nardi, this volume, as well as Robey, Raymond, and Anderson, this volume), we thus need to realize that materiality relates to what is *relevant* or *pertinent* to a given situation, that is, "the relation [of something] to the matter at hand" (Webster's Dictionary). We will speak, for instance, of "facts material to the study" both to mean not only what bodies or physical artifacts were mobilized to study something but also what facts were relevant or pertinent to a given study. All these meanings point to the fact that speaking in terms of matter or materiality when referring to a given situation or thing amounts to questioning what this latter depends on[2] in order to be or exist, that is, what stands under it (substance), what pertains to it (pertinence), or what causes it (reasons). In other words, it enjoins us to adopt a performative approach to matter, which leads, for instance, Barad (2007) to speak in terms of "materialization" (see also Scott and Orlikowski, Chapter 6).

[1] Of course, saying "what's the matter with you?" can also be used to mark a form of reprobation vis-à-vis what a person might have done or is about to do, but such an indirection can only function if the literal meaning is also conveyed (see Searle, 1979 as well as Cooren, 2010).

[2] Relevance comes from the Latin relevare, which means to raise up, to relieve (see Cooren and Sanders, 2002). When X is deemed relevant to Y, it therefore means that X brings about or explains Y. See also Ekbia and Nardi (Chapter 8).

From a CCO perspective, a way to investigate the question of materiality in discourse and interaction thus consists of studying the different ways by which materiality expresses itself when people communicate with each other, that is, how what appears to *matter* or be relevant to them gets itself materialized in their conversation, what Pentland and Singh (this volume) would also call "materiality in practice."[3] Echoing Latour (2008), one could also note that what matters to someone in a given situation can also be called a *matter of concern*, that is, what matters to her is, by definition, what concerns, preoccupies, interests, or *animates* her. In keeping with our relational ontology, we thus see that there is no divide between the world of materiality and the world of discourse and communication. *Any* form of communication is indeed animated by a form of concern, preoccupation, or interest, however minimal it may be.

Investigating the question of materiality from a CCO perspective thus amounts to noticing what appears to move, animate, concern, or preoccupy the people in interaction. For instance, if someone appears to be concerned about a specific situation, it means, *by definition*, that this situation *preoccupies* her, a source of preoccupation that can be evoked, conveyed, or translated in what she says or does. Instead of separating the material world from the discursive/interactional world, we thus see that human interactants are not simply and only enacting a given situation (although they certainly do this, see Weick, 1979), but that they are *also* reacting to it, *marking the agency* of what appears to matter to them, that is, what animates them.[4]

In keeping with premise (*d*) of the CCO perspective, it will be thus crucial to remain, analytically speaking, as open as possible about *who* or *what* appears to be acting or active in a given situation. Although identifying what people do in interaction (and how they do it) remains a key aspect of our analyses, it is noteworthy that these people are themselves *inhabited* by specific concerns or preoccupations, which will mark what appears to strike, affect, interest, touch, or even disturb them in a given situation. If such an analytical move could be accused of falling into psychologism, the good news is that interactants often stage these effects of animation in their own talk and conduct (Cooren, 2010). Indeed, how else could experts or analysts precisely infer the effects of many materialities otherwise (e.g., whether a gun is a weapon, historical artifact, or trophy)?

[3] As they write, "Materiality is not about artifacts, people, ideas, or any thing. Or rather, it is about all of them, but they only become material when they influence a particular course of actions or events that we value. Materiality is all about actions, values, and consequences in context" (Pentland and Singh, Chapter 14: 294).

[4] In some respects, connections could be made with Activity Theory and its materialistic perspective, especially through the principle of object-orientedness (Nardi, 1996). However, we tend to depart from this theory, which is still, for us, too human-centered (see Latour, 1996, as well as Groleau, 2006).

In terms of co-construction and co-orientation (premise *c*), human interactants are certainly active in co-constructing and co-orienting to a given situation, but it will be also crucial to recognize that *they are not alone on the construction site*, so to speak. In keeping with Orlikowski's idea of entanglement (2007), what appears to *mat(t)er(ialize itself)*[5] in a given discussion should also be deemed as what *participates* in the co-construction of the situation. For instance, when someone points to an artifact during an interaction, an analyst could say that she is orienting to it, which marks a form of preoccupation on her part, but one could also point out that this artifact might have as well *caught her attention or eye*, meaning that it appears to matter to her and marking what this artifact is literally *doing* in this situation.

In keeping with Latour's position (1996), there is no origin to action, but a series of translations from one contribution/action to another, whether these contributions are considered human or nonhuman. An artifact or situation that suddenly appears to *matter* in a given discussion might, for instance, influence or even *dictate* how the interaction unfolds because it shows, demonstrates, or illustrates something to the human interactants, something that might be highly consequential regarding the situation at hand.

One could retort that such contributions—showing, demonstrating, or illustrating—are nothing without the human interactants' *interpretation and its discursivity*. Our point is not to question such evidence, but to recognize, on the contrary, that interpreting something precisely consists of *recognizing what it tells us*. We could even contend that our analytical positioning consists of taking what the interactants are saying or doing very seriously (a move that we borrow from ethnomethodology even if ethnomethodologists themselves would certainly disagree with our positions about material agency). For instance, if X says, "Look, I'm right" by pointing to an artifact, it means that one way for us, as analysts, to interpret the situation is to conclude that this artifact, which X is pointing to, is supposed to *tell* her interlocutor that X is right about something that she was holding as true.

As we see, recognizing material forms of agency does not amount to questioning human beings' capacity to make a difference, that is, what we tend to call their agency (Giddens, 1984). On the contrary, we contend that it *reinforces* it by showing that, when human beings do or say something, it is also many other things or beings that potentially express themselves, giving

[5] By this, we want to convey that noticing that something *materializes itself* in a given discussion also means that this thing appears to *matter* to at least one of the participants. For instance, if someone brings out a specific issue (the poor state of the environment, for instance) in a conversation, it also means *eo ipso* that this thing (the poor state of the environment) is supposed to matter to him or her. When this question materializes itself in a discussion, it thus also means that it matters to at least one person who starts to speak on its behalf, making this issue present in the conversation.

them—the human beings—a form of legitimate power or authority (Benoit-Barné and Cooren, 2009; Taylor and Van Every, 2011). When X says, "Look, I'm right" by pointing to an object, it is not only she who is saying that but also the object whose presence or form is supposed to confirm or prove her position. We could state this otherwise by suggesting that human beings are not "free agents" as much as they are hybrids—"enabled agents" and "constrained agents" vis-à-vis material forms of agency. To suggest otherwise is to ignore, following Weick (1979), the materially based *bracketing of their conditions*, which humans routinely perform in accounting for their worlds (see also Kallinikos, 2004).

In keeping with premises (*a*) and (*e*), then our analytical move thus consists of studying the realm of communicational events, knowing that many things *get communicated* when people interact with each other (in French, one would say "*Ça communique,*" an expression that is almost impossible to translate but that conveys the idea that many different things get communicated when people communicate: ideas, emotions, reflections, knowledges, concerns, realities, situations, etc.). What remains to be shown, however, is how the analytical productivity of such a positioning can be illustrated empirically. This is what we propose to do in the next section.

An Illustration

One of the strengths of the CCO approach lies in its capacity to illustrate empirically its theoretical positioning (Robichaud et al., 2004; Cooren et al., 2006; Fairhurst, 2007; Benoit-Barné and Cooren, 2009; Brummans et al., 2009; Fairhurst and Cooren, 2009; Cooren, 2010; Chaput et al., 2011; Taylor and Van Every, 2011). Given that the CCO analyses tend to be exclusively focused on action and communication, the empirical challenge thus consists of unfolding the various forms of agency that might be expressing themselves in a given interaction.

This is what we propose to do with the following episode by focusing especially on the question of materiality and its various forms of expression. The scene is taking place in January 2003 in a sixty-story commercial skyscraper located in Manhattan, New York. The first author, armed with a video camera, has been following Denis, the building manager of this building, throughout his working day. Denis is in charge, among other things, of supervising all the renovation work that is currently under way in his building. One of his main tasks consists, in particular, of telling subcontractors what needs to be done in each office space that is being renovated.

When the episode starts, we just arrived in an office space that Denis is restructuring for a tenant (all the tenants of this building are organizations and firms). He is followed by Robert and Alberto, who are, respectively, vice-president and technician of a company specialized in the installation of air conditioning for commercial purposes. Because of the renovation under way, Denis has to install a new air conditioning system. He asked Robert and Alberto to join him so that they can tell him what should/could be done about that.

Here is how the sequence begins:

1 ((They get out of the elevator, all laughing and joking. As they walk into a
2 corridor to get to the client's premises, we see that construction work is
3 going on. A stepladder is lying on the floor.))
4 Denis: All right. Robert, Ran- uhm (.) They're expanding. They're taking-
5 (2.5) ((while talking, Denis is walking from the hallway to the office space,
6 followed by Alberto and Robert. Denis then briefly stops walking, turns over
7 and points to something in the back when he says "They're expanding."
8 Robert then looks back in the direction Denis was pointing to))
9 Robert: All that?
0 ((Robert makes a small gesture with his left hand, marking the space that is
1 supposed to be "taken." When he turns back, he realizes the others went in
2 and walks towards them. Denis continues talking.))
3 Denis: They're taking all of this. (1.0) I'm putting uh a demising wall ((common
4 wall)) up here. (1.0) So uhm ((They all look around the premises. A worker
5 comes from an adjacent room.)) Sabor, Crancik ((tenants' names)) is going
6 to be taking this whole place ((they turn around a corner in the premises))
7 Alberto: Straight across.
8 Denis: We're turning this into two offices (0.5) okay? This is going to be their tech
9 room (.) something like this. ((Walks towards his right, pointing in front of
0 him with his hand, as to mark where the walls of the tech room will be.
1 Alberto walks toward the space that Denis just defined with his arm)). With
2 all of their equipment. I'll need a diffuser in there.
3 ((While placing his first order, Denis walks toward Robert who is looking at
4 the ceiling of an adjacent room. Robert point towards the ceiling and
5 makes a motion with his hand.))
6 Robert: You've got two diffusers up here ((waving upward toward the ceiling))
7 (1.0)
8 Denis: Oh, we're good here.
9 Robert: Yeah, you're fine here.
0 Denis: ((Points in front of him.)) I'm gonna need something in this tech room.
1 Robert: Okay
2 Denis: And a full return or whatever because they've got ((points in different direction))
3 some- they're not generating a lot of heat but they're adding a few things to
4 [it.
5 Robert: [Right

Although it is difficult to recreate the general scene from a simple transcript, whoever watches the video will realize that Denis, Robert, and Alberto are completely surrounded by materiality as they move about the office space: walls, windows, doors, of course, but also the various tools, equipment, and construction materials found on the floor or along the walls. Interestingly, we also see a change of vocal tone or register as they are getting out of the elevator and entering the office space. While they were making jokes and laughing in the elevator, Denis, the building manager, marks the beginning of a new sequence by saying "All right. Robert, Ran- uhm (.) They're expanding. They're taking-" (line 4).

This change of register (from joking to business talk) unsurprisingly coincides with their progressive arrival in the office space. Robert and Alberto are basically here to know what Denis wants them to do in this space, which means that their arrival in the said space also corresponds with their visual/physical encounter with what is supposed to concern them, professionally speaking. Such preoccupation is marked by the fact that Alberto and Robert have almost immediately started to look around them (lines 8, 14), especially to where Denis points. While we could say that it was mainly Denis, Alberto, and Robert who were joking in the elevator, it is now the building manager and the air-conditioning professionals who are also and especially interacting from now on.

Acting and speaking *as* a building manager indeed means that specific preoccupations or concerns have to be voiced in connection with the renovation that is under way, while acting or speaking *as* air-conditioning professionals means, among other things, that one is attending to what the building manager will signal, that is, the various items and problems he will speak *to* during this visit. For instance, we see, in line 22, how Denis places his first order by saying "I'll need a diffuser in there," but it is also what he says and shows to Robert and Alberto that is of interest to us.

Indeed, we observe, from lines 18 to 22, how Denis says, "This is going to be their tech room" (lines 18–19), information he further materializes by adding "something like this" (line 19), while making an invisible square with his hand in order to mark where the walls of the tech room will actually be. As Alberto is walking toward the space that Denis just defined, the latter then says, "With all of their equipment. I'll need a diffuser in there" (lines 21–22). Interestingly, we see how his placing this order is substantiated by what he not only just said but also *showed* to Alberto and Robert.

As a building manager, Denis indeed knows (as do his interlocutors) that tech rooms have to be cooled down because of the heat they generate, which explains why a diffuser appears necessary. By informing his interlocutors that this space he is showing them is going to be a tech room "with all of their equipment," he is therefore substantiating his request, that is, providing the reasons why a diffuser has to be installed there. Interestingly, we see that the order is placed by saying, "I'll need a diffuser in there," marking the fact that he, Denis, is the one who needs a diffuser. However, what *stands under*, that is, substantiates such need is (the fact) that a tech room is about to be installed there, as specified and shown by Denis himself.

Although material agency is, per se, made *invisible* in what Denis is saying, that is, it is bracketed out, we can note that it is *made visible* by the square he is making with his arms. Furthermore, anybody who knows something about air-conditioning is supposed to realize that the situation he just described and defined *dictates* that a diffuser be installed. In other words, everything happens as though it were not only he who was dictating that a diffuser be installed but also *the situation itself*, which is displayed in what he is saying and showing. What is relevant or pertinent to this situation, that is, what pertains to it and is supposed to matter to Denis, is that there is going to be a tech room with a lot of equipment. This equipment is going to generate heat, something that obviously concerns or preoccupies him, a source of preoccupation that is implicitly evoked and translated in what he says or does.

We thus see that Denis is not simply *co-enacting* or *co-producing* a given situation, a situation to which Robert and Alberto, of course, contribute. He is *also*, to a certain extent, reacting to it, *marking the agency* of what appears to matter to him. By showing and explaining to Robert and Alberto what is about to be built, he is also justifying why he needs a diffuser. In terms of agency, a series of translations appear to be operating: a space that is about to be built (the tech room) happens to matter to him because of what he knows of this kind of room and the heat they generate. In turn, this preoccupation gets translated through what he is saying to Robert and Alberto, who can then understand what dictates, in this specific case, the installation of diffusers, which is what is *supposed* to concern them at this point.

As we see, *there is no gap between materiality and discursivity*, as what Denis is saying is supposed to translate his concerns in this situation at this moment. Relationally speaking, we could then observe that preoccupations, concerns, or worries fill the gap because of their hybrid nature (physical, mental, and discursive). They can be pointed to (this is what Denis is doing when he marks with his hand where the tech room will be). They inhabit the person who experiences them (Denis says that *he* needs a diffuser, marking a form of appropriation about what needs to be done, that is, what the situation dictates), and they can be expressed discursively (when he says that he needs a diffuser, we also understand that this is a preoccupation that expresses itself at this point, a preoccupation that is supposed to translate what, according to him, the situation calls for).

If we go back to the interaction, we then see Robert pointing to the ceiling of an adjacent room while making a motion with his hand (lines 23–25). He then says, "You've got two diffusers up here" (line 26). In terms of material agency, we could claim that these two diffusers that he is pointing must have, by definition, *caught his attention*, an attention or consideration that is supposed to get translated in what he is now saying to Denis. "You've got two diffusers up here" can then be understood as an indirect way for him to ask Denis what or if something should be done about them (otherwise, why would he be saying that?). Interestingly, we then see Denis moving into the adjacent room and then saying, after a 1-second-long silence, "Oh, we're good here" (line 28), so as to mark the irrelevance of preoccupation/concern/worry for them.

As we see, material agency is not a phenomenon that the analysts arbitrarily identify, but something that interactants implicitly or explicitly orient to. Indeed, another way to translate what Robert is doing in saying "You've got two diffusers" (line 26) could consist of noticing that he is also asking what the presence of these two diffusers *dictate* in terms of intervention on their part. In other words, what does the presence of these two diffusers tells them (in terms of preoccupation and intervention)? In this respect, Denis's response—"Oh we're good here" (line 28)—implicitly points to the absence of concerns on his part, an absence of worries that he extends to a "we," which could be referring to Robert, Alberto, and himself or to Denis's own construction team. In saying, "Oh we're good here," it is therefore as if he were implicitly saying that they—whoever *they* are—should not be worried about these two diffusers. "Being good" implicitly means that since these two diffusers are good where they are, the "we" in question should also be good here, marking an interesting translation between matters of facts and matters of concerns.

In terms of materiality, we could also notice, through this example, that what appears to matter at a specific point (the two diffusers Robert is pointing to) can lose a part of its materiality because interactants come to the conclusion that what could have been a matter of concern for them precisely does not happen to be so. This is what is happening when Robert responds, "Yeah, you're fine here" (line 29), which, among other things, contribute to the closing of the discussion about these two diffusers he was pointing to. In the rest of the sequence, these two diffusers will not matter(ialize themselves) in the interaction anymore, which means that they will not animate or stand under, that is, substantiate, any concern or preoccupation on the part of Denis and his interlocutors.

Our point is not to say that these two diffusers suddenly vanish from the top of the ceiling—they keep a part of their materiality to the extent that they remain *simply there* because of their parts that *stand under* them, making them what they are—but they will not matter in the discussion anymore, at least in this sequence. Not mattering anymore means, as pointed out previously, that they will not substantiate any concern or preoccupation, but also that they will precisely lose their materiality *as* (objects of) concern and preoccupation.

In other words, *as* diffusers, they do not lose their materiality, but *as* concerns or preoccupations, they do, precisely because they do not matter anymore.[6]

Going back one last time to the interaction, we see how Denis quickly goes back to what he needs once the discussion on the diffusers has been closed ("I'm gonna need something in this tech room" (line 30)). This turn-at-talk can be interpreted as a way to bring Robert back to what should matter to him because it matters to Denis, that is, the diffuser for the tech room that is about to be built (and not the ones that Robert was pointing to). Interestingly, we can notice how something that does not exist yet, at least physically in the office space—the tech room—can display some agency, to the extent that its upcoming existence has already started to matter or count to Denis, occupying and inhabiting his mind, that is, preoccupying him, a preoccupation that he is conveying again to Robert.

As shown in this latest analysis, our point is not to defend a materialistic position, but to show, on the contrary, that what matters does not have to be physically material per se. Indeed, we saw that two diffusers that are physically present can lose their materiality *as preoccupations*, while a tech room, which is physically absent, can definitely matter as a preoccupation, precisely because it materializes itself in people's minds and discussions. Our world thus constantly navigates between these effects of presence/absence where certain questions, issues, beings, realities come to materialize themselves in specific circumstances, while losing (a part of) their materiality in others (see also Scott and Orlikowski, this volume, on stabilizing and destabilizing).

As pointed out by Étienne Souriau (1956) more than fifty years ago, we should think in terms of "the existential incompletion of each thing" (p. 5). That is, we should not adopt an *either/or* logic when we start thinking of the (in)existence of anything, but rather a *both/and* thinking (see also Cooren et al., 2006, esp. p. 3). What does it mean concretely in the situation(s) here and now? It means, as pointed out by Souriau himself, that things have *various modes of existence* (or materiality) and that simply thinking in terms of their mere existence versus inexistence (or materiality vs. immateriality) begs the question of *the type of description* (Descombes, 1996) that

[6] Just to be clear, we are NOT saying or implying that something would completely and absolutely lose its materiality when it would not matter anymore to people who are engaged in a debate or discussion. It is *as a form of preoccupation or concern* that it can stop mattering to people. But this does not mean, of course, that, *as a thing*, it would suddenly vanish from our world. This would amount to holding a solipsist position, which would go against our relational ontology. For instance, and as we all know too well, things that apparently do not matter to some people can start to strike back, as Weick (Weick, 1990, 1993; Weick and Sutcliffe, 2001) and Latour (2004) brilliantly showed us on several occasions. Striking back precisely means that people can sometimes be almost forced to realize that what did not initially matter to them actually does matter because it materializes itself through various forms (dangers, outrage, deaths, etc.). What is crucial is to keep thinking and speaking relationally while acknowledging the multiple forms by which certain things start to express themselves.

is adopted when such existence or inexistence (materiality or immateriality) is proclaimed.

For instance, the tech room is physically inexistent even if its agency can definitely be felt in the discussion we just analyzed. Why is it so? Precisely because this tech room has at least two modes of existence: *As* a preoccupation or matter of concern, it has already started to exist in Denis's mind and discourse even if *as* a room per se, it does not exist yet (even if its contours can be shown by Denis to Robert and Alberto).[7] As pointed out previously, the key point is to acknowledge the *relational* ontology of our world (Robichaud, 2006), that is, to recognize that something like a tech room can be (described) (as) not only a physical location but also a preoccupation, a sketch on a blue print, a space defined by Denis's arms, or even a simple word ("tech room").

As we see, such a relational ontology is possible only if we do not operate what Alfred North Whitehead (1920) denounces as the *bifurcation of nature*, that is, the flawed separation of the physical dimensions of a given object, that is, its form, matter, solidity, etc. (what John Locke (1690/1959), for instance, would have called its "primary qualities") from its experience and comprehension by human beings, that is, its color, taste, attractiveness, or even beauty (what Locke would have called "secondary qualities"). All these qualities are relational, that is, acknowledged under a specific description, which is precisely what is illustrated by the excerpt we analyzed.

Conclusion

Thinking about the question of material agency does not mean that human beings are suddenly dispossessed of their own agency, that is, their own capacity to make a difference or "to act otherwise," as Giddens (1984) would say. On the contrary, it allows analysts to take the question of their agency, but also their *authority* and even *power* extremely seriously. As long as the only vocabulary at our disposal was one of structure (read, "conditions of production") and (human) action (read, "the productions themselves"), the key question of power and authority could be bracketed out—as often denounced by critical theorists when they refer of the work of conversation analysts (see, e.g., the debate between Schegloff (1997) and Wetherell (1998)). Or, on the contrary, power and authority questions are so pervasive that analyzing an interaction could almost be deemed completely futile (see,

[7] Although we do not have the space to address this question, it is noteworthy that Charles Sanders Peirce (1897) developed the same viewpoint in his logic of relations (see Descombes, 1996, as well as Rorty, 1967), as well as Jacques Derrida (1992, 1994) and what could be called his *hauntological* vision of our world.

e.g., Bourdieu, 1991, for an extreme version of such a position, but also Deetz et al., 2007).

However, if analysts start to adopt a relational ontology, they are, we contend, in a very good position to speak interactionally of (human) agency, power, and authority. For instance, we may acknowledge, analytically speaking, that when Denis is speaking at a specific moment, it is not only he, for instance, who is mentioning that something should be done but also—*and through his conduct and turns of talk*—the situation here and now he is referring to, a situation that, according to him, *dictates* that specific actions be carried out. Such effects, which can also be associated with a form of ventriloquism (Cooren, 2008, 2010), can interestingly be extended to other forms of agency, which have been alluded to during our analysis. For instance, as he was entering the office space, we also saw how he started to speak *as* a building manager, a building manager who is himself supposed to speak *on behalf of* or *for* the (interests of the) real estate agency he is representing.

To all these voices we can also add his expertise and experience, which can also be felt and heard when he implicitly shows that he *knows* that a room full of equipment *calls for* a diffuser. When he speaks, it is therefore also his experience and expertise that speak, an experience and expertise that is also displayed or incarnated through Robert and Alberto's conducts and questions. Speaking of material agency thus means that one can begin to acknowledge, analytically speaking, all the things that appear to *matter* in a given interaction. Whether these things are forms of knowledge, situations, expertise, emotions, concerns, interests, or even organizations (just to name a few of them), they can potentially make the human participants more authoritative and powerful because of the capacity these latter have to ventriloquize them.

As we tried to show in this chapter, the question of materiality should be understood and studied *relationally*, that is, we should always ask ourselves, "material to whom or what?" instead of simply asking what is material or not material, a question that, we contend, can never be answered in and by itself. If something is material, it means that this thing is the ground, reason, or cause for something else, that is, it stands under, brings about, materializes, or explains something else. For example, something can be said to be material because people can touch, smell, see, hear, or taste it, which means that this thing stands under, explains, or causes what people feel and say when they enter into relationship with it through their senses (see, e.g., Fayard, Chapter 9).

But, as we saw, the question of materiality does not need to be reduced to these forms of sensory experiences and can be extended to objects or situations that are deemed to be less immediately present. We are especially thinking of things like expertise and experience, but also preoccupations, concerns, worries, interests, and, through them, realities and situations, which can be felt and unfolded in a conversation. All these things can express themselves in any interaction, translating what is supposed to matter in a given situation, whether it is the fate of our planet, the rumor of an upcoming earthquake (see Burrell, this volume), or a simple diffuser that needs to be installed.

As long as only human beings are considered to be the sole actors in the interactional scene, discourse and materiality bifurcate, thus falsely separating the world of communication from the so-called "material world," an expression that, as we saw, does not mean anything in and of itself (see also Kallinikos, this volume, as well as Searle, 1995, for a similar position, but based on different premises and leading to different conclusions). Now that we showed, both theoretically and empirically, that many other things do contribute to the ongoing development of any interactional event, we are, we think, better equipped to demonstrate why matters always matters in communication, which is another way to demonstrate why communication always matters in matters of concerns.

References

Ashcraft, K. L., Kuhn, T., and Cooren, F. (2009). Constitutional amendments. *The Academy of Management Annals*, 3(1), 1–64.

Austin, J. C. (1962). *How to do things with words*. Cambridge, MA: Harvard University Press.

Barad, K. M. (2007). *Meeting the universe halfway*: Durham, NC: Duke University Press.

Beahky, B. A. (2003). Sharing meaning across occupational communities: *Organizational Science*, 14, 312–30.

Bencherki, N. and Cooren, F. (2011). Having to be: The possessive constitution of organization. *Human Relations*, 64(12), 1579–607.

Benoit-Barné, C. and Cooren, F. (2009). The accomplishment of authority through presentification. *Management Communication Quarterly*, 23(1), 5–31.

Boden, D. (1994). *The business of talk: Organizations in action*. Cambridge Polity Press.

Bourdieu, P. (1991). *Language and symbolic power* (G. R. M. Adamson, Trans.). Cambridge: Polity Press.

Brummans, B. H. J. M., Cooren, F., and Chaput, M. (2009). Discourse, communication, and organisational ontology. In F. Bargiela-Chiappini (ed.), *The handbook of business discourse* (pp. 53–65). Edinburgh: Edinburgh University Press.

Chaput, M., Brummans, B. H. J. M., and Cooren, F. (2011). The role of organizational identification in the communicative constitution of an organization. *Management Communication Quarterly*, 25(2), 252–82.

Cooren, F. (2000). *The organizing property of communication*. Amsterdam: John Benjamins.

—— Taylor J. R., and van Every, E. J. (eds.) (2006). *Communication as Organizing*. Mahwah, NJ: Lawrence Erlbaum.

—— (2008). The selection of agency as a rhetorical device. In E. Weigand (ed.), *Dialogue and rhetoric* (pp. 23–37). Amsterdam: John Benjamins.

—— (2010). *Action and agency in dialogue*. Amsterdam: John Benjamins.

—— Bencherki, N. (2010). How things do things with words.

—— Fairhurst, G. T. (2009). Dislocation and stabilization. In L. L. Putnam and A. M. Nicotera (eds.), *The communicative constitution of organization* (pp. 117–52). Mahwah, NJ: Lawrence Erlbaum Associates.

—— Matte, F. (2010). For a constitutive pragmatics. *Pragmatcs and Society*, 1(1), 9–31.

—— Sanders, R. E. (2002). Implicatures: A schematic approach. *Journal of Pragmatics*, 34, 1045–67.

—— Taylor J. R., and Van Every, E. J. (eds.) (2006). *Communication as organizing*: Mahwah NJ: Lawrence Erlbaum.

—— Kuhn, T., Cornelissen, J. P., and Clark, T. (2011). Communication, organizing and organization. *Organization Studies*, 32(9), 1149–70.

D'Adderio, L. (2011). Artifacts at the centre of routines. *Journal of Institutional Economics*, 7(2), 197–230.

Deetz, S. (2003). Reclaiming the legacy of the linguistic turn. *Organization*, 10: 421–9.

—— Heath, R., and MacDonald, J. (2007). On talking to not make decisions. In F. Cooren (ed.), *Interacting and organizing* (pp. 225–44). Mahwah, NJ: Lawrence Erlbaum.

Derrida, J. (1992). Force of law. In D. Cornell, M. Rosenfeld and D. C. Carlson (eds.), *Deconstruction and the possibility of justice*. New York: Routledge.

—— (1994). *Specters of Marx*. New York: Routledge.

Descombes, V. (1996). *Les institutions du sens*. Paris: Éditions de Minuit.

Eisenberg, E. (2006). *Strategic ambiguities: Essays on communication, organization, and identity*. Thousand Oaks, CA: Sage.

Fairhurst, G. (1993). The leader–member exchange patterns of women leaders in industry: A discourse analysis. *Communication Monographs*, 60, 321–51.

—— (2007). *Discursive leadership*. Thousand Oaks, CA: Sage.

—— Cooren, F. (2009). Leadership as the hybrid production of presence(s). *Leadership*, 5 (4), 469–90.

—— Putnam, L. L. (2004). Organizations as discursive constructions. *Communication Theory*, 14(1), 5–26.

Fayard, A.-L. and Weeks, J. (2007). Photocopiers and water-coolers. *Organization Studies*, 28, 605–34.

Garfinkel, H. (1967). *Studies in ethnomethodology*. Englewood Cliffs, NJ: Prentice Hall.

—— (2002). *Ethnomethodology's program*. Lanham, MD: Rowman & Littlefield Publishers.

Giddens, A. (1984). *The constitution of society*. Cambridge: Polity Press.

Groleau, C. (2006). One phenomenon, two lenses. In F. Cooren, J. R. Taylor, and E. J. Van Every (eds.), *Communication as organizing* (pp. 157–77). Mahwah, NJ: Lawrence Erlbaum.

Heritage, J. (1984). *Garfinkel and ethnomethodology*. Cambridge: Polity Press.

Hutchins, E. (1995). *Cognition in the Wild*. Cambridge, MA: MIT Press.

Kallinikos, J. (2004). Farewell to constructivism. In C. Avgerou, C. Ciborra, and F. Land (eds.) *The social study of information and communication technology* (pp. 140–61) Oxford: Oxford University Press.

Kuhn, T. (2008). A communicative theory of the firm: *Organization Studies*, 29(8–9), 1227–54.

Latour, B. (1996). On interobjectivity: *Mind, Culture, and Activity*, 3(4), 228–45.

—— (2004). *Politics of nature*. Cambridge, MA: Harvard University Press.

—— (2005). *Reassembling the social*. Oxford: Oxford University Press.

—— (2008). *What is the style of matters of concern?* Amsterdam: Van Gorcum.

—— (2011). Reflections on Étienne Souriau's Les différents modes d'existence. In B. Levi, N. Srnicek, and G. Harman (eds.), *The speculative turn* (pp. 304–33). Melbourne: Re.press.

Law, J. and Mol, A. (1995). Notes on materiality and sociality. *The Sociological Review*, 43, 274–94.

Leonardi, P. M. and Barley, S. R. (2008). Materiality and change. *Information and Organization*, 18(3), 159–76.

Locke, J. (1690/1959). *An essay concerning human understanding* (Vols. 1 and 2). New York: Dover Publications.

McPhee, R. D. and Zang, P. (2000). The communicative constitution of organizations. *The Electronic Journal of Communication*, 10(1/2), 1–16.

Nardi, B. A. (1996). Studying context. In B. A. Nardi (ed.), Activity theory as a potential framework for human–computer interaction research (pp. 69–102). Cambridge, MA: MIT Press.

Orlikowski, W. J. (2007). Sociomaterial practices. *Organization Studies*, 28(9), 1435–48.

—— Scott, S. V. (2008) Sociomateriality. *The Academy of Management Annals*, 2, 433–74.

Peirce, C. S. (1897). The logic of relatives. *The Mouist*, 7(2), 161–217.

Pickering, A. (1995). *The mangle of practice*. Chicago: University of Chicago Press.

Putnam, L. L. and Nicotera, A. M. (eds.) (2009). *The communicative constitution of organization*. Mahwah, NJ: Lawrence Erlbaum Associates.

Robichaud, D. (2006). Steps toward a relational view of agency. In F. Couren, J. R. Taylor, and E. J. Van Every (eds.) *Communication as organizing* (pp. 101–14). Mahwah, NJ: Lawrence Erlbaum.

—— Giroux, H., and Taylor, J. R. (2004). The meta-conversation. *Academy of Management Review*, 29(4), 617–34.

Rorty, R. (1967). Relations, internal and external. In P. Edwards (ed.), *The encyclopedia of philosophy* (Vol. 7, pp. 125–33). New York: Macmillan.

Schegloff, E. A. (1997). Whose text? Whose context? *Discourse & Society*, 8(2), 165–87.

Searle, J. R. (1979). *Expression and meaning*. Cambridge: Cambridge University Press.

—— (1995). *The construction of social Reality*. New York: Free Press.

Souriau, É. (1956). Du mode d'existence de l'oeuvre à faire. *Bulletin de la société française de philosophie*, 25, 4–44.

Taylor, J. R. (1988). *The organisation h'est qu'un tissu de commuication*. Montreal: Cahiers de recherches en communication.

—— (1993). *Rethinking the theory of organizational communication*. Norwood, NJ: Ablex.

—— Van Every, E. J. (2000). *The emergent organization*. Mahwah, NJ: Lawrence Erlbaum.

—— Cooren, F. Giroux, N., and Robichand, D. (1996). The communication basis of organization. *Communication theory*, 6(1), 1–39.

—— —— (2011). *The situated organization*. New York: Routledge.

Weick, K. E. (1979). *The social psychology of organizing*. New York: Random House.

—— (1990). The vulnerable system. *Journal of Management*, 16, 571–93.

—— (1993). The collapse of sensemaking in organizations. *Administrative Science Quarterly*, 38(4), 628–52.

—— Sutcliffe, K. M. (2001). *Managing the unexpected*. San Francisco, CA: Jossey Bass.

Wetherell, M. (1998). Positioning and interpretative repertoires. *Discourse & Society*, 9, 387–412.

Whitehead, A. N. (1920). *Concept of nature*. Cambridge: Cambridge University Press.

16

The Materiality of Rumor

Jenna Burrell

This chapter explores a circulating rumor and its process of materialization, considering the more general question of how the nonmaterial comes into material being. In examining this process, I mark a distinction between rumors material *aspects* and material *effects*.[1] The effort to distinguish between what is part of (an aspect) versus what follows from (an effect) raises the question of units of analysis, of defining the boundaries of the "stuff" under examination. A broadly encompassing notion of the material is most appropriate to the current case and in working out these distinctions. By contrast to other chapters in this collection, I work with a notion of the material that extends beyond the man-made or technological, including the natural world and the corporeal as material domains. This broader stance is aligned with material culture studies where complementary insights about matter and materiality have been explored in parallel with the recent work in organizational theory and in science and technology studies scholarship (Miller, 1998, 2005). The materiality of rumor specifically is linked to the body and the production of speech through the vocalizing organs, and the functioning of human memory. These are critical in constituting rumor's material aspects.

Studying rumor from a materialist standpoint helps to underline the point that *matter is not the only thing that matters* (Leonardi, 2010). In other words, the nonmaterial can also be hugely consequential via its materialization. What this consequentiality might mean will depend upon where one draws the line between what is material and what is nonmaterial. To be consequential is to manifest in some way as a diverting force. Thus, a critical question is how something nonmaterial is ultimately enacted or manifested. I wish to add

[1] I thank Paul Leonardi for pointing out this distinction at the workshop.

to this initial assertion a caveat—*without matter, nothing matters*. These two claims may seem to be contradictory and indeed are when a definition of material and nonmaterial rests on their mutual exclusivity—that is, that there are wholly material "things" and then there are wholly nonmaterial things. What I wish to establish instead is the interdependency and inextricability of material and nonmaterial. I argue this in opposition to what Faulkner and Runde claim for an independent ontological status of nonmaterial objects as separable from their material "bearer" (Faulkner and Runde, 2011). Yet, when it comes to the nonmaterial, to be able to experience and speak about it requires some form of externalization and in this way must engage with the material.

A study of rumor, in particular, presents an intriguing challenge to materialist accounting because of the fundamental nonmateriality of the rumor itself as a circulating tale of an event or happening which often proves illusory. The imaginary of such tales and their underpinnings in shared but unarticulated concerns and anxieties is similarly nonmaterial. However, this belies the powerful material force frequently generated in the wake of rumor, for example, in the mass movement of populations, collective boycotts, even riots or physical attacks that sometimes follow. Rumors show a capacity to generate an emergent organizing. Furthermore, a rumor, in and of itself, through its continual reperformance endures well beyond more ordinary and everyday acts of speech. A study of rumor as a special spoken genre with such peculiar properties lends itself to a broader reconsideration of speech and representation and the tie between materiality and discursivity, as is argued elsewhere in this collection (Cooren et al., Chapter 15).

The materialist turn in social theory has typically been positioned as a critique and alternative to analytical approaches that position language and discourse as principally constituting the social world. For example, the materialist argument of actor network theory (ANT) offers a relational materiality performed through the assemblage of material elements, rather than as emanating from fixed properties of things (Law, 1999). This is explained by John Law, Bruno Latour, Madeleine Akrich, and others through an analogy that relocates the principles of a *linguistic* semiotics (concerned with meaning produced in the relationship among words) to a *material* semiotics (concerned with the material effects of relationships between objects/entities) (Akrich, 1992; Akrich and Latour, 1992; Latour, 1993). However, one outcome of this analytical move is that language, words, and the work of representation have been left with an ambiguous and underdefined role in ANT studies. This stance has yielded a concern with apparent material *entities* rather than material *continuities* and, in general, a tilt toward an object-centrism evident in the early fascination of ANT accounts of sailing ships, electric cars, door-closers, scallops, etc.

A new direction in materialist accounting has begun to explore instead, not those cases of sharply bounded physicality, the tangible and substantial, but the conceptual edges where matter or substance is not so evidently massed as an apparent "object." This is most evident in recent work engaging questions of digital materiality (Kallinikos, Chapter 4; Yoo, Chapter 7; Ekbia and Nardi, Chapter 8) (Kallinikos, Lanzara et al., 2010; Faulkner and Runde, 2011; Sunden, 2003). Likewise, a return to language and in particular to performed speech, attending to its distinctive and varied materiality serves to complement and extend this kind of work. Verran (2001) in an early effort pursued this by exploring the materiality of numeracy through the bodily materiality of fingers and toes—in the base-10 system of counting in English versus the multibase system of 20, 10, and 5 in the Yoruba language. The present case of rumor is also a reminder that an expanded analysis of materiality does not depend upon the emergence of novel phenomena or material inventions. Pervasive digitalization is a fascinating turn in contemporary life in many parts of the world, but is certainly not the only way to consider materiality beyond the physical object.

As speech and the human voice are the typical and traditional format for producing instances of rumor, we may begin by considering how speech (and more generally the corporeal) has been situated explicitly or implicitly in materialist theory, specifically in early ANT work. The efforts of actor network theorists made the case for a social theory that accounted for the material world by emphasizing the limits of the human body (including voice) as a mechanism of social ordering. They turned analytical attention to the significance of structures, machines, and texts and their role in social constitution. Primatologist Shirley Strum with anthropologist Bruno Latour pointed out that baboon societies rely upon their bodies alone and that this limited the scope of their social ordering (Strum and Latour, 1987). In human societies, the great surplus of things, an ever-expanding object world, was how we were able to extend society beyond our primate neighbors. The point was not that the corporeal was nonmaterial, but that there were limitations in the degree and the scope of the body as an enacting force.

Law makes a series of assertions in *Organizing Modernity* that summarize this point noting, "...some materials last better than others. And some travel better than others. Voices don't last for long, and they don't travel very far. If social ordering depended on voices alone, it would be a very local affair. Bodies travel better than voices and they tend to last longer. But they can only reach so far—and once they are out of your sight you can't be sure that they will do what you have told them...machines, though they vary, may be mobile and last for longer than people. Texts also have their drawbacks. They can be burned, lost or misinterpreted. On the other hand, they tend to travel well and they last well if they are properly looked after" (Law, 1994:

102). Law constructs this provisional hierarchy (in terms of a *tendency* toward durability—from more durable to less):

Texts/machines/buildings
↓
Bodies
↓
Voices

However, to prevent a reification of this ordering (which would contradict the very principle of a relational materiality) Law uses the language of, "may have," "is liable," or "tends to." He goes on to explain these qualifications asserting that, "it sounds as if I am saying that mobility and durability are properties given in nature. But this is wrong. Mobility and durability—materiality—are themselves relational effects." In a 2007 update of ANT he restates the tendencies in durable ordering but again qualifies that, "stability does not inhere in the materials themselves" (Law, 2007). What I attempt in the analysis that follows is to offer a compelling counterexample to show how an entity (an instance of rumor) produced largely through voice, when caught up in the right sort of relations, may in fact be quite durable yielding a social reordering across geographic distance counter to Law's hierarchy. In this way, we might examine more thoroughly the idea of durability as a relational effect. An examination of rumor in the case considered below is meant to do precisely that.

While any given instance of speech may be ephemeral, once we add to that a more relational consideration looking at the contents of what is spoken and the larger unifying phenomenon (i.e., patterns in format, what motivates the speech in the first place, and what is being attempted through speech), what becomes apparent is the diverse materiality (both in terms of material aspects and material effects) of different formats of spoken exchange. Certain oral forms such as rumors, jokes, aphorisms, or songs can be contrasted to more mundane, unmemorable everyday utterances that make up the greater proportion of spoken exchange. These forms work along the vocalizing organs and the brain and memory in different ways. There is consequently a diversity to the particular materiality in the way they are enacted, both between formats (rumors vs. jokes) and even between instances of a format (different examples of rumor). This diversity is easily overlooked where materialist accounts treat speech/voice/language performance as the homogenously weak other to the durability of physical artifacts. This chapter looks in particular at rumor, a speech form that challenges assumptions about the homogeneous ephemerality of the spoken. Rumors manage to spread far and wide, evading the efforts of official sources to refute and diminish them. Rumors can persist for years or decades. They endure in a way that recent materialist analysis has tended to consider only in built structures, machines, or texts.

Matter, Human Scale, and Tangibility

To argue for the materiality of rumor, it is necessary to clarify further what I treat as matter (meaning substance) relating this to the way I have framed the issue around consequentiality, that is, of matter as a redirecting force. General definition work has sometimes used the measure of the sensing and receiving human body as the judgment of what is or is not matter. Leonardi in his suggestion for how we might come to understand digital (as opposed to physical) artifacts rests the issue on tangibility, that sense of touch is the final arbiter on what is matter (Leonardi, 2010). One might ask why, for example, nano-sized manufactured items should be excluded from the material realm because our skin receptors are not fine-grained enough to feel them. Why should a virus, for example, because we cannot pick up and grasp it, therefore not be considered as material. A magnified image of this virus printed onto paper showing it as an apparent and bounded object is material by such a judgment, while whatever is depicted is not. A similar line of thought seems to guide Faulker and Runde (2011) in their definition of non-material digital objects (more specifically, bitstrings) in a way that largely overlooks an inescapable and fundamental constitution in the materiality of the integrated circuit. The etching on silicon yielding the microprocessor is another microscale process whose results are boxed up in the computer case and thus removed from direct human apprehension. These examples show a certain arbitrariness in registering as matter only what is human-accessible and specifically human scale.

In addition to broadening a consideration of the material in relation to scale, I also extend consideration beyond the sole sensory channel of touch. Particles, displaced sound waves, light patterns and disturbances, all of these are received primarily through other sensory channels, and yet can certainly initiate consequences in their own distinctive way. Since such displacements and disturbances are external to the one who senses them, beyond the recipient's impetus and control, they may be considered matter/substance. In this broader definition, one can see that anything at all that is experienced has its material component. This is precisely the point necessary to the argument that *without matter nothing matters*. This definition work prioritizes processes of materialization (rather than the work of categorizing material and nonmaterial). While an oral form (such as rumor) is not in its essence a material thing, we must still ask the important question of how it is materialized and how this compares to other formats of language production and, more narrowly, of speech acts.

The definition of matter/substance calls reciprocally for a clear definition of rumor with reference to the material. I define rumor here as an account of some event the teller believes to have taken place in the real world, but that is received second-hand (rather than observed directly) by the teller. This is a different definition of rumor than the one that prevails in public understanding typically equating rumor to *false* belief. The problem with a fixation on truth in rumor is that it looks past the possibility of rumors' constitutive role and focuses instead on the rationality and credulity of those who receive and tell such stories. This perspective can be critiqued as an example of what Barad calls the "trap of representationalism," the normative notion that language performance is *supposed to be* a mirror of reality (with divergences such as false rumors a problem or failure of this normative order) (Barad, 2003). As a comprehensive definition of rumor, "falseness" also falls short since rumor scholars studying such tales in circulation have found they sometimes prove to be true (Kapferer, 1990). That said, that rumors are frequently false does help to illuminate how such a speech form constitutes something in and of itself rather than merely indexing the apparent physical/material world. On the whole, rumor defined as a second-hand account is the more consistent aspect of the definition and is what is carried through in the following analysis.

"The Day the Nation was Fooled"

On January 12, 2010, a 7.0 magnitude earthquake struck the country of Haiti centered only 10 miles from the capital Port Au Prince. Shortly afterwards, once the global media outlets managed to land correspondents in country, the world began to see broadcasts of the grim devastation: imploded buildings, slabs of concrete leaning precariously, the bodies of unfortunate victims obscured beneath piles of rubble, and massive tent cities filled with distressed and dispossessed survivors.

Nearly 5,000 miles away in the West African nation of Ghana, the populace began to receive these mass-mediated images of the disaster on TV, on the radio, and in reports in the local newspapers. Then on the night of January 18, six days after Haiti's earthquake, this distant event became suddenly personalized as a warning describing an impending earthquake due to strike Ghana at any moment began to spread. It moved from region to region in Ghana through mobile phone calls and text messages. Ghanaians called family and friends urging them to leave the homes where only moments before they had peacefully slept, these often concrete-brick structures were much like the buildings in Port Au Prince. People fled into the streets, gathering in open spaces, some carrying their most valued possessions—a television, important documents, jewelry, or cash. By shouting and pounding on doors, concerned neighbors attempted to rouse those unreachable by phone. Throughout the night, these displaced citizens prayed, listened to the radio, called one another until the phone networks failed from overuse, and eventually tried to catch a bit of uncomfortable sleep in the nighttime chill of harmattan season. By early morning, as the local news shows grappled with this event, seeking confirmation or clarification, it became apparent that no earthquake had taken place and that none was imminently due, that no official body had issued this prediction, and instead that a powerful bit of misinformation had swept through the telecommunication networks, from citizen to citizen, generating this dramatic outcome. Authorities from the Ministry of Information, the Geological Survey Department (GSD), and the National Disaster Management Organisation (NADMO) gave interviews to the newspapers and on the radio, refuting the rumor and urging citizens to return to their normal routines.

My research assistant Kobby referred to the event with a sense of humor as, "*the day the nation was fooled.*" A feature article from the Ghana News Agency recalling the incident a couple of weeks afterwards was titled, "the text message that robbed Ghanaians of sleep." Having encountered a report of this event while browsing Ghana's major news website Ghanaweb, I decided as part of a planned return to Accra, Ghana's urban capital city, that I would try to collect some first-hand accounts of the event as it was experienced by Ghanaian citizens. This extended an interest in story-telling and technological sense-making sparked by rumors about "big gains" from Internet scams that I had been told years before by Internet café users in Accra (Burrell, 2011). The interviews carried out four months after the incident captured the event from the perspectives of a diverse cross-section of Ghanaian society. The group of interviewees ranged in educational attainment though skewed toward the more educated and English-fluent. They included both men and women, held a range of different religious beliefs (animist, protestant, evangelical, and Muslim), and ranged from affluent, well-connected professionals to low-income and unemployed individuals.

Among those I spoke with, all but two had heard the rumor that night giving some sense of its pervasiveness in the well-connected urban capital. It was, however, received with varying degrees of skepticism and belief. Among the highly skeptical, the rumor was not passed along. In fact, some made attempts to counteract its spread. For example, there was Farouk, a young leader from Mamobi, a low-income settlement in the heart of Accra, a local boy-made-good who had recently graduated from University. He received many calls the night of the rumor from people seeking to verify its truth, "people were calling to find out. And because of, you know, I read papers. I get more information and people will want to find out from me, is it true, is it true? In our house, if there's any news, people want to hear from me whether that is the news... so people called and I tell them that 'oh, you should forget about this, it's not true.'" However, those who accepted some small possibility that the warning might be true passed it along in light of the messages life-or-death implications. Freeman, a policeman, stated, "I was not thinking it will happen," yet despite his doubts he was moved by his sense of obligation and concern noting, "as a citizen of the country you show love to others. I can't wait for my colleagues to also die... I won't be happy if I should lose a friend or a family member. Especially a family member whose responsibility is on me." Consequently, as soon as he had received the warning, he worked through his mobile phone address book calling people one by one to make sure his entire social network had been informed.

The actions undertaken by those who heard the earthquake rumor to alert others stemmed from this sense of responsibility, but also required a basic sense that the tale was credible. Ghanaians mentioned several reasons why they took this particular earthquake prediction as likely to be true. The rumor itself was often passed along with a reference to an authoritative source which contributed to its credibility. Rumor tellers referred to reliable media (the radio, the Internet) as well as official institutions such as NADMO and certain professions presumed to be authorities on the matter. In particular, meteorologists, geographers, geologists, and astronomers were mentioned. Another contribution to credibility was the precedent set, not just by the Haiti earthquake, but by smaller earthquakes that had been reported in Ghana, including one that had occurred in the mid-1990s. One young woman, Joyce, also noted that the actions and retelling by others confirmed for her that the story had some truth to it. She noted, "I summoned the people in my house and lo and behold someone too has heard it. So it's like, there was evidence and when we came out, we saw people outside." Belief that stems from the fact that others tell and believe a tale generates a "social truth" typical in the circulation of rumors (White, 2000). As it is retold and diffuses more widely, it therefore gains credibility and becomes further strengthened.

Material Aspects of Rumor

From a materialist stance, what is apparent in rumor is its enactment and constitution in a pattern of continual reperformance through face-to-face speech (and, as of late, in other interpersonal modes such as phone calls, text message, email, etc.). This reproduction depends also upon the way it lodges in the memories of those who hear it. Yet it is not just that such a story is remembered, but that it compels retelling. Rumor's material aspects are formulated in human speech and memory, both critical to how a rumor comes to be known, experienced, and socially consequential. Rumor is materialized again and again with each spoken instance. Rumor is defined not only by the disembodied symbolic elements of the story but also by this particular pattern of widespread reproduction. Thus, rumor is not nonmaterial, existing apart from the way it is borne materially, rather the nonmaterial and material interdependency is what constitutes *rumor* as opposed to a more ephemeral recounted story.

While accepting that rumor is not just a type of story (distinct from its materialization), we may still ask what role is played by the contents of the rumor itself, the sequence of words, in accomplishing this durability. One might consider whether rumor's impact is through its indexicality, the way it points to material things in the world positing a relationship among that which it indexes. This would bring the motivating force underlying the spread of rumor back around to a materiality outside of it—in this case concrete structures, the movement of the earth, and the threat this poses to fragile human bodies. However, that this particular rumor is *false* and is eventually acknowledged as such is how we can see that rumor is rather a shared imaginary and in this aspect nonmaterial. There is not an actual NADMO prediction the rumor stems from, no actual earthquake that hit within a reasonably immediate timeframe,[2] rather these exist only in the rumor itself. Certainly, to have the kind of impact it does, rumor leverages material relations drawing from the kinds of patterns experienced or witnessed in the world (i.e., in the coverage around Haiti's earthquake), but is also surplus to that.

Another materiality to account for in this instance of rumor is the critical role played by the mobile phone network in Ghana providing 24-hour connectivity. The availability of cheap phones and widespread phone ownership had taken place only within the last ten years in Ghana, becoming nearly ubiquitous only in the past few. Prior to the mobile phone, phone access was generally limited to urban areas and residential phones were uncommon. For most Ghanaians, phone calls were limited to the daytime hours when public phone kiosks or communications centers were open for business.

Beyond the infrastructure of cell towers and phone handsets, it was not just the new convenience of calling and reaching someone directly day or night, but additionally the social conventions around phone use in Ghana, as well as competition between networks and promotional pricing schemes that played into the rapid spread of this rumor. Ghanaians typically kept their phones on and at the ready overnight. I found that there was little social approbation

[2] Though it should be noted that some interviewees did seem to believe that an earthquake had hit at some point after the rumor began to spread, but a smaller one, and in another region of Ghana.

against late night phone calls; at the same time it was common for callers to take a little offense when they called and the call was not picked up. Furthermore, with the many phone network providers in Ghana competing to win customers, given the price sensitivity of this customer base, each network heavily promoted pricing schemes including discounted hours, special in-network rates, and bulk air time (phone credit) buying discounts. Ghanaians were frequently experts at these schemes, keeping multiple SIM cards on hand for different phone networks in order to optimize calling rates. Given the time of night when the rumor spread, typically a time when phone traffic was light, calling rates were heavily discounted. As Freeman the policeman noted, "it was 'free nights' when you call, [the phone network provider will] give you 99 percent [discount]. So I called more than 30 people. When I call you I will also tell you to call a different person whom you know. Call him and get him informed." Further contributing to the credibility of the rumor, Ghana's national phone network became overloaded and calls stopped being connected. This breakdown of the phone network also confirmed a sense of the out-of-the-ordinary, of breakdown,[3] and of a vast, collective response underlining the social truth to this rumor.

One might ask why rumor is the focus of this materialist analysis rather than the seemingly more significant materiality of this novel technological infrastructure? Undeniably, changes in communications infrastructure brought about by the mobile phone were key in the peculiarly rapid and forceful materialization of this event. Many Ghanaians themselves mentioned mobile phones in realizing this nighttime exodus. Fauzia, a young unmarried woman who the night of the rumor was residing in Tarkwa, a mountainous area about 120 miles from Accra, said she heard about the earthquake warning in person from a neighbor. When asked about what made such a scenario possible, she responded that it is, "the way the world is these days." She recalled a prior reliance on the postal system and the written letters she remembers receiving many years ago from a sister who had traveled abroad. These days, as she noted, "news spread with the help of the mobile phones." Jacob, a University-educated bank worker living in a suburb of Accra, began to wonder following the event about the value of mobile phones in Ghana. He noted, "the message spread very fast even through the hinterland...," and though in this case it was a false tale, "we realized that [this] electronic way of communication... if something should have happened... a lot of people would be alerted and they would have saved a lot of lives." This concern is also reflected in a news item

[3] Larkin (2008) points to breakdowns as a key materiality to consider in African studies. His study of media infrastructures in Nigeria demonstrates how breakdowns reflect matters trajectories beyond what human intentions invest in them.

on the Ghanaweb site titled "Ghanaians ready for early warning systems"[4] noting the rapidity of the message's diffusion with the new telephone infrastructure in the country and the apparent responsiveness of Ghanaians to such a warning. The inclination was to blame or credit the mobile phone (and its network) as a kind of responsible party to the event. Yet, certainly none would admit to the mobile phone having a capacity to write and spread such a message autonomously. Instead Ghanaians pointed to a deceptive human source, an unknown rumor-monger, and called for the authorities to trace, identify, and hold this person accountable. Thus, motive was located exclusively in a human source and a critical materiality principally in the phone. What I have attempted to show, instead, by bringing the many pieces of this story together is that motive and materiality were diffused across various human and nonhuman components in what was an emergent unfolding scenario.

Ultimately, despite being referred to as "the text message that robbed Ghanaians of sleep,"[5] the common denominator in the earthquake rumor event in Ghana was the rumor itself, which, in practice, circulated in several formats—voice calls, text messages, and face-to-face retelling. The rumor provided the impetus for this compressed and rapid message dissemination that the phone was in place to facilitate. Thus, the rumor itself (and underlying it certain anxieties and notions of plausibility) served as the initiating force and as the constant in the event while the phone was its amplifier. Of course, the rumor was no more essential to the particular way this event unfolded than the phone. It would not have taken place with quite this scope or rapidity without either.

Material Effects of Rumor

Apart from the material aspects of this rumor, there are also its material effects. The most visible effect was this large population moving *en masse* from their homes into the streets as a result of the rumor's circulation. One can imagine how astounding the figure would be if we were to calculate all the kinetic energy expended seemingly from thin air. An immediacy and urgency in the response is evident, for example, in the way Joyce recalls her response to hearing the rumor. She noted (laughing at the memory of it) that she fled into the street without even being fully dressed, "a friend of mine called me around 3 o'clock... I wake up, take my cloth [wrap]. I wasn't even wearing

[4] "Ghanaians Ready for Early Warning System" (January 18, 2010), Ghana news Agency, http://www.ghanaweb.com/GhanaHomePage/regional/artikel.php?ID=175315

[5] "The Text Message that Robbed Ghanaians of Sleep" (February 5, 2010), Ghana Broadcasting Corporation, http://www.ghanaweb.com/GhanaHomePage/NewsArchive/artikel.php?ID=176127

pants! oh... I was afraid!" This reflects a measure of the rumor's force, the way that the issued warning: "NADMO has predicted an earthquake is about to strike here in Ghana" in combination with a spoken command—"leave your room, go outside"—yielded almost total conformity.

This particular rumor generated a kind of emergent organizing, facilitating a leaderless movement tapping into the wisdom or the madness of the crowds. The earthquake rumor and its aftermath in Ghana, as some observers noted, served unintentionally as a dress-rehearsal for an emergency response system. In the comments emerging after this rumor event, there was a certain awareness and celebration of how the collective population handled the circumstances—the consideration Ghanaians expressed not only for kin, but for acquaintances and neighbors, for more vulnerable groups such as the elderly, and the rapidity of the alert and response. "We thank God that we have community," Fauzia noted.

Officials of NADMO and other agencies refuted having ever issued such a warning. This left the populace of Ghana to sort through alternative explanations to try to make sense of the rumor and how it got started. The aftermath of the rumor generated other sorts of rumors. Meta-level analysis of the event yielded new rumors intended to make sense of the original earthquake rumor. There was nearly universal consensus, as noted above, that the rumor was a willful act of deceit. It was "somebody's lie" as Joyce described it. There were calls from political leaders, journalists, and the general public to identify the responsible party or parties. Those considered to be the most likely candidates for this act of deceit included attention-seeking radio DJs, entrepreneurial Christian preachers, armed robbers attempting to create chaos and expel people from homes in order to loot, or politicians wishing to generate an incident that would distract from political campaign problems.

A young man named Nana who was living in the low-income slum of Mamobi had the most detailed theory around the rumors possible political motivation. His explanation turned first toward an analysis of the radio stations in Ghana, what political parties they supported, who owned the station, and their political affiliations. The National Democratic Congress (NDC) were the party in power at that time, with the NDC president John Atta-Mills taking office in January 2009. Nana asserted that the rumor, "was political because we heard, a day before, NDC had come to Tamale to elect their leaders and the information I got from a friend in Tamale was... that some people [there] started texting [the rumor]." Having situated the rumors origins in a particular political event in the major northern town of Tamale, Nana then explained the motive for such a tactic, "a lot of people think there was going to be chaos at the election... it was not going to be successful. They know the media was going to talk about it. So they just brought that thing for people to put their attention to the earthquake." Nana, in resolving the rumor, situated the tale within the broader social terrain and the circumstances of media and politics in Ghana. In referencing these key figures, he drew upon a shared set of expectations about how such figures operate in and upon society as well as the most likely suspects who would desire this sort of influence and be capable of effecting it.

Alternately, in the literature on rumor, scholars generally argue that false narrative details in rumors likely stem from misunderstandings and guesswork under circumstances of incomplete information (Allport and Postman, 1965; Shibutani, 1966; Turner, 1993) not the puppet-mastering of malicious individuals or groups. To identify what narrative details can produce the momentum of rumor and the right social situation for such a circulation appears difficult (Turner, 1993).[6] Supporting the notion that rumors are essentially sourceless and emergent, no person or institution was ever officially identified as the source of Ghana's earthquake rumor, just as with prior documented studies of rumor.[7] Rumor to some degree is always beyond the initiation or control of the long chain of individuals who receive or retell it. This suggests that sequences of action following from the origins of a phenomenon in the nonmaterial can effect a kind of "inverse instrumentality" as Ekbia and Nardi argue (this volume) when they describe material systems that position human users in ways that are not clearly voluntaristic. In the present case, humans are instruments of a rumor's mobility, whereas Ekbia and Nardi consider a networked game (World of Warcraft) and electronic medical records in this capacity to demand humans for their completion.

In these material effects, we may glimpse a sense of why societies have and even why they may, perhaps, *need* rumors. Clearly, this efficient, populist mechanism of message dissemination communicates information rapidly when information is desired. It helps to look at the patterns documented in scholarly work on rumor to further illuminate the role they typically fill. Certainly, not every second-hand narrative becomes a rumor. We may think of enduring rumors as the survivors of a filtering process among all of the everyday spoken exchanges between people. They offer the most memorable of tales that resonate with the fears and imagination of a population. Bodily threat is a recurring and compelling theme. One common type of rumor involves a claim that powerful persons, institutions, or governments are intentionally spreading disease among certain undesirable populations or

[6] For example, British Knights shoes were the target of a widely circulating rumor among African-American populations that the corporation that sold the shoes supported the KKK following from a subtle connection interpreted from the brand name and logo.

[7] The most carefully documented example of rumor-mongering (apart from psychology experiments) is in corporate rumors about "tropical fantasy" a drink that targeted African-American consumers and was thought to be made to cause sterility in those who drank it. Turner (1993) documents how the corporation invested a great deal of time and energy into tracing this profit-impacting story, finding that business adversaries participated consciously in spreading the rumor, but they found no evidence that it was these business adversaries that started the rumor.

rendering these populations infertile (Turner, 1993), or stealing their organs (Scheper-Hughes, 1992) or blood (White, 2000). Luise White considers enduring rumors that circulated in East Africa for decades that suggested colonial authorities were killing and extracting blood from colonial subjects (vampire-like) to render health or economic advantages.

On the whole, rumors often take a more vague sense of opposition, animosity, or threat and attach it to concrete events and actions, pinning institutional responsibility, fixing a date and time, pointing to the method of the threat, and sometimes offering mitigating practices to manage it—in this instance, fleeing into open spaces for safety. Rumors thus channel and diffuse ambiguity and anxiety (Turner, 1993: 29–30). They may offer a release, a sense of purposeful action taken, of a threat having been handled. Thus, it seems that materialization in and of itself is perhaps cathartic, releasing a population from the burdensome isolation of an unarticulated worry. In the best case scenario, this process of materialization can be essentially harmless as in the current example, though darker scenarios of violence against scapegoats stemming from what turns out to be false information is also possible.[8]

Conclusion

The turn to questions of materiality—in understanding organizational forms—need not be a wholesale abandonment of language. Rather, a materialist stance can serve as a grounding for the consideration of how language is performed and the broader material effects spun off from these performances. There is much to be gained from studying speech acts, the construction of narratives and the relationship among words in and of themselves, but connecting that to how they come to be diversely materialized. There is a flexibility in language (and the way it is performed orally or in writing) that must also be acknowledged. Specifically, the possibility of speaking about something that does not otherwise exist, to speak in the future tense, or to knowingly tell a fictional tale, to speak of and share abstract and non-literal concepts reflects its inescapable tie to the nonmaterial despite its apparent material aspects.

Upon careful examination it is clear that speech acts exhibit diverse materiality with some formats (such as rumor), demonstrating remarkable durability in the sense of traversing across social networks or enduring through time.

[8] Here I am thinking specifically of rumors after the World Trade Center bombings in New York that circulated in some countries of the Middle East that the bombing was a Western conspiracy and that all Jews who worked in the building were warned in advance and stayed home on 9/11. This functioned as grounds for continued ethnic and religious opposition. Rumors after Japan's 1923 Kanto earthquake that Koreans were poisoning wells, committing arson and robbery, led to mob violence and the murder of thousands, it is estimated, of Koreans, Chinese, and others.

Here I have argued for this materiality as twofold: one, in the uttering of the rumor itself, as a product of the body and its speech capabilities and of human memory. The compulsion not only to retain but to retell the rumor is the mechanism by which rumors travel far and wide distinguishing them from more ordinary day-to-day speech acts. A second aspect of the materiality of rumor was its capacity to motivate a population to act quickly and decisively in response to what ultimately proved to be false information. This material consequentiality of a "mere" speech act is a sharp contrast to the way spoken commands and voice in general has been propped up in some materialist theorizing as generally and typically ephemeral and as less impactful than other material modes. This chapter further cautioned against overstating the significance of a novel technological artifact (in this case, the new role of mobile phones play in spreading a rumor) in materialist accounting. The novel or most apparently material element is not necessarily the driving force or unifying element in a studied phenomenon.

The earthquake in Haiti itself was a powerful and sudden inversion of the hierarchies of durability that, as John Law notes, "tend to" hold: the terror of buildings crumbling and turning to rubble, of what was previously silent and immobile, a haven of protection and security crashing down upon its unfortunate inhabitants. In Haiti there was also a rapid sweep of rumors in the wake of the earthquake, specifically rumors of a tsunami threat that motivated the panicked flight of many Haitians from their tent camps. Furthermore, rumors emerged in distant lands, specifically in Ghana where such a news item resonated in a deeply personal way.

A final note on the time I have spent in urban Ghana that I have come to see as an especially verbose society, a society that relishes wit, the brilliant turn of phrase, where popular musicians brag of the authenticity of their training in proverbs,[9] and where Internet scammers, a population I have previously studied, often attribute their own success or the success of others to "powerful words." It is therefore appropriate in such a context to restore to this discussion of the material—language performances and broader nonmaterial realms of affect, myth, and metaphor, or even theory and ideology considering the interdependency between nonmaterial and material. My claim here, though, is that the materiality of language performances matter everywhere, *not* that they constitute a distinctively Ghanaian condition. No population is inherently susceptible to rumor and it is very much a phenomenon of circumstance appearing out of anxiety and uncertainty. The highly educated (Scheibel, 1999), public officials, and journalists (Tierney, Bevc et al., 2006) are all sus-

[9] For example, there is this line from a popular song "Borga, Borga" by Sarkodie, who in a boast directed at young music industry upstarts he says, "When you people [i.e. competing musicians] started I was in the village speaking proverbs."

ceptible to false rumors and, at any rate, rumors do not always prove to be false. Words matter, but not as merely a floating realm of signifiers. We will benefit in a broad variety of research contexts from considering materiality in a way that avoids excluding language and its production as an unfortunate slip into the rejected realm of pure symbolism.

References

Akrich, M. (1992). The de-scription of technical objects. In W. E. Bijker and J. Law (eds.), *Shaping technology/building society: Studies in sociotechnical change* (pp. 205–24). Cambridge, MA: MIT Press.

——Latour, B. (1992). A summary of a convenient vocabulary for the semiotics of human and nonhuman assemblies. In W. E. Bijker and J. Law (eds.), *Shaping technology/building society: Studies in sociotechnical change* (pp. 259–64). Cambridge, MA: MIT Press.

Allport, G. W. and Postman, L. (1965). *The psychology of rumor*. New York: Russell & Russell.

Barad, K. (2003). Posthumanist performativity: Toward an understanding of how matter comes to matter. *Signs: Journal of Women in Culture and Society*, 28(3), 801–31.

Burrell, J. (2011). User agency in the middle range: Rumors and the reinvention of the Internet in Accra, Ghana. *Science, Technology and Human Values*, 36(2), 139–59.

Faulkner, P. and Runde, J. (2011). The social, the material, and the ontology of non-material technological objects. *Unpublished*.

Kallinikos, J., Lanzara, G. F., et al. (2010). Introduction to the special issue. *First Monday*, 15(6–7).

Kapferer, J.-N. (1990). *Rumors: Uses, interpretations, and images*. New Brunswick: Transaction Publishers.

Larkin, B. (2008). *Signal and noise: Media, infrastructure and urban culture in Nigeria*. Durham, NC: Duke University Press.

Latour, B. (1993). *We have never been modern*. New York: Harvester Wheatsheaf.

Law, J. (1994). *Organizing modernity*. Oxford: Blackwell.

——(1999). After ANT: Complexity, naming and topology. In J. Law and J. Hassard (eds.), *Actor network theory and after* (pp. 2–14). Oxford: Blackwell Publishers.

——(2007). Actor network theory and material semiotics. http://www.heterogeneities.net/publications/law2007ANTandMaterialSemiotics.pdf

Leonardi, P. M. (2010). Digital materiality? How artifacts without matter, matter. *First Monday*, 15(6–7).

Miller, D. (Ed.) (1998). *Material cultures: Why some things matter*. Chicago: University of Chicago Press.

——(2005). *Materiality*. Durham, NC: Duke University Press.

Scheibel, D. (1999). 'If your roommate dies, you get a 4.0': Reclaiming rumor with burke and organizational culture. *Western Journal of Communication*, 63(2), 168–92.

Scheper-Hughes, N. (1992). *Death without weeping: The violence of everyday life in Brazil.* Berkeley, CA: University of California Press.
Shibutani, T. (1966). *Improvised news: A sociological study of rumor.* Indianapolis: The Bobbs-Merrill Company, Inc.
Strum, S. and Latour, B. (1987). Redefining the social link: From baboons to humans. *Social Science Information*, 26(4), 783–802.
Sunden, J. (2003). *Material virtualities: Approaching online textual embodiment.* New York: Peter Lang.
Tierney, K., Bevc, C., et al. (2006). Metaphors matter: Disaster myths, media frames, and their consequences in hurricane Katrina. *The Annals of the American Academy of Political and Social Science*, 604, 57–81.
Turner, P. A. (1993). *I heard it through the grapevine: Rumor in African-American culture.* Berkeley: University of California Press.
Verran, H. (2001). *Science and an African logic.* Chicago, IL: University of Chicago Press.
White, L. (2000). *Speaking with vampires: Rumor and history in colonial Africa.* Berkeley: University of California Press.

VII

Epilogue

17

Matter Matters: Materiality in Philosophy, Physics, and Technology

Albert Borgmann

"Matter matters" is meant neither as a tautology nor as a pun, but as the suggestion that what is truly important must be essentially material; it must be tangible, have weight, and occupy a definite place and time.[1] Nothing really matters unless it is material. Accordingly, materiality has become a concern, I'd like to suggest, because a certain kind of immaterial object is growing alarmingly in number and importance—programs, data, computer files, all the objects, in short, that come under the heading of technological information (Borgmann, 1999: 123–212). But there is also a broader apprehension that the sturdy moral framework of old and the reliable heft of things have disappeared. The material solidity of the world seems to be slipping over the horizon, and we are worrying about the trace it has left, namely, materiality—a second-order concern as when you are worried about literacy rather than literature or legality rather than the law. My proposal is that the historically evolving perspective of philosophy, physics, and technology will reveal a way from materiality to what matters today.

Surprisingly perhaps, matter was already an issue 2,500 years ago in Western philosophy. Though philosophy has survived to this day, it has surrendered the decisive concern with matter to physics in the seventeenth century CE. And although physics too has survived into the third millennium CE, the crucial treatment of matter has been handed over to technology toward the end of the twentieth century. But to say that the leading concern with matter has passed from one discipline to another is not to say that an ancestral discipline was made irrelevant by its offspring. On the contrary, my hope is that drawing on all three we can clarify what really matters.

[1] I am indebted for help to my colleague Armond Duwell.

In pretheoretical circumstances, both historically as well as today in the natural circumstances of everyday life, there is no distinction between matter and form. As a historical example, consider the beginning of James Welch's *Fools Crow*, the story of a Blackfeet band at its first encounter with the European invaders. Welch describes how the protagonist of the story, initially called White Man's Dog, "watched Cold Maker gather his forces. The black clouds moved in the north in circles, their dance a slow deliberate fury" (Welch, 1987: 3). Is Cold Maker winter itself or is he the spirit who moves the stuff of the threatening clouds and bone-chilling air? Perhaps he was both, sometimes winter and at other times a spirit who would talk to a warrior in a dream. Zeus, as we usually think of him, is not lightning itself any longer, but the god who hurls thunderbolts. Poseidon is the god who lives in the sea and on occasion stirs it up with his trident. But for the ancient Greeks, Helios was still the name of the sun and the god, and the course of the god's chariot was the course of the celestial body.

In Ancient Greece, the general tendency, however, was for divinity to detach itself from material things and assume the nature of a person who inhabits and governs the thing. But once divinity had become personalized, it was humanized as well. The gods became human, all too human. The unpredictability of the weather became the whims or the wrath of a god. Divine grace became the favor of a goddess. Natural fertility became the philandering of a god. When the gods first had become persons in their own right and then lost their credibility as gods, tangible reality became an issue in its own right as well. Piety was displaced by curiosity, and the rule of the gods was replaced by the questions: What is at the bottom of reality? What is the world made of? What is the basis of the variety of things?

In its first appearances, godless matter was still one with form, *hyle* with *morphe*, a position that has come to be known as hylomorphism; alternatively, matter was still one with life, *hyle* with *zoe*, the view that historians of philosophy call hylozoism. But beginning in the Pre-Socratic period and clearly in the classical era, the question as to the basis of reality became an explicit concern, and two different answers emerged. One answer was: The basis of everything is structureless matter. The other said: The basis of everything is immaterial structure.

The latter answer was inspired by the Pythagoreans who claimed that everything at the bottom was number. Plato in the *Timaeus* gave the claim a more precise and plausible answer, arguing that four of the five regular solids constituted the traditional four elements. Tetrahedrons constituted fire, octahedrons air, icosahedrons water, and cubes earth. The fifth, the dodecahedron, presumably was the template for the cosmos (Plato, 1952: 130–5).

The other answer, that the world is ultimately made of unstructured matter, was Aristotle's in book Z of *Metaphysics* (Aristotle, 1961: 316–19). In his time, the natural world may still have been pervaded by gods and spirits, but it was at any event, as it is today, hylomorphic and hylozoic. In a lion or a pine, we do not distinguish between matter and form. A tree's form is one with its matter, and when alternative matter was used and the question "What's Wrong with Plastic Trees?" was famously answered with "nothing," controversy ensued (Krieger, 1973).

In the world of artifacts, however, then as now, the distinction between matter and form is commonsense. The form of the Greek temples betrays the wooden ancestry of the lithic ones that have come down to us. To take a more recent example, Michelangelo's Pieta is made of marble. But Carlos Slim's museum in Mexico City houses a copy cast in bronze—one form (the Pieta), two kinds of matter (marble and bronze). Michelangelo's David is of marble as well—two forms (Pieta and David), one kind of matter (marble).

Yet the material that the Pieta and David are made of has its own form. The particular marble has a color, a hardness, a specific weight, etc. So what is marble made of? It is made of limestone that has been exposed to heat and pressure. But limestone in turn has a form. So what is *it* made of? What Aristotle did is to push a commonsensical distinction to its unimaginable conclusion. What would totally formless matter look like? Whatever goo or soup you imagine, it will have some shape and color, however blurry.

Though Aristotle and Plato (in the *Timaeus*) differed on the nature of reality at its most basic, they agreed that a divine principle was needed to explain how the world at large was constructed on its ultimate basis. In the *Timaeus*, it is the demiurge, the divine craftsman, who arranges the elements according to the eternal and immutable patterns Plato called ideas. For Aristotle, the origin of all formation is the prime mover whose pure form was articulated and transmitted to the earthly realm by the intelligences that we know as the planets.

In the alchemy of the Christian middle ages, we get a first glimpse of the cultural forces that grew out of the difference that distinguished Plato's view from Aristotle's. Though he did not use or coin the term, historians of philosophy call Aristotle's structureless matter "prime matter." In alchemy, it figures as *prima materia*, the stuff that was thought to underlie all elements and, so it was hoped, could be transmuted from one form to another by the philosopher's stone, say from the form of lead to the form of gold. The cultural complexion that was foreshadowed by *prima materia* and the philosopher's stone is this: Structureless matter represents unlimited possibility, immaterial structure represents unlimited power.

Alchemy was the wayward offspring of philosophy and the fumbling ancestor of science. It was Newton's *Principia Mathematica* of 1687 that established physics as the rightful explanation of the material world. In the seventeenth century, physics was still called natural philosophy as the full name of

Newton's treatise shows: *The Mathematical Principles of Natural Philosophy* (Newton, 1999). But that was already a convention. Physics had left and displaced philosophy as the authority regarding the basis and structure of the world.

Matter was crucial to Newton's physics, not as an actor but as the stage. More precisely, the stipulation that matter was conserved in all changes of force, velocity, and acceleration was the precondition of the three laws of motion. Frank Wilczek has called the principle of the conservation of matter Newton's "zeroth law," the one that has to come first for the other three to work (Wilczek, 2008: 11–17). Matter does not figure directly in the three laws. What shows up, particularly in the second, is what Newton called "quantity of matter," namely mass (Newton, 1999: 403–4). In Newton's universe, things no longer had their own weight. What something weighs depends on its relation to something else. An astronaut weighs more in relation to earth than in relation to the moon. Matter, thought of as mass, is unearthly. For Newton, it is invariant—the same anywhere in the universe.

Although Newton universalized matter, his assumption that its "quantity" is conserved turned out to be parochial. When in Newtonian physics you let two elastic balls collide head-on and measure the velocity of each before and after the collision, the ratio of the difference between the before and after velocity of each is the ratio of the difference of their masses. These ratios consistently reflect the mass of a particular ball, no matter how you vary its velocity or the mass or velocity of the opposing ball. Mass is conserved. But, to use Wilczek's example, if you let an electron and a positron collide at speeds close to that of light, the result typically is ten pions plus one proton plus one antiproton. But mass is not conserved. The total mass of the after-collision particles is about 30,000 times the mass of the before-collision particles (Wilczek, 2008: 16).

The background to the experiment, whose outcome would have been so shocking to a Newtonian scientist, is the rise of relativity and quantum theory. Whether a lay person expects mass to be conserved, I am not sure. But what has to be unexpected and unsettling to the ordinary person is the disappearance of a plausible mode of analyzing matter and the dissolution of the framework we take for granted. As for analysis, we are not surprised that the things we know can be divided into smaller and smaller parts. We may not have preconceived notions as to what the smallest current particles look like and whether they are the ultimate ones or are further analyzable in the future. But we would expect that at any one moment they would have a definite location, that they would have a here and a now in space and time.

That is no longer so. Special relativity theory tells us that what is now for you may be later for me, and what is separated by a foot for me may be separated by two feet for you. Quantum theory tells us that an electron as best anyone can know before we measure its position is as much here as it is there. Matter had lost its own weight in Newtonian physics. Now it has lost its traditional identity as well, that is, the possibility of being uniquely identified at any moment by reference to its place and its time. Nothing definite seems to matter anymore.

You cannot be unsettled by what you do not know. Most people in America are ignorant of the physics that has dissolved the traditional substance and setting of matter. Thus, culturally the teachings of physics may be neither here nor there. People do not know physics. Could they understand physics to begin with? The rudiments could certainly be taught in college or even in high school. Two-dimensional space–time diagrams can convey both the conceptual and the mathematical basis of special relativity. Vector diagrams can do the same for the strange state an electron is in, its superposition, and the way measurement resolves a superposition.

Assuming that general education in colleges and universities were equal to its obligations, would what students are taught inform their conception of matter? The nightmare that haunts professors is the question of the half-life of their teachings. How long does it take typical students to forget half of what I taught them this semester? A couple of months? It could be less, but let us assume it is two months. Within a year, the students would remember a sixty-fourth of what they once knew. Such precise prognoses are whimsical at best, but they gloss what many of us know—information that does not become regular working knowledge is evanescent.

Are most people in America, to take one example, left then with a hylomorphic and hylozoic conception of living things and with a commonsense understanding of matter and form in artifacts? Has contemporary physics left the ordinary understanding of matter untouched? Before I answer the question, I must point out that the question of matter is far from settled in physics. There are in fact three open questions. In order of increasing difficulty they are:

(1) Do the Higgs boson and the Higgs field that supposedly lend particles mass in fact exist?

(2) The kind of mass and matter we understand constitutes only about five percent of all the matter and energy there is. The rest is dark, that is, mostly unknown. What are dark matter and dark energy made of?

(3) Quantum theory tells us how matter behaves in the small and relativity theory how it behaves in the large. But the two theories are incompatible with one another. What will a unified and fully intelligible theory of matter look like?

Still, I am confident that answers will be forthcoming and, more important, that what relativity and quantum theory tell us will essentially be incorporated rather than refuted by those answers. After all, just as Newtonian physics has proven itself successful in the range of velocities we are intuitively familiar with, so quantum and relativity theory have been successfully applied in technology. Here then we come to the transition from physics to technology. In one regard this is another transition from one scholarly discipline to the next, from the science of physics to engineering science. The latter is a simplified and more readily usable version of the former and has its own terms, rules, and models (Moriarty, 2009).

In another sense, the transition is from theory to reality. Technology is a realization of physics. Of course, if the laws of contemporary physics represent the most basic and regular aspects of reality, as I am convinced they do, then, broadly put, everything is, was, and will be a realization of physics. In particular, all cultures—the Homeric, the classical Greek, the western medieval, and the early modern—have been realizations of contemporary physics. The question at hand is whether the discovery, knowledge, and conscious application of modern and especially contemporary physics have highlighted features that, although always latently present, have now come to be visibly mirrored in contemporary culture. Obviously, these features have no compelling cultural force; otherwise all cultures would have exhibited them. The fact they now have cultural parallels is a contingent historical fact, and the most we may be able to say about the explanation of the origin of such parallels is this: The defining event of the modern era, the Enlightenment, was a liberation of the human spirit that opened up both the vistas of modern science and the space of their technological application.

There is a closely related feature of contemporary physics that bears upon our comprehension of reality. It is culturally invariant in one sense and, in another, a significant cultural background condition for what matters in this chapter. All discernible objects in the universe, including all cultural objects whatsoever, are traceable to their physical micro and macro structures. Every star and every flower can be analyzed as to their chemical structure and gravitational field. And thus everything can be illuminated as never before. But it matters culturally whether an object is uniquely or indifferently reducible and traceable to its immediate context and beyond. A real rose is uniquely traceable to its stem that is part of this bush, which is nourished by these particular roots, embedded in this certain slope, and so on and on. A rose on an iPhone, to the contrary, is immediately traceable to an LCD screen, to wiring, to computer chips, and so on, all of which are indifferently related to the rose, that is, no more uniquely than the text that follows the rose on the screen of the iPhone (Borgmann, 2011*a*).

Before I turn to the particulars of the parallels between physics and culture, I must note popular versions of those parallels that make you hesitate to trace alternative ones because these popular versions are so mistaken in their physics and misleading in their culture. Thus, it is occasionally said relativity theory has demonstrated that there are no longer absolutes or constants, that everything is relative and variable. But that is just wrong. Similarly, it is sometimes said quantum theory has proven that observers inevitably construct or disturb what they observe. But this claim rests on an interpretation of quantum physics that is controversial and burdened with theoretical difficulties. Even on its own terms, the observer-induced part of quantum measurements is both strictly constrained in its possibilities and subject to uncontrollable randomness.

There are more considerable parallels, drawn by well-trained physicists, parallels between relativity or quantum theory and traditional world views, such as eastern mysticism or the world view of the Blackfeet (Peat, 2005; Capra, 2010). I will set aside the question of just how compelling and illuminating these parallels are. In any case, the parallels that are of concern to my chapter are those that might shed light on the destiny of materiality and matter in contemporary culture. They would obtain between phenomena both of which are without precedent. My conjecture, at any rate, is that there are two such kinds of parallels between contemporary physics and the prevailing culture of the advanced industrial societies, what I have been calling "contemporary culture" for short (Borgmann, 1984). The first kind of parallel reveals unprecedented and all but irreversible features that contemporary physics and contemporary culture have in common. These features, however, have been elaborated in contemporary culture in ways that are morally dubious and, though of course compatible, not positively consonant with certain features of physics. The second kind of parallel goes counter to paradigmatic contemporary culture, but is consonant with a new conception of what matters in the good life.

Special relativity and contemporary culture jointly show how the framework of matter has changed. Both suggest that there is an indispensable universal structure of reality and an aspect of individual difference or choice. In relativity theory, the inescapable constant is the speed of light. It is absolute in two ways: (*a*) It is the absolutely greatest speed there is. (*b*) The speed of light is absolute in being relative to nothing. Whether you are stationary relative to the source of light or moving with the source of light or against it, it does not matter. However, you measure the speed of light, you find it is always the same, one foot per nanosecond or 186,282 miles per second. The other indispensable structure is a metric space of four dimensions, three of space and one of time. A helpful approximation is a two-dimensional space of two coordinates, one of space and one of time.

But in that space, there is no universal now, and there are no absolute spatial distances. Everyone's here and now is the center or origin of his or her coordinate system. Everyone who, and everything that, is moving in relation to you generally has not only a here that differs from yours but also a different now and distances that are shorter for you than for her, him, or it. And you can increase such temporal and spatial differences by accelerating your velocity relative to that other person or thing. If you do, you also increase your (relativistic) mass relative to that person or thing.

The indispensable common structure in the culture of technology is the system of utilities, the transportation links, the electrical grid, the Internet, the water supply, and more. No one in the industrial countries could survive for long without some support of that structure. Most of us, moreover, are tied to that structure through work. But within that structure and most of the time, each of us is the center of a coordinate system. Each of us decides where to live and with whom, what to eat, and how to entertain ourselves. Distances are elastic, depending on our means of transportation. The markers and boundaries of time that once were firm and shared are beginning to shift and to dissolve.

Most people are ignorant of the trajectory of dematerialization that departs from mythology, runs through philosophy and physics, and ends in technology. But people seem to have a rough and intuitive understanding of what philosophy and physics would tell them explicitly and precisely: There is no absolute framework of matter; there are infinitely many frames of reference, lawfully based on an underlying structure. There are no impenetrable material substances; everything can be decomposed and transformed, and the rigidity of distances is penultimate and subject to shrinking or cancellation. This half-aware grasp of dematerialization may be a starting point, but cannot be a substitute, for knowledge of philosophy and physics.

With all the analogies between physical space and cultural space, there is an important disanalogy. The properties of physical space are crisply all or nothing, finite or infinite, conserved or variable, absolute or relative. But in cultural space, there is a continuum from all to nothing; properties are more or less robust. In the technological culture, there is still mass and weight, a here and a now. But the tendencies are toward less resistance and persistence and toward greater ubiquity and control over time.

The lack of absolute frameworks of time and space and the absence of immutably massive things are features that physics and contemporary culture inevitably share, or such at any rate is my suggestion (Taylor, 2007). But contemporary culture in addition exhibits features that are not paralleled by physics and are of dubious cultural value. As Aristotle and Plato had indicated, when we push the question of what things are made of to its deepest level, we arrive at one of two answers—structureless matter or immaterial structure. The solid and substantial world gets dematerialized either way. The cultural counterpart of structureless matter is realized in the cultural space of ever open possibilities. It is not a perspicuous and homogeneous world, however. It rests on a structure that for its part is well ordered, thoroughly articulated, and to most of us opaque. In fact, the substructure of structureless possibility fits the model of the alternative basis of reality—immaterial structure. At bottom, the world does not consist of unstructured goo, but of something more like Plato's triangles that he claimed to be the elementary particles of the four elements and that have their best contemporary counterpart in the electron. Like the equilateral triangles that make up the tetrahedron (the element of fire), electrons are indistinguishable from one another and, like an equilateral triangle, strictly and completely represented by their mathematically determinable properties—charge, spin, mass, energy state. The electron is also, at least as of now, the paradigm of irreducibility. It was the first elementary particle to be discovered, and it has remained elementary to this day.

Most important for our purposes, the electron for better or worse has served as the exemplar of the unlimited transformative power that seems to issue from the knowledge of how matter is structured at the bottom. It is a power that unites determinacy and limitlessness. Although an electron in superposition is neither here nor there, we definitely, on resolving and measuring its position, find it with equal probability either here or there. But while measurement appears to fix it in space and time, an electron can also be thought of as defying space and time (Vedral, 2011: 43). If you run electron twins (a so-called Cooper pair) through a superconductor, separate them at a T-junction, run them through a nanowire in opposite directions, measure a particular spin of one of them, and find it to be, say, clockwise, you know that when the same spin of its twin is measured, it will turn out to be counterclockwise. The spins, when measured, will always and instantly be opposite to one another whether the electrons are separated by a foot or a galaxy. Such separated electron twins are said to be entangled, and entanglement seems basically to reside in a distanceless realm.

In metric space, electronic effects travel at a speed close to that of light. The swiftness, determinacy, and minuteness of electrons make them powerful carriers of information. Thus, the technology that is named after electrons, electronics, is now controlling everything—engines, utilities, manufacturing, finances, traffic, weapons systems, and of course information and entertainment. Immaterial structures control the material world. Nominally, the ultimate control is ours. But large parts of the total electronic structure are now so far-flung and complex that no person, no team, no nation can truly understand and direct them any longer. Nor could any government regain control by shutting them down and starting over. We are all dependent on their continuing functions.

Just as there is a realm of limitless possibility within the structure of utilities, so there seems to be a realm of limitless power within the electronic structures. It is a realm we often call cyberspace. It parallels the distanceless realm of quantum entanglement. In cyberspace, the intervals of space and time have been displaced by ubiquity and instantaneity. Nothing is distant in space or remote in time. Material space beyond cyberspace is becoming pliable; distances can be made to shrink by improvements in transportation and production.

Brute matter has not altogether lost its brutality, far from it. It assaults people in earthquakes, tsunamis, hurricanes, tornadoes, floods, and in the deprivations of hunger and illness. But we think of such brutal assaults as technological challenges rather than as mythic forces, and rightly so. But the common notion that all kinds of burdens, even such tolerable burdens as dark moods and homely appearance, are unacceptable and temporary is morally troubling. Similarly, the life of unlimited possibilities still collides on occasion with massive resistance—mass is resistance to force. But the structure of utilities at times offers questionable evasions where collisions are looming. Everything actual is thought to be relative to possibilities, alternatives, and replacements.

Does that mean that the rightful force of matter is irretrievable? My proposal is that physics furnishes not only parallels that are inevitably consonant with physics and dubiously elaborated in contemporary culture but also contains features that parallel new ways the good life can materialize today. To begin with quantum interpretation, one of the great puzzles in making sense of measurements is that, although quantum physics is a total theory of reality, measurement as conventionally understood falls outside of quantum mechanics. You cannot make quantum-mechanical sense of it. In particular, there is no account in the traditional conception of measurement of what happens to the possibilities that were not realized. Quantum mechanics is strange, and if your interpretation eliminates one strange feature, you are saddled with another. But the smoothest and most consistent interpretation says that a measurement is an event within the quantum world, and when it occurs, it resolves a superposition into two previously probable but now actual states, one being realized in the world you and I inhabit, and the other in another world. This is the "many worlds" interpretation. You have a successor in every one of the worlds that have been realized, and every successor is you as much as you. The philosopher David Albert in a conversation with the physicist Sean Carroll has pondered what bearing this may have on personal identity, and his point, I want to suggest, has a deeper and revealing significance (Carroll and Albert, 2008).

Albert imagines that we have a million particles before us, all spinning counterclockwise in the vertical direction. Particles spin in all directions at once, including the horizontal. (Quantum mechanics is strange.) If you measure the horizontal spin of one of these vertically spinning particles, it will be randomly either clockwise or counterclockwise. How many possible combinations of the million spins are there? Answer: Two to the millionth power, an unimaginably large number. Now Albert imagines that on measuring the million different horizontal spins he discovers that all of them have turned out to be counterclockwise. He finds it surprising, not that *someone* has found the spins all the same; after all, the world has branched into two to the millionth worlds, and in one of those worlds, all the spins had to be counterclockwise. He wonders why *he* is that someone.

We can further sharpen Albert's question this way. I measure the spins and get an apparently random sequence of clockwise and counterclockwise spins. Just out of curiosity I convert all the clockwise spins into ones and the counterclockwise spins into zeroes. Then I apply the ASCII notation that converts one eight digit sequence of ones and zeros after another into a letter or punctuation mark, and lo and behold my ASCII conversion has yielded the first thirty chapters of Genesis in the New Revised Standard Version of the Bible. Again, someone had to get the sequence of spins that translates via ASCII into Genesis. But why me?

The cultural analog, I suggest, is the poignancy of personal identity in a universe of countless alternatives. There is no solid and encompassing order that assigns me a place from birth till death; and what singles me out seems to be a random event, one that, before it occurred, was no more likely than innumerable others. My being the person that I am in fact seems to be a brute given. But that fact has another aspect as well. What is brute is also forceful, and what is given I have to come to terms with.

What centers your life and lends it its identity is, let us imagine, a moment when you fell and were badly injured and the occasion when you first came upon the person you love more than life. You do not recall location plus times plus contents. You remember events. An event constitutes and occupies its own place and time. But an event could not matter if it were not material through and through. What matters needs to have depth. It must be grounded without rupture and traceable without loss of meaning. Life is lived out in the interval between such events. That interval is the spine of your identity, and like the space–time interval between events in special relativity, it remains no matter your changing frames of reference. Of course, the defining events of your life are not the extensionless points of special relativity where four coordinates intersect. But they are compact nevertheless and give point to your life. They tell you what matters, and the manifold of reality unfolds from there. They give you a standpoint from which to comprehend and take responsibility for the several and expanding contexts of your life—your family, your town, your country, the global community, and at length the universe and its lawful structure. Life is not always eventful in this sense. We often mistake ourselves for alchemist and demiurges, in possession of unlimited possibilities and powers, when in fact we are drifting with the currents of pointless choices and decisions.

To sum up and conclude, physics can serve as the foil, and offer parallels, for the good life. As a foil it illuminates and renders transparent the deep and wide structure of events, things, and persons. As a source of parallels, it suggests that the world of endless possibilities gets closure in events that we have to accept and respond to and that define our identity and the identity of things that become landmarks of life. To make this a little more concrete, consider two illustrations.

First imagine a mountain peak in the Rockies. Physics can illuminate it layer by layer, without rupture or cognitive dissonance, telling us that it has been thrust up by the collision of tectonic plates and piled up in part of granite, composed in part of quartz, consisting in turn of silicon–oxygen tetrahedrons (we are down to Plato's elements of fire). But no expertise in physics could have foretold you that someday you would tumble down a steep and snowy slope, end up badly battered, would be taken on a toboggan to the ambulance, your distraught daughter skiing alongside while you are begging for oxygen. Nor could anyone know that one day you and your family would arduously climb a peak on foot and be rewarded at the top with a vista of the valley where you live. You live your life in the interval between the event that taught you mortality and the event that revealed your world to you. Through these events a peak became a landmark, and the landmark helped to define the identity of your beloved.

And then imagine a violin that your daughter's violin teacher recommended you buy because it had a fine sound. Physics can tell you how to trace its material structure from the wood to the organic polymers it is composed of and how to trace the melodies that strike you as the sounds of strings to the mechanics of vibrations and the compressions of air. But no physics could prepare you for what you discovered when at home you peered inside the violin—a certificate that read: "Johann Georg Neiner Lauten u. Geigenmacher in St. Petersburg 1821." An event of nearly two centuries ago opened up, a violin maker from Mittenwald in Germany moving to St. Petersburg, drawn perhaps by the enterprising Catherine the Great.

Neither could you have known, though you may have hoped, that your daughter at sixteen would play the solo part of the first movement of Vivaldi's concerto in A minor; and that a decade and half later, you and your daughter would take the violin to the formidable René A. Morel in New York City to have the neck reset, and a block, the post, and the bridge replaced. Those events made the violin a landmark in your life and in part made your daughter who she is.

Persons and things would not matter and could not contain so much meaning if they did not consist of tangible and infinitely complex matter and call up a consistently traceable context. Nor could they materialize so splendidly were they dependent on the sanction of some absolute framework. Whether a divine principle is needed to explain all this as Plato and Aristotle thought is no longer a question that can count on a common answer. But all good people will agree that the beings that truly matter are sacred (Borgmann, 2011*b*).

References

Aristotle (1961). *Metaphysics*. Trans. H. Tredennick. Cambridge, MA: Harvard University Press.

Borgmann, A. (1984). *Technology and the character of contemporary life*. Chicago: University of Chicago Press.

——(1999). *Holding on to reality*. Chicago: University of Chicago Press.

——(2011a). Intelligence and the limits of codes. In T. Bartscherer and R. Coover. (Eds.), *Switching codes*. Chicago: University of Chicago Press.

——(2011b). The sacred and the person. *Inquiry*, 54, 183–94.

Capra, F. (2010). *The tau of physics*. Boston, MA: Shambhala (originally published in 1975).

Carroll, S. and Albert, D. (2008). bloggingheads.tv (July 22). http://bloggingheads.tv/diavlogs/13487retrieved June 18, 2011.

Krieger, M. H. (1973). What is wrong with plastic trees? *Science*, 179(February 2), 446–55.

Moriarty, G. (2009) (originally published in 2007). *The engineering project*. College Park, PA: Pennsylvania State University Press.

Newton, I. (1999) (originally published in 1687). *The principia*. Trans. I. B. Cohen and A. Whitman. Berkeley, CA: University of California Press.

Peat, F. D. (2005). *Blackfoot physics*. Boston, MA: Weiser Books (originally published in 2002).

Plato (1952). *Timaeus*. Trans. R. G. Bury. Cambridge, MA: Harvard University Press.

Taylor, C. (2007). *A secular age*. Cambridge, MA: Harvard University Press.

Vedral, V. (2011). Living in a quantum world. *Scientific American*, June, 304(6), 38–43.

Welch, J. (1987) (originally published in 1986). *Fools crow*. New York: Penguin Books.

Wilczek, F. (2008). *The lightness of being*. New York: Basic Books.

Author Index

Note: references to footnotes are indicated by the suffix 'n' followed by the note number, e.g. 310n7.

Subject Index

Note: page numbers in *italics* refer to figures and tables. References to footnotes are indicated by the suffix 'n' followed by the note number, e.g. 302n4

Subject Index

Made in the USA
San Bernardino, CA
16 October 2013